797,885 Books

are available to read at

Forgotten Books

www.ForgottenBooks.com

Forgotten Books' App
Available for mobile, tablet & eReader

ISBN 978-1-331-93274-1
PIBN 10256337

This book is a reproduction of an important historical work. Forgotten Books uses
state-of-the-art technology to digitally reconstruct the work, preserving the original format
whilst repairing imperfections present in the aged copy. In rare cases, an imperfection in
the original, such as a blemish or missing page, may be replicated in our edition. We do,
however, repair the vast majority of imperfections successfully; any imperfections that
remain are intentionally left to preserve the state of such historical works.

Forgotten Books is a registered trademark of FB &c Ltd.
Copyright © 2015 FB &c Ltd.
FB &c Ltd, Dalton House, 60 Windsor Avenue, London, SW19 2RR.
Company number 08720141. Registered in England and Wales.

For support please visit www.forgottenbooks.com

1 MONTH OF
FREE
READING
at
www.ForgottenBooks.com

By purchasing this book you are eligible for one month membership to ForgottenBooks.com, giving you unlimited access to our entire collection of over 700,000 titles via our web site and mobile apps.

To claim your free month visit:
www.forgottenbooks.com/free256337

* Offer is valid for 45 days from date of purchase. Terms and conditions apply.

Similar Books Are Available from
www.forgottenbooks.com

General Biology
by Leonas Lancelot Burlingame

A Manual of Bacteriology
by Herbert U. Williams

The Human Body
A Popular Account of the Functions of the Human Body, by Andrew Wilson

Practical Biology
by W. M. Smallwood

Manual of Practical Anatomy, Vol. 1
Upper Limb; Lower Limb; Abdomen, by D. J. Cunningham

Problems of Biology
by George Sandeman

Science and Human Affairs from the Viewpoint of Biology
by Winterton C. Curtis

History of Biology
by L. C. Miall

The Bacteria
by Antoine Magnin

A Manual and Atlas of Dissection
by Simon Menno Yutzy

Practical Microscopy
by Maurice Norton Miller

Elements of the Comparative Anatomy of Vertebrates
by W. Newton Parker

The Elements of Animal Biology
by S. J. Holmes

Handbook of Plant Dissection
by J. C. Arthur

Biology and Its Makers
With Portraits and Other Illustrations, by William A. Locy

An Introduction to Vertebrate Embryology, Based on the Study of the Frog, Chick, and Mammal
by A. M. Reese

Elementary Biology
Plant, Animal, Human, by James Edward Peabody

Scientific Method in Biology
by Elizabeth Blackwell

The Biology of Birds
by J. Arthur Thomson

Fathers of Biology
by Charles McRae

THE ANATOMY OF THE HUMAN SKELETON

THE
ANATOMY OF THE HUMAN SKELETON

BY

J. ERNEST FRAZER, F.R.C.S.Eng.

PROFESSOR OF ANATOMY IN THE UNIVERSITY OF LONDON AND LECTURER IN THE MEDICAL SCHOOL OF ST. MARY'S HOSPITAL ; EXAMINER IN ANATOMY FOR THE UNIVERSITY OF LONDON; EXAMINER IN ANATOMY FOR THE PRIMARY FELLOWSHIP OF THE ROYAL COLLEGE OF SURGEONS OF ENGLAND; FORMERLY EXAMINER IN ANATOMY FOR THE CONJOINT BOARD OF THE ROYAL COLLEGES OF PHYSICIANS AND SURGEONS

SECOND EDITION

WITH 219 ILLUSTRATIONS

LONDON
J. & A. CHURCHILL
7, GREAT MARLBOROUGH STREET
1920

PREFACE

It is not necessary to lay emphasis on the importance of a knowledge of the skeleton as an integral part of the study of human anatomy, and, in the literature bearing upon the subject, we find masterly accounts of the constituent bones which rank as classics in the education of the student. In this book I have ventured to wander in some degree from the well-trodden road and to lead the reader by other ways to the comprehension of his subject. My intention has been to induce him to think of the bones as they exist in the body rather than as they lie on the table before him, and to do this I have laid stress—because he must use the prepared specimens—on the meaning of small details and on the relations of the bone, and have relegated the pure description of the dry bone to a secondary place : in other words, each part of the skeleton has been used as a peg on which to hang a consideration of the neighbouring structures, in the hope that this may afford a new point of view to the reader and enable him to grasp the intimate connection between them.

Such a way of regarding the skeleton opens up a very extensive field of description, and within the limits of a student's hand-book it is only possible to deal with some out of the many points which offer themselves for development, but I hope that those of which I have treated in this volume may be of value to the student and may lead him to think of the skeleton as something more than a dry subject for study, and to search for reasons for the hundred and one abstract and concrete qualities which his own observation will prove any particular bone to possess. If it has this effect, one of my objects in writing the book will have been attained.

The majority of the illustrations, which the generosity of Messrs. J. & A. Churchill has enabled me to insert, are intended merely to help the student to apply the descriptions in the text to the actual specimens : if, in spite of their many artistic imperfections, they are of use in this respect, I shall be content. They have been drawn from specimens in my possession or in the Anatomical Department in the School of this Hospital.

It is a pleasure to acknowledge my indebtedness to my colleague, Dr. R. H. Robbins, for his careful reading of the proofs, to Mr. R. M. Handfield-Jones for the same service in a part of the work, and to my wife for help in preparing the book for the press.

J. ERNEST FRAZER.

Medical School,
St. Mary's Hospital, W.

PREFACE TO SECOND EDITION

THE kindness extended by reviewers to this book on the occasion of its first appearance, and the steady demand for it since the War, give me reason to hope that this edition will still find a sphere of usefulness among students of anatomy.

A number of reviewers expressed the opinion that the bones of the skull had not been presented so fully "dressed" as were the other constituents of the skeleton. I was quite conscious of this at the time of writing the book, and had put them forward in that way after some reflection, and deliberately. I have thought over the matter again in preparing this edition, and have concluded that the old reasons, those which influenced me before, are at least as potent now, so that I have refrained from an attempt to alter the plan on which the description of the skull was based.

The whole text has been submitted to careful revision, and a few new illustrations added.

J. ERNEST FRAZER.

CONTENTS

CHAPTER I.—PRELIMINARY.

	PAGE
Constitution of Skeleton	1
Varieties of Bones and names of Parts	2
Formation and Condition of Bones	3
Study of Skeleton and Factors in Moulding Bones	5
Plan of Description used in the Book	7

CHAPTER II.—VERTEBRAL COLUMN.

FUNCTIONS: TYPICAL VERTEBRA	9
REGIONS OF COLUMN	11
COMPLETE COLUMN	13
CERVICAL VERTEBRÆ: General Description	15
Articulation with Cranium	15
Detailed Consideration of Cervical Vertebræ	20
DORSAL VERTEBRÆ: General	26
Detailed Consideration	28
LUMBAR VERTEBRÆ: General	32
Detailed Consideration	33
VERTEBRAL DEVELOPMENT	36
SACRUM: General Description	38
Detailed Account	40
Ossification	43
COCCYX	44

CHAPTER III.—THORAX.

RIBS: General Description	46
COSTAL CARTILAGES: General	47
COMPLETE THORAX	48
Peculiar Ribs	50
Detailed Account of Ribs	51
COSTAL CARTILAGES: Details	55
DEVELOPMENT of Ribs and Cartilages	57
STERNUM: General Account	57
Detailed Account	58
Development	60
Ossification	61

CHAPTER IV.—LIMBS.

LIMB GIRDLES	62
Hypothetical Development of Limbs	62
Relations between Limbs	63
UPPER LIMB: General Account of CLAVICLE	65
Sterno-clavicular Joint	66
Details of Clavicle	67
Development of Clavicle	69

CHAPTER IV.—continued.

	PAGE
SCAPULA: General Account	69
Detailed Consideration	71
Movements	77
Development	79
HUMERUS: General Account	80
Shoulder-joint	82
Detailed Consideration of Humerus	83
Ossification	89
ULNA AND RADIUS	89
ULNA: General Account	90
RADIUS: General Account	90
Attachments between Ulna and Radius	91
Elbow-joint	93
ULNA: Detailed Consideration	94
RADIUS: Detailed Account	97
Ossification of both Bones	100
HAND	101
CARPUS: General Account	101
Carpal and Wrist Joints	103
METACARPUS and PHALANGES	105
Carpus: Separate Bones	106
Metacarpus: Separate Bones	110
Phalanges	115
Development of Hand Bones	116

CHAPTER V.—LOWER EXTREMITY AND PELVIS.

OS INNOMINATUM: General Description	118
PELVIS as a whole	121
Detailed Consideration of Os Innominatum	124
Its Development	137
General Account of FEMUR	138
Hip-joint	140
Details of Femoral Relations	142
Ossification of Femur	151
PATELLA	152
Development	154
TIBIA AND FIBULA	155
TIBIA: General Account	155
FIBULA: General Account	156
Knee-joint	158
Detailed Description of Tibia	161
Detailed Description of Fibula	165
Ossification of both Bones	168
FOOT. TARSUS	169
METATARSUS and PHALANGES	170

viii Contents

CHAPTER V.—*continued.*

	PAGE
FOOT as a whole	170
Arches of the Foot	172
Separate Tarsal Bones	175
Metatarsal Bones	183
Phalanges	185
Development of Bones of Foot	186

CHAPTER VI.—SKULL AND HYOID.

SKULL : General Description	188
Skeleton of CRANIUM	189
FACIAL Skeleton : General Account	196
Separate Bones. FRONTAL	198
Development	202
PARIETAL Bone	202
Development	204
OCCIPITAL Bone	204
Development	208
TEMPORAL	209
Development	218
SPHENOID	219
Development	226
SPHENOIDAL TURBINATES	227
MAXILLA	228
Development	231

CHAPTER VI.—*continued.*

	PAGE
PALATE Bone	233
Development	234
HARD PALATE	235
SPHENO-MAXILLARY FOSSA	236
ETHMOID	237
Development	240
LACHRYMAL Bone	240
VOMER	241
INFERIOR TURBINATE	242
NASAL Bone	243
NASAL FOSSÆ	244
Septum	244
Outer Wall	244
Cells	247
Air Sinuses	249
MALAR	251
Development	252
ORBIT	253
MANDIBLE	253
Development	261
DEVELOPMENT OF SKULL as a whole	261
Skull at Birth	266
Growth of Skull	269
Skull Measurements	269
HYOID	271
Development of Hyoid	274

THE ANATOMY OF THE HUMAN SKELETON

Chapter I

PRELIMINARY

The **SKELETON** constitutes the framework on which the soft tissues of the body are supported, enabling them to retain their definite position in the body and, in the case of the limbs, affording a strong system of levers by means of which the muscles may change the situation of the body as a whole or of its various parts. To serve such functions the skeleton must be strong yet elastic, and must permit of movement without effort yet also without any serious lessening of its strength. These ends are attained by dividing up the skeleton into numerous constituent pieces, by making these of very strong and hard yet somewhat elastic material (*bones* and *cartilages*), and by connecting these bones and cartilages by means of *joints* or *articulations*, at which the firm structures can move on one another without resistance and yet securely owing to the fact that they are appropriately attached to each other by strong *ligaments*. If, then, we wish to study the skeleton, we should not confine our attention to the bones, but should also consider the cartilages, ligaments, and joints that are concerned in maintaining the form of the body and in enabling it to move about.

Cartilages * are tough, but elastic and compressible to a considerable extent. In the adult skeleton they are found completing the skeleton of the thorax (*costal cartilages*), holding bones firmly together (as in the vertebral column), or filling up intervals between bones (as in certain parts of the skull), coating the articular surfaces of bones (*articular cartilages*), or interposed between two articular surfaces (*inter-articular cartilages*). Cartilage occurring in the body is in general of the hyaline variety when covering articular surfaces of bone, but has the structure of one of the varieties of fibro-cartilage in other situations, save when it is taking the place of a bone in the skeleton.

Articulations can, as a whole, be divided into *synarthrodial* or immovable and *diarthrodial* or movable. Diarthrodial joints possess a synovial cavity to permit of easy movement, and this cavity may be wholly or partly divided by an inter-articular cartilage or *meniscus* into two parts. Synarthrodial joints are firmly fixed by interlocking of the bones concerned or by interposition of a thin but strong layer of fibro-cartilage over a large area on each of the bones. A subdivision of this class, where a certain very small amount of motion is permitted between the bones owing to the intervention of a thicker pad of fibro-cartilage, is usually termed an *amphiarthrosis*.

Ligaments are bands of fibrous tissue that pass over a joint to connect the bones ; naturally they are best developed over diarthrodial joints, where the articulating bones have no connecting medium between them, but depend on outside agents for security. It is evident that, if the ligaments are not to be a hindrance to free movement, they must be so disposed with regard to the joint surfaces, and these must be so shaped with reference to the ligamentous attachments, that any given ligament will either be tense and effective at one part of the movement or will remain tense throughout it ; in the latter case the movement of the articulation can only occur in one

---

* For detailed histology and for accounts of varieties of tissues consult special works on histology.

Anatomy of Skeleton

plane, and the ligament must be attached practically in the axis round which the movement takes place.

The *bones* constitute the larger part of the skeleton. Bone is extremely hard, but at the same time exhibits a small amount of elasticity and toughness. It differs from mere calcification of cartilage, in that it has a definite organised structure (Fig. 1), but it can be looked on as composed of animal or organic matter impregnated with earthy salts : analysis gives (roughly) about one-third or less of animal matter and the rest as mineral matter. The former gives toughness and the latter hardness to the composite result.

Fresh bone possesses a pinkish hue on the surface, which is covered by a fibrous *periosteum* of variable thickness. Bone exists in two states in the body, termed, according to the appearance it presents, *compact* and *cancellous* ; examples of both sorts of bone can be seen in a section along a bone of any size. Complete Haversian systems only occur, of course, in the compact bone, which is made up of a series of the systems (Fig. 2).

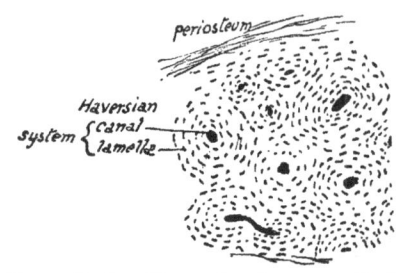

FIG. 1.—Section of bone under a low power. Each Haversian system consists of a central canal containing vessels and surrounded by a number of concentric lamellæ of bony tissue, between which are minute lacunæ which communicate by very fine channels and hold cells (bone-corpuscles). Compact bone is composed of an aggregation of such systems.

The shaft of a long limb bone is composed of compact bone which makes a shell that is hollow internally : this is the *medullary cavity*, and it is walled directly by the compact shaft-wall in the middle of the bone, but at each end cancellous tissue appears and occupies the cavity. The medullary cavity and the interstices of the cancellous bone are filled with very vascular *marrow*, of which two distinct varieties are found—(a) *yellow marrow*, found in the shafts of long bones, and containing about 96 per cent. of fat, and (b) *red marrow*, occurring in the articular ends of the long bones and elsewhere and containing 75 per cent. of water, with very little fat and a considerable amount of organic solids.

The skeleton comprises an *appendicular skeleton* of the limbs as well as an *axial skeleton* of the trunk.

The various bones of these parts are divided, according to their shape and appearance, into *short, long, flat,* and *irregular*.

Many and various names are given to the different markings and irregularities seen on bones : thus a hole in a bone may be termed a *foramen, canal,* or, in certain cases, a *meatus, hiatus,* or *(aquae) ductus*. Prominences projecting more or less from the general level are called *processes, trochanters, tuberosities, protuberances, tubercles,* and

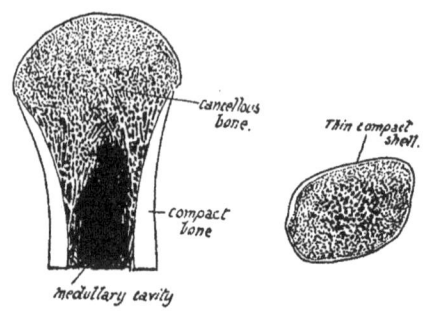

FIG. 2.—Diagram of a section of an end of a long bone (a) and a short bone (b) to show the arrangement of cancellous and compact tissue in them.

spines, or, if more linear in disposition, *ridges, spines, crests,* or *lines*. Depressions on the surfaces of bones may be *fossæ, cavities, fossettes,* or *foveæ* ; if more linear in direction, *grooves* or *sulci* ; if a large cavity exists in a bone it may be described as a *sinus, cell,* or *antrum*.

Preliminary

All the names just given appear to be used more or less indiscriminately, but in a very general way the order in which they are placed above corresponds with the diminishing size of the different structural characters to which they refer.

A projecting articular process on a bone is frequently referred to as the *head*, its narrowed attachment to the rest of the bone as the *neck*, and the remainder constitutes the *body*, or, in a long bone, the *shaft*. A *condyle* is a protruding mass carrying an articular surface, and a *ramus* is a broad arm or process of bone projecting from the main body.

All the terms given above, with numerous others of more special usage, will be better understood both in their meaning and application by the student after a short time spent in the study of the individual bones, and no more need be said about them now.

Bones do not take shape as such *ab initio*, but are preformed in the embryo as condensations of mesenchyme, which in some cases become cartilaginous before ossification commences in them, but in others remain unchondrified: in the former case the bones are said to be formed *in cartilage*, and in the latter they are formed *in membrane*. These terms simply mean that the bones have replaced cartilage or non-cartilaginous "membrane" as the case may be. The process of ossification is essentially similar in both varieties of formation, save that in chondral ossification the cartilage is calcified first and then absorbed and replaced by the true bony formation. Certain large cells, called *osteoblasts*, have the power of depositing or forming bone round themselves: they exist in the covering tissues of the developing bone (periosteum or perichondrium), and, in the case of chondral ossification, grow into the cartilage and occupy the spaces made in that structure by the confluence of the cell-spaces that goes with calcification. The early bone thus formed is removed by the action of other cells known as *osteoclasts*, and in this way a medullary cavity is provided in long bones while additional bone is being laid down on the surface under the periosteum, so that the bone increases in thickness. Thus there is no direct ossification of cartilage, but a replacement of it by bone that is made in the same way as in membrane bones.

The greater part of the skeleton is composed of cartilage bones, including that of the limbs, trunk, and base of skull; the bones of the face and vault of the skull are formed in membrane. The chondral skeleton is the modified *endoskeleton*, and the bones formed in membrane are usually considered to represent dermal bones, structure formed originally as part of the *exoskeleton*, which in the course of evolution have sunk to a deeper position and joined the other skeletal structures.

The invasion of cartilage by osteoblasts leads to an ossification commencing in the cartilage and termed *endochondral*, to distinguish it from that taking place under the periosteum and known as *ectochondral*: it is evident that ectochondral ossification, as indeed is the endochondral process really, is simply ossification in membrane occurring *on* cartilage.

The ossification commences constantly at one spot and at a fairly constant time in each individual bone, making a *centre of ossification;* the process extends from this centre. But a bone may have more than one centre—in fact, most bones have more than one and some have several. Such centres can be divided into *primary* and *secondary*, and as a rule the primary centres appear before birth and the secondary ones after that event. The smaller secondary centres form what are termed *epiphyses:* an epiphysis is therefore a part of a bone which is developed from a secondary centre, is at first separated from the main bone by a connecting area of unossified cartilage, and joins the main ossification at a later date to make the adult

4   Anatomy of Skeleton

bone. Epiphyses can be seen well in a long bone: Fig. 3 shows the condition of things in the femur, where the primary centre forms the shaft of the bone, and epiphyses lie on this as bony caps joined by plates of unossified cartilage to it. Such a shaft is known as a *diaphysis*. Epiphyses can be classed as *traction, pressure*, or *atavistic :* in the figure the epiphyses of the head and lower end, being in the line of weight-transmission, are pressure epiphyses, while those at the muscle attachment are traction epiphyses. Atavistic epiphyses are not found on the femur, and when they occur they are supposed to represent some past condition or process of the bone which has now no apparent function in the human skeleton. It is not impossible, however, that some of the other epiphyses are modified earlier processes and have therefore certain atavistic values.

The *epiphysial cartilage* that lies between the bony tissue of the epiphysis and diaphysis has an important function. We have seen that the shaft increases in thickness from the deposit of bone under the periosteum, but it is clear that this cannot lengthen the shaft, and a special arrangement is necessary if growth in length is to go on except by the slow process of interstitial increase: such an arrangement is provided by the growth in depth of the epiphysial cartilages, which just keeps ahead, so to speak, of the extension of diaphysial ossification toward the end of the bone. So the bone can grow in length as long as the cartilage remains unossified, and the growth ceases when the epiphysis joins the main bone. It follows from this that growth goes on longest at the end where the epiphysis is the last to join, and such an end is termed the *growing end ;* in all long bones except the fibula the centre which forms the epiphysis at the growing end, and therefore joins last, is the first to appear.

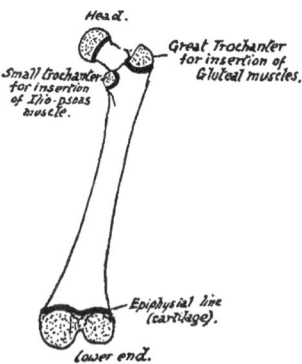

FIG. 3. — Diagram of femur. The stippled areas represent the epiphyses which are formed from centres of ossification distinct from that of the main shaft; each of these epiphyses is separated from the shaft by a layer of cartilage, indicated by the thick black lines, and the adult consolidation of the bone is brought about by the ossification of these cartilaginous plates. Their presence is thus a sign of immaturity.

A knowledge of the "growing ends" is necessary for the surgeon who may have to deal with injuries affecting them, with subsequent premature junction with the shaft and, as a consequence, shortening and deformity of the bone concerned. The deformity, of course, would not be so great, although present, if the injury affected the epiphysis at the other end of the bone. The limb bones have their growing ends at the shoulder and wrist in the arm and forearm respectively, and at the knee ends of the bones in the lower limb.

The vascular supply of bones is obtained from the arteries in the neighbourhood: in long bones many small vessels enter near the ends and in the epiphyses, and their tracks are seen as vascular foramina in the dried bone; but the vessels also enter the shaft, and one or two larger vessels in particular make what are termed the *nutrient foramina* in the body of the bone. In long bones the direction of the nutrient foramen is in a direction away from the "growing end" of the bone: the reason for this disposition of the vessels is obscure. The larger vascular foramina are occupied by veins as a rule, and nerves and lymphatics accompany the vessels into the bone. The results of certain diseases and injuries of bones is explained by reference to the way in which their blood supply reaches them ; thus, whereas the stripping of the shaft of a long bone will result in necrosis of the exposed osseous tissue, the bones of the cranial vault may be laid quite bare without such result, because their blood supply is almost wholly derived from the meningeal vessels on their deep surface.

Before the student starts on the study of the details of the skeleton he should have some knowledge of the factors that are responsible, in each individual bone, for the appearance it presents in the adult state. The following considerations ought therefore to be understood, for on the comprehension and appreciation of their bearing on any particular bone will depend the student's ultimate grasp of the bone and its possible modifications : he must, moreover, bear in mind that a knowledge of bones and muscles and intermuscular planes is essential to a proper understanding of the anatomy of the body, that the planes of muscles cannot be properly grasped without study of the skeleton, that the bones are only incompletely studied without the muscles and soft parts in relation with them and attached to them, and that many of the smaller details of the structure of the body that puzzle the student can be solved at once by an intelligent appeal to the skeleton. The principal factors responsible for the appearance of the adult bones are most easily illustrated in the long bones of the limbs ; but the student must remember that they are also operative in the other bones, and as his studies proceed he should endeavour to see how they have come into action in these other instances.

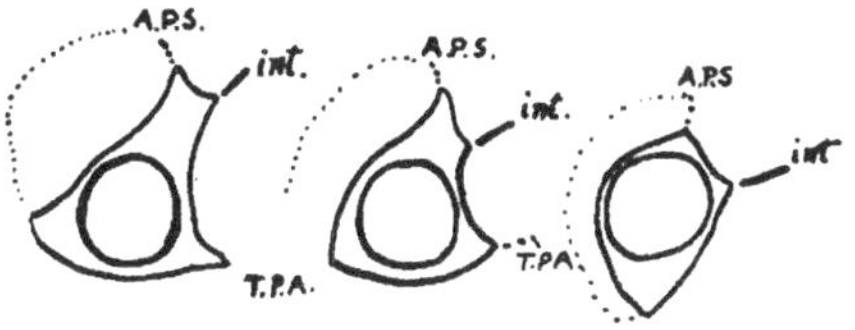

1. The shape and moulding of the surfaces of a bone depend on two main factors—(a) its primary "build" for performing certain functions and to resist certain strains, and (b) secondary moulding resulting from the arrangement and "pressure" of the surrounding structures.

FIG. 4.—Schematic sections through the fibula at various levels from above down, to show how the primary cylindrical bar of the bone is overlaid and hidden by the mouldings due to the muscles applied to it. The dotted line covers the Peronei. *A.P.S.* is the anterior peroneal septum, *int.* the interosseous membrane, and *T.P.A.* is the aponeurosis covering Tibialis posticus.

Thus in the fibula there is a strong bar of bone running throughout its length that is of a primary nature and remains in itself unaltered by the various structures attached to it, but it is covered in and partly hidden by secondary surfaces moulded over it and shaped by the covering muscles, and it can only be recognised here and there through some of these surfaces, making a convexity where there would otherwise be flatness or a concavity : again, in the lower jaw, we have the angled shape as a necessity of function, to bring the teeth into opposition, and combined with this as other primary constructions there are (see Fig. 210) the thick splenial portion to support the down-push of the alveolar part that carries the teeth, and the existence of a thick bar to counteract the tendency of the angle to open out ; while on these and not affecting them we find the secondary moulding shown, for example, in the presence of glandular fossæ and a coronoid process. Or in the scapula a thick bar (see p. 70) extends from the lower angle to the glenoid process, and its relation to the axillary border varies individually in detail, because the border is a secondary one which varies with the development of the Subscapularis that makes it, whereas the bony bar is quite unaffected by the muscles that lie over it. Considerations such as these will enable the student who endeavours to understand their bearing to comprehend the meaning of the existence and appearance of the various surfaces seen on a bone, to appreciate the way the bone lies in relation to surrounding structures, and to think of its different parts in terms of these as they exist in the living body.

2. In addition to the secondary mouldings just mentioned there are *secondary markings* to be distinguished on bones. These are made by ossification spreading a little way into the attachment of fibrous tissue to the periosteum, and become evident after birth, as a rule about puberty. They *indicate fibrous attachment only*, be it ligament, tendon, aponeurosis, or fascia, and are never made as a result of attachment of purely muscular fibre. They are manifestly surface elevations or markings and easily distinguished from primary ridges in the bone. An excellent example is seen in the oblique or vertical line on the tibia or the deltoid impression on the humerus. Sometimes the *secondary marking* is placed on *a primary ridge*, as in the linea aspera of the femur : at birth there is only the *primary* posterior border visible on the shaft, into which the aponeuroses are inserted, but later the bony growth invades these fibrous structures and thus the *secondary* linea aspera is added to the primary posterior border. Secondary markings are not always so apparent, and in some cases may be hardly visible, but they can always be felt by the finger, and it is a good thing for the student to acquire the habit of examining all bony surfaces with the finger, which, with a little training, will be found more useful than the eye for certain purposes.

Pressure and saving of space are at the bottom of tendon-formation, and thus in the body, where the mass of the packed muscles exceeds that of the skeleton, the muscles narrow to their attachments on the bones as a rule, *i.e.*, they have their muscle fibres ending partly or altogether in less bulky tendons : this is particularly noticeable where they pass over joints, to allow for movement without compression of other structures. Thus it comes about that secondary markings are present to indicate the attachments of most of the muscles, but the marking is only for the fibrous tissue, and therefore may not by any means correspond with the whole area taken up by the actual attachment. A purely muscular insertion must have practically the same area as the section of the muscle, whereas a tendon can thin down to a very small size, whence we reach the conclusion that the amount of secondary marking for any muscle will be inversely as the amount of pure muscle reaching the bone, or, in other words, *caeteris paribus*, the larger the area the smaller the amount of secondary marking in general. Used with knowledge and proper care these markings are of the utmost value in obtaining an accurate notion of attachments and consequently of relations of a bone, and the understanding eye can learn from them other facts about other structures : for instance, we know without further investigation that Tibialis posticus has one or two intramuscular tendons, from the fact that its area of fibular origin is crossed by one or two secondary "oblique lines," or the absence of a secondary capsular marking on the back of the neck of the femur might lead us to infer that there are no transverse fibres on the back of the capsule and presumably therefore only circular fibres.

But although muscles make no secondary markings, yet their areas of attachment can occasionally be made out by a careful and close inspection, revealing in a fine bone a slight change in what might be called the "surface texture" of the bone : this, however, is not a secondary marking in the sense used above. It is better to confine the use of the term to a spread of ossifying process and thus not to include such c   n es or the impressions produced by nerves or vessels, etc., in contact with the bhaeg

We might sum up this part of the subject by saying that the student should study the groups of muscles and other structures that lie against the bone and thus see how the different surfaces come into being, that he should see how these surfaces may have shape and level altered by the primary " build " of the bone, and should notice the secondary markings on the surfaces, remembering that they indicate fibrous attachment, and making use of them to mark out the accurate and detailed relations of the various aspects of the bone.

The key to many things that puzzle the student of anatomy is to be found in a study of the bones of a part, and it should be a rule for every dissector that he must undertake his dissection of a part with the skeleton of the region beside him, so that he may constantly refer from one to the other and back again ; in this way he will begin to understand how and why certain structures are found in certain situations, and to look at them from new points of view. It is a common error to imagine that any explanation of some anatomical fact or relation can only be sought in its development—it will repay the student to seek for mere mechanical and physical reasons for what he observes, for these can be brought in directly in all sorts and conditions of anatomical matters : then, if he wishes to pursue the subject further, he will have a *point d'appui* from which to direct his embryological or other researches. He should bear in mind that the disposition of the soft parts has its effect on the structure of the bones, and he should look for this as much as for the effect that the bones produce in the arrangement of the tissues. In this way he will obtain an idea of the skeleton as it exists in the body, with all its attachments and relations—a much more interesting study than that of the dry bones lying on the table, and one that will be of everyday use to him in future practice.

To study the skeleton, however, it is necessary to use the dry bones, but if the student makes a point of trying to find reasons for the various things he observes on the specimen, he will not only add to the interest of his occupation but will be surprised to find how much descriptive anatomy can be appreciated from such a study.

When considering bones or reading descriptions of them—such, for example, as are given in this book—the student should take care to have the specimen before him ; a description that is not verified is of little permanent value, and the reader ought to make a point of referring every statement made in the description to the test of observation on the specimen. For this reason he should be provided with several examples of the particular part of the skeleton he is studying, for comparison between many specimens is the best way of impressing the main or fundamental points of any structure on the mind, in spite of individual variations. Also he ought, whenever it is possible, to have the various appropriate dissections and preparations that enable him to appreciate the bones *in situ*, and constant reference to these will be of the utmost benefit. Finally, he should always see how much of the skeleton can be felt and studied on his own or some other living body.

The particular bone that he studies, and with which the other specimens are compared, should if possible be one from an adult male in which the secondary markings are well developed, but not excessively so.

In the account of the skeleton that follows each bone is at first described shortly, and a reader approaching such a study for the first time ought to go carefully through the description with the bones beside him, until he is acquainted with the general build and form of the particular bone and the names, position, and nature of all its principal parts, for he is not able to understand further details until he has mastered these elementary matters.

A longer consideration of things concerning the bone follows. Such a consideration could not be understood unless the reader had dissected the part concerned, so it is assumed that he has done so and is more or less familiar with it. It is in this part particularly of the description that the bones for comparison and the dissections ought to be used for constant reference, and it must not be forgotten that the descriptions are intended for application to the specimens and not to the drawings : the majority of figures in the following pages are introduced as guides to facilitate the recognition of the various parts on the actual skeleton. There can be no other object in having drawings of bones, and Anatomy cannot be learnt from pictures.

Chapter II

VERTEBRAL COLUMN

The human spinal column is made up of some thirty-three segments or *vertebræ* : the length of the column of bones is much increased by *intervertebral discs* of fibro-cartilage placed between the segments.

The column has two main functions—(*a*) support of the trunk, etc., and transmission of its weight to the pelvis and lower extremities ; (*b*) protection of the spinal cord and its membranes. The structure of the whole column accords with the necessities of these functions, and thus the vertebræ of which it is composed show their indi

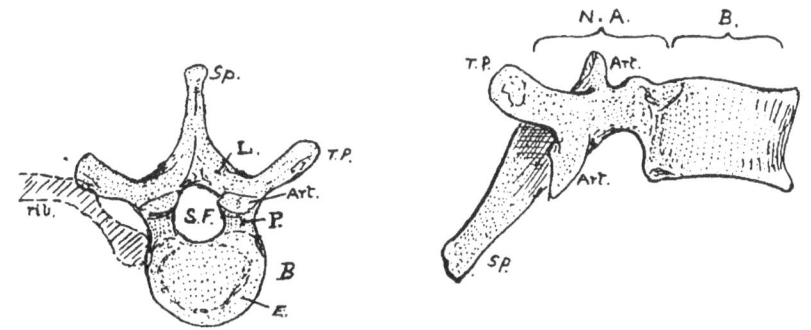

Fig. 5.—A " typical " vertebra (mid-dorsal). From above and from the side. *B.* body ; *N.A.* neural arch ; *S.F.* spinal foramen ; *Sp.* spinous process ; *T.P.* transverse process ; *L.* lamina ; *P.* pedicle ; *Art.* articular process. A rib is also indicated in position on one side. *E.* is the " epiphysial plate " or ring on the body.

vidual agreement with the general arrangement of the whole : they are modified in details according to their position, but they are all built on the same general principle.

This principle may be followed on a vertebra taken from the centre of the column and used as an illustration of the positions of the chief parts of a " typical " vertebra.

Typical Vertebra (Fig. 5).—In front a strong *body* carries and transmits weight, while the *neural arch* behind this covers in the spinal cord : in conformity with this we find the strong pads of the intervertebral discs placed between the bodies of contiguous vertebræ, whereas the arches are connected by ligaments and tend to overlap one another.

In the articulated column the successive arches and ligaments, with the backs of the bodies, enclose a *neural* or *spinal canal* for the cord and membranes, and the portion of the canal that is enclosed in the neural arch of each separate bone constitutes the *spinal foramen* of that vertebra.

The neural arch has *spinous* and *transverse* processes projecting from its back and sides respectively, for the attachment of muscles and ligaments : the transverse processes also help to support ribs in their attachment to the column. Each half of the neural arch is divided by the position of the transverse process into a posterior part between this process and the spine, termed the *lamina*, and a portion in front between the process and the body, which constitutes the *pedicle*. These two parts are under somewhat different conditions: the pedicles are thick and somewhat rounded bony bars transmitting to the body the weight of the body-wall carried by the rib to the transverse process, whereas the laminæ are not concerned in this, but are flattened and sloped to cover in the spinal canal. It can be seen, however, that thickened portions extend in the laminæ between the spines and the pedicles, better marked as we proceed down the column; these bear the strain of muscles acting on the spines. The arch articulates by means of *articular processes* with the arches above and below it, so that it carries two pairs of such processes, *superior* and *inferior*. They are placed in a general way above and below the base of the transverse processes—that is, the upper ones are always at the junction of the processes and pedicles, but the lower are usually further back and even on the lower borders of the laminæ.

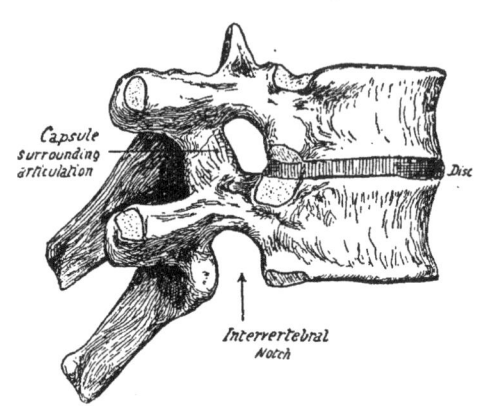

FIG. 6.—Two dorsal vertebræ articulated to show the disc in position and the formation of the intervertebral foramen.

Turning now to the pedicles, we observe that in vertical depth they are not equal to the bodies to which they are attached. In this way "notches" lie above and below the pedicles, and, because the pedicle is attached nearer the top than the bottom of the body, the inferior *intervertebral notch* is much the deeper of the two. The notches are bounded behind by the articular processes, and when the vertebræ are in apposition it is seen (Fig. 6) that the upper and lower notches of contiguous vertebræ are combined to form an *intervertebral foramen*, which is bounded above and below by pedicles, behind by the joined articular processes, and in front by the intervertebral disc and the lower part of the body above it. The spinal nerves escape through the intervertebral foramina, and small branches of the segmental arteries enter the canal through them.

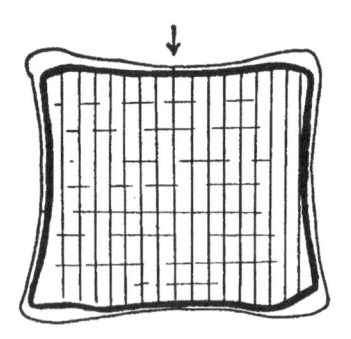

FIG. 7.—A scheme to show the direction of the lamellæ of cancellous bone in the body of a vertebra; the main direction is vertical, in the line of pressure as indicated by the arrow, but weaker lamellæ bind these together transversely.

The body, being the weight-carrying part of the vertebra, is built accordingly. A certain amount of elasticity in the column is obtained by the formation of the body almost entirely from cancellous tissue, and the strongest lamellæ of this run in the direction of pressure from above downwards, being held together by weaker horizontal lamellæ (Fig. 7).

The **intervertebral discs** add greatly to the elasticity of the column, at the same time connecting the bodies strongly together, so that very little movement is possible between adjacent bones, although considerable power of motion of the column as a whole results from the presence of a number of segments. Each disc consists of a peripheral *annulus fibrosus* of fibrous tissue more or less concentrically arranged, surrounding a softer and more elastic *nucleus pulposus*. The disc is fastened to layers of hyaline cartilage which cover the upper and lower surfaces of the vertebral bodies : the hyaline cartilage ossifies in part as an epiphysis to the body, so that these surfaces of the body present an *epiphysial plate* in the shape of an incomplete ring (Fig. 5).

In the foregoing general account of a vertebra the terms " arch " and " body " have been used in their ordinary descriptive sense. It is necessary, however, to say that the term " centrum " is frequently used in the same sense as " body," but that it does not mean quite the same, being really a more strictly accurate term for nearly the whole of the " body." To understand the difference it is necessary to go a little way into the development of a vertebra. The bone is laid down in cartilage which afterwards ossifies, and looking at it in its early stages it can be appreciated at once that there are two main parts in it—the " centrum " and the " neural arch " (in two pieces). But when the fusion occurs the distinction is not so evident, yet it can be understood if it is remembered that the rib is only carried by the neural arch and does not articulate with the centrum. This being so, the junction of the two parts—the *neuro-central suture*, as it is termed—must lie in front of the articulation with the head of the rib, and yet the part that carries this head belongs in a varying degree to the descriptive " body," as will be seen when dealing with the several vertebræ.

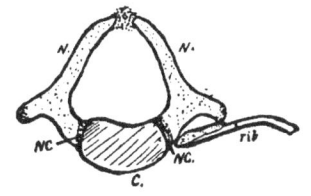

FIG. 8.—A dorsal vertebra from a new-born child. *N.N.* are the two halves of the neural arch, joining the centrum, *C.*, at the neuro-central suture *N.C.* Observe that the rib only articulates with *N.* yet the front portion of *N.* may be included in the descriptive ' body."

The difference between the values of the terms is not great from the point of view of description, but in the interests of accuracy it should be borne in mind, and the word " centrum " ought only to be used when the morphological entity which lies between the neuro-central sutures is definitely meant ; on the other hand, the word " body " is conveniently applied to the whole mass that lies in front of the free pedicles.

Regions of the Column.

The vertebral column is divided into regions in which the vertebræ show characteristic modifications. The most nearly typical region is that which carries the movable ribs, and is called *dorsal :* the dorsal vertebræ are twelve in number.

Above this, in the neck and supporting the skull, are seven *cervical* vertebræ, while between the dorsal region and the pelvis are five *lumbar* segments. Below these are five *sacral* vertebræ, fused together to form the *sacrum* and firmly fixed between the bony side-walls of the pelvis, and the sacrum rapidly tapers below to give place to the degenerated *coccyx :* the coccygeal vertebræ are usually four in number. The presacral segments are often termed the " true vertebræ," the sacrum and coccyx being then called " false vertebræ." The column can thus be analysed as follows

    7 Cervical V.⎫
   12 Dorsal   V.⎬24 presacral V. + 5 sacral V. + 4 coccygeal V. = 33 V.
    5 Lumbar V.⎭

Of the *presacral* vertebræ, the dorsal form less than half, the lumbar less than a third, and the cervical a fifth or more in the recent column.

Anatomy of Skeleton

This is. the usual division, but occasionally exceptions may be encountered: thus there may be twenty-three or twenty-five presacral vertebræ, consequent on forward or backward shifting of the pelvis, and this may be correlated with change in the number of sacral vertebræ, or the sacral may be altered in number without an alteration in the presacral number. Again, the number of dorsal vertebræ may appear to be altered by the occurrence of a rib on the last cervical segment, or on the first lumbar, or by partial suppression of the first or last of its own series, but these are not as a rule really alterations in number, but only in appearance. The coccygeal vertebræ are inconstant in number, as they are in a state of degeneration, representing

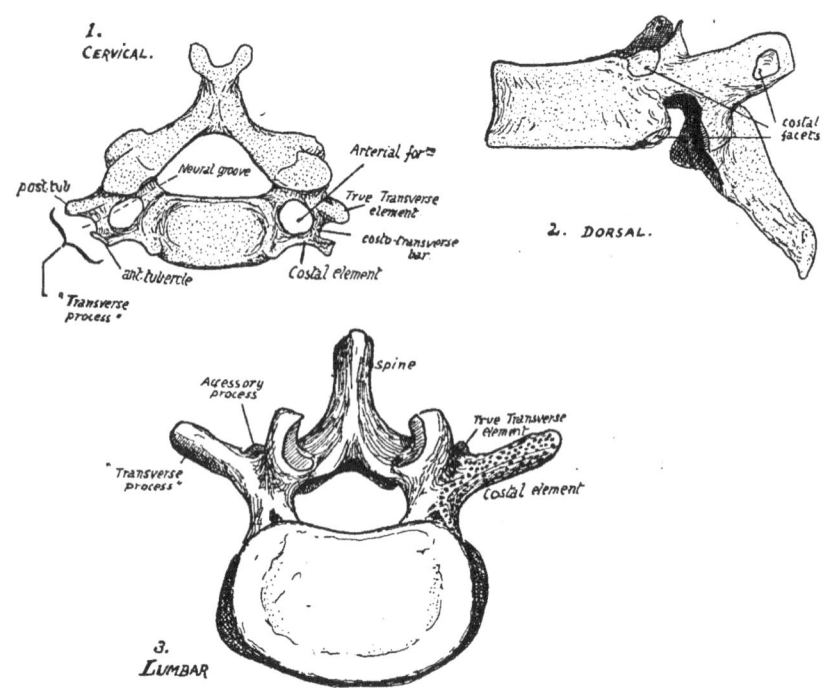

FIG. 9.—Vertebræ taken from the three presacral regions to show the characters of these regions. Cervical and lumbar vertebræ seen from above, dorsal from the side.

the bony skeleton of the tail: in the embryo there are six of these rudiments to be found.

In the various regions the true vertebræ exist under different physical conditions and accordingly exhibit distinguishing modifications, and the distinction can be even carried so far as to enable the observer to recognise and place the individual vertebræ in their proper order in most cases.

The different groups can be known at once by their distinguishing characters. Thus the cervical vertebræ possess an arterial *foramen* in their "transverse processes," the dorsal group carry free ribs and therefore present *costal facets* on their bodies and (in the majority) on their transverse processes, while the lumbar vertebræ have neither

Vertebral Column

foramen nor facet, but are distinguished by the massive size of their bodies and arches and the thickness and square cut of their spines. In the figure (Fig. 9) these differences are exemplified by drawings of vertebræ from the middle of their respective regions. At the same time it is seen that a *costal element* must be considered as entering into the composition of most of the vertebræ. This forms a free rib in the dorsal series, but is included in the so-called transverse process in the other regions, and in the sacrum. In the cervical series the costal element forms the front wall of the *vertebrarterial foramen*, so that the " transverse process " really is a compound of a true transverse element and a costal element, separated by the foramen and joined by a *costo-transverse bar*. In the dorsal region the transverse process is a true one, and not compound, but in the lumbar region it is really a costal process, the true transverse element being reduced and its tip represented by the small *accessory process* that is visible behind the base of the costal element.

In the sacrum the different morphological constituents of the bone can also be recognised, as will be pointed out at a later stage.

Before examining the special characters of the individual vertebræ the articulated column should be considered as a whole. The discs account for more than one-fifth of the total length of the column. A striking character of the **complete column** is the presence of curves : there are four of these in the sagittal plane, convex forward in the cervical and lumbar regions, concave forward in the dorsal and sacral. The dorsal curve is primary, and can be looked on as the persisting curve of the embryonic axis. The sacral curve is referred to as primary also. The other two curves are secondary and compensatory, the cervical bend appearing towards the end of fœtal life to enable the head to be raised from the thorax, while the lumbar curve results from the assumption of the upright sitting posture and extension of the legs, therefore appearing after birth. The sharp bend between the lumbar and sacral regions is a modification in the general curve that begins to appear in the second half of intrauterine life.

The primary curves have corresponding differences between the front and back heights of the bodies of their respective vertebræ : in the acquired and compensatory curves the bend is accounted for mainly by corresponding differences in height in the intervertebral discs. This is specially marked in the lumbar convexity, where the curve is altogether due to the ventral thickness of the discs in lower races : in higher races the sum of the front measurements of the bones is higher than that of the posterior surface of the bodies, so that the curve is not altogether due to the discs. The lumbo-sacral disc is about two and a half times as thick in front as behind.

The greater depth of the fronts of the lumbar bodies in Europeans is really confined to the lower two or three members of the series. In lower races the bodies, except the last, are shallower in front.

The curves of the column become more accentuated up to puberty, and the dorsal curve appears to increase somewhat in length and depth after general growth has ceased.

The most anterior part of the cervical convexity is the front of the body of the fourth cervical vertebra, and the curve ends below about the second dorsal : the body of the seventh or eighth dorsal usually forms the point of greatest posterior projection in the dorsal curve, which ends usually below the first lumbar vertebra.

Compare male and female columns, and observe that the sacrum in the woman is turned back to a greater extent. This would make a more prominent lumbo-sacral angle, but it is modified by the lumbar curve becoming more pronounced : thus in women the fourth lumbar is usually more prominent forward than in men.

The presence of curves in the column above the sacrum adds to its elasticity, while the number of curves gives it a higher resistance to weight than would be afforded by a single curve; weight is transmitted to the top of the sacrum and from thence to the innominate bones, so that the sacral curve is merely an adaptation to the contents of the pelvis, not concerned in weight-transmission, and differs in the sexes.

A slight lateral curve is sometimes seen in the line of bodies, concave to the left in the dorsal region and probably due to use of the right arm, with some compensatory deviation above and below.

The line of the spines shows frequent irregularities without any deviation in the line of the weight-bearing bodies.

On each side of the line of spines the *vertebral groove* runs down the length of the column, floored by the laminæ and backs of the articular and transverse processes · it is occupied by the deep and intermediate layers of the post-vertebral muscles, the superficial mass lying on these and extending further out.

Observe the width of the transverse processes. More or less regular in the cervical region, but great in the first and last members of the series, it decreases from above down in the dorsal vertebræ, and increases again in the lumbar to the third or fourth.

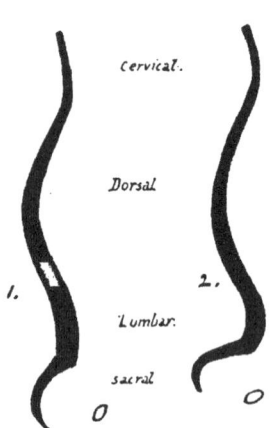

The number of vertebræ permits a considerable amount of movement in the complete column without calling for more than a very small amount between the individual segments, thus avoiding the weakening that would result if the segments were fewer and longer. The mechanical stability of the column under all normal movements is assured by making the axis of rotation pass through the bodies and not through the arches, so that the bodies are not displaced from one another in movement. To this end the discs join the bodies together, and the thicker the disc the greater the amount of movement between the bodies. Thus we find that the discs are thickest in the lumbar and cervical regions, where movement is freest, and shallow in the dorsal series, where naturally the amount of motion is much restricted.

FIG. 10.—Schemes to show the differences between (1) male and (2) female curves. Observe the relatively greater prominence forward of the fourth lumbar segment in the female column.

Where the strong and more flexible lumbar column joins the weaker and more restricted dorsal spine, *i.e.*, about the dorso-lumbar region, is the weakest part of the column surgically.

In addition to the discs, the bodies are attached to each other and to the discs by the anterior common ligament, a strong band increasing in width and thickness from above downwards and fastened to the lips of the bodies and the front surfaces of the discs. The posterior common ligament is attached in the same way to the back aspects of the bodies and discs, but is a much weaker and narrower band that decreases in size from above but widens opposite each disc, consequently having a denticulated appearance.

On the sides of the bodies are other lateral bands or sheets of fibrous tissue, thinner than the anterior ligament, and connected with the attachment of muscles to the column. Thus the whole series of bodies may be said to form a flexible column with the discs, ensheathed in a fibrous covering; the fibrous sheath does not interfere with movements, being on the discs, but adds to the strength of the whole.

It is evident that a forward bending of a part of the column will lead to separation between the corresponding laminæ and spines, and thus we find that the laminæ are connected with those above and below by *interlaminar ligaments* (*ligamenta subflava*) composed of almost pure elastic tissue ; these are prolonged back between the spines as *interspinous ligaments*, best developed in the lumbar region and only represented higher up by a fibrous or areolar web that makes markings on the bones. At the top of the column, where the movements of the head demand a greater range of motion between the first two vertebræ and the skull, the interlaminar ligaments become modified and represented by thickened areolar tissue : in this case the functions of ligaments are performed by the short muscles that connect the top of the column and the skull. The interspinous ligaments below the level of the 2nd c.v. may be looked on as simply the thickened areolar sheaths of the interspinous muscles, which also act as extensible ligaments limiting the excursus of the spines.

In the same way the separation of the transverse processes in lateral flexion of the column is met by intertransverse muscles modified partly into tendinous fibres ; but true intertransverse ligaments other than these are not easily found, save in a part of the dorsal region in which some fibres of the costo-transverse series appear to be also in part intertransverse.

Looking again at the back of the column, it is seen that the imbricated laminæ and spines hide the contained spinal canal to such an extent that it is usually only possible to see into it easily in the lower two or three lumbar intervals. It is evident, therefore, that here the ligamenta subflava are most visible, completing the covering of the canal, and this region is chosen of necessity for the operation of inserting a needle into the membranes of the cord to draw off cerebro-spinal fluid : the proceeding is helped by forward flexion of the spine, which is a movement most easily carried out in this region.

The movements between the parts of the different vertebræ may be shortly summed up : the bodies, being in and round the centre of motion, move least, and the various processes move more, in proportion as they are far away from the body in the plane of the particular movement.

The spinal cord and its membranes, lying deep to the neural arch which moves away from or to its neighbours in flexion or extension, obtains the utmost possible amount of freedom from movement by having no attachment to the arches but being held to the back of the column of bodies, both by the irregular and slight adhesion of the dura mater to the posterior common ligament and by the attachment of its nerves to the pedicles and discs through their sheaths of dura mater.

CERVICAL VERTEBRÆ.

The drawing of a cervical vertebra in Fig. 9 would represent, with minor modifications, the general appearance of any member of this series from the third to the sixth inclusive ; but the type has been distinctly altered in the upper and lower vertebræ of the region.

In the first two vertebræ the alteration is remarkable, in association with the necessity for providing free movement of the head on the column, but it is slighter in the seventh segment and connected with the change from the physical conditions of the cervical to those of the dorsal region.

The first vertebra is termed the *Atlas*, because it supports the globe of the skull : the second is called the *Axis*, because it forms the pivot round which the atlas turns and carries the skull.

Anatomy of Skeleton

Cervical Characters (third to sixth).—The characters of a typical cervical vertebra are :—*Body* small, shallow from above downwards, concave transversely above and sagittally below, so that it overlaps the next body in front, broader transversely than antero-posteriorly in proportion 3 : 2. *Pedicles* rounded and near mid-level of the body, so that upper and lower intervertebral notches are nearly equally deep. *Laminæ* flat, long, meeting nearly at a right angle, and overlapping those below. *Transverse process* contains (vertebrarterial) foramen for the vertebral vessels, separating true transverse and costal elements, these being joined by a costo-transverse bar, and beyond this projecting as *posterior* and *anterior tubercles* respectively : the tubercles separated by a *neural groove* for issuing spinal nerve lying on the " transverse process." *Articular* processes on strong columns of bone : upper looks backwards and rather upwards, lower looks forwards and rather downwards. *Spinous process* bifid, ending in two angled projections. *Spinal foramen* large and more or less triangular in shape.

SEVENTH CERVICAL VERTEBRA.

This differs from those just described mainly in its long, strong, and non-bifid spine, and in the smallness of its arterial foramen. The condition of the spine is associated with the attachment of the ligamentum nuchæ and certain strong muscles, and it can be felt in the living neck, whence this segment is sometimes termed " vertebra prominens." The costo-transverse foramen is small because it does not transmit the vertebral artery.

Other modifications are seen in the smallness of the anterior and marked size of the posterior tubercles on the " transverse process," with corresponding differences in the constituent elements : occasionally, however, the costal element may be enlarged and become free, forming a *cervical rib*. Occasionally, on the other hand, it may be absent. The body may sometimes present a mark, though hardly a facet, showing the situation of the head of the first rib, or the attachment of its ligaments on the lower and lateral margin.

ATLAS (Fig. 11).

Easily recognised : it has no body attached to it, this part having joined the Axis. It presents an *anterior arch*, and a *posterior arch*, with a thick *lateral mass* on each side joining these arches : lateral mass carries the upper and lower articular processes, of which the former is markedly concave for reception of occipital condyle and shows a tendency to be subdivided at junction of the two arches.

A prominent median *anterior tubercle* is on the front of the anterior arch, and on its back aspect a smooth articular facet for the odontoid process of the axis. A small rough *posterior tubercle* represents the spinous process on the posterior arch.

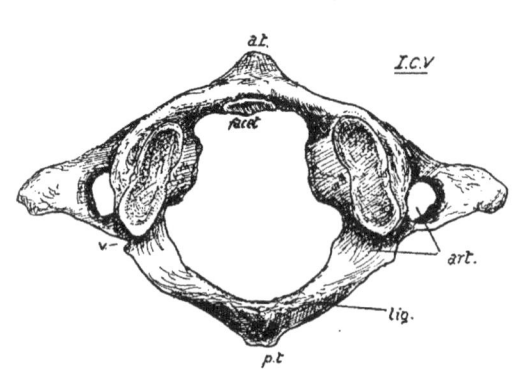

FIG. 11.—First cervical vertebra, from above : *art*. groove and foramen for vertebral artery ; *a.t*. anterior tubercle ; *p.t*. posterior tubercle ; *lig*. ridge for posterior occipito-atloid ligaments ; *V*. notch for small vein.

Posterior arch shows on its upper surface a broad groove for the vertebral artery and suboccipital nerve, just behind the articular mass and leading from the foramen in the transverse process ; behind the groove, a ridge for the posterior occipito-atlantoid ligament. Transverse process projects from the lateral mass and has a large foramen : its front bar may show a small terminal tubercle, but the end of the process corresponds with the *posterior* tubercles of the lower vertebræ. The large hole enclosed by the arches has a front part into which the odontoid process projects from below, and a larger posterior part which is the proper spinal foramen. Lower articular processes, being broader than the upper ones, encroach somewhat on the front part of the space by their inner margins, and on these are vascular foramina and tuberculated roughnesses for the attachment of the transverse ligament that passes behind the odontoid process, and of the accessory alto-axoid ligament.

AXIS (Fig. 12).

Remarkable owing to the presence of the *odontoid process*, or *dens*, which projects upwards : this is really the centrum of the Atlas, separated from that bone and fixed to the upper surface of body of Axis, and it articulates by its front surface with the facet on the anterior arch of the Atlas. Dens has an extremity roughened by attachment of occipito-odontoid ligaments, and a slight constriction behind for the transverse ligament, a bursa intervening ; the true *body of the Axis* is below the base of the process.

Upper articular surfaces look upwards and outwards and are placed a little distance outside the base of odontoid, but the lower processes are further back, corresponding with the position of the process in the lower cervical segments : the second nerve makes its exit *behind* the upper process, and the third *in front* of the lower, as in the vertebræ below the Axis.

Transverse process short, ending in a single (posterior) tubercle : arterial foramen is a canal which, passing upwards, reaches under part of upper articular mass and turns outwards sharply. Internal to the arterial canal a thick column of bone transmits the weight of the Atlas and head from the upper articular process to the body.

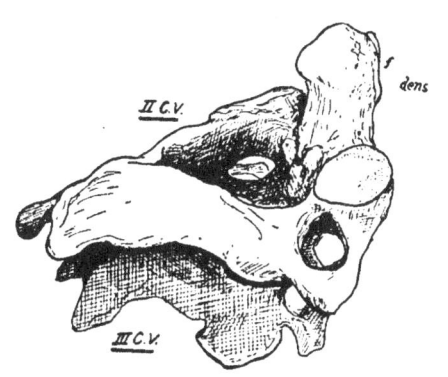

FIG. 12.—The second cervical vertebra shown resting on the third vertebra ; seen from the right and above. *f*, facet on the front of odontoid process. Observe that only the lower articular process is in line with those of the vertebræ below.

The laminæ are strong, and a broad, strong, elongated bifid spine is continued back from them.

The pedicles can be seen from below, and recognised as being in their usual positions, but their upper surfaces are hidden by the upper articular processes

The nodding movements of the skull (flexion and extension) take place between the Atlas and Occipital, while rotation is carried out between the Atlas and Axis. To permit and control these movements there are, in addition to the capsules of the articular processes, two important systems of ligaments. The first is a strong band thrown across from one lateral mass of the Atlas to the other, behind the odontoid

Anatomy of Skeleton

process, thus holding the process up against the front arch of the Atlas, and completing a collar round the process that grasps it tightly, with intervening bursæ: evidently with such a band, the Atlas cannot move away from the odontoid, but can rotate round it as a centre. This *transverse ligament* is connected by *upper and lower crura* of vertical fibres with the occiput and back of the body of the Axis respectively, but these bands are comparatively weak and do not seem to be of much importance in limiting movement: the whole structure is sometimes referred to as the *cruciform ligament* (Fig. 13).

The second system consists of strong and thick bands connecting the upper end of the odontoid process with the margin of the foramen magnum in the occipital bone: there is one of these on each side, *lateral occipito-odontoid* or *check ligaments*, and a feeble median one, between the process and the front margin of the foramen, termed

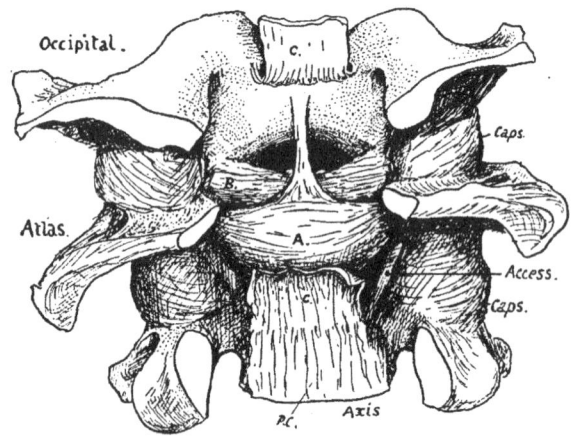

FIG. 13.—The ligaments of the occipito-atlo-axoid articulation exposed by removal of the neural arches and contents of the canal. The posterior common lig. is continuous with the membrana tectoria, which has been divided and turned back at C.C. to expose the cruciform lig. A. This covers the back of the odontoid, but the check ligaments B. can be seen passing to the upper end of the process from the margin of the foramen magnum. *Access.* is the accessory atlo-axoid lig. The median "suspensory" lig. is hidden by the upper crus of the cruciform lig.

the *middle odontoid* or suspensory ligament. The middle band has no mechanical value, being only the continuation into the skull of the perinotochordal tissues, but the lateral bands are of prime importance. Flexion and extension between the Atlas and Axis is effectively prevented by the close grip obtained by the anterior arch and transverse ligament on the odontoid process. On the other hand, rotation between the skull and the Atlas is checked as a result of the deep concavity of the upper articular surfaces of the Atlas: any attempt at rotation of the Occipital tends at once to lift the bone up on to the raised front and back margins of these surfaces and thus immediately tightens the check ligaments and anterior occipito-atloid, and is moreover resisted by all the short muscles passing to the neighbouring skull base from the column.

Thus, for practical purposes, we may look on the skull and Atlas as one mass rotating on the Axis, and on the Atlas and Axis as one mass on which the skull can

Vertebral Column

perform nodding movements: so the skull and the Axis are on opposite sides in both these forms of movement, and the occipito-odontoidal ligaments become the most important bands associated with limitations of these movements.

Flexion of the head is checked by tightening of these ligaments: this is seen in Fig. 14, A, where it is shown that when the skull is brought down from the level AA to BB the condyle rotates in the articular cup of the Atlas, and the upper attachment of the check ligament is carried backwards and upwards (from a to b) on the condyle, so tightening the band and checking the movement.

The opposite movement is checked by the tightening of the anterior occipito-atloid ligament (aoa) and the front fibres of the articular capsule.

Rotation is also limited by the lateral occipito-odontoid ligaments. These are shown in Fig. 13, and consist of a thick mass of fibres on each side, the posterior and deeper ones showing a tendency to pass forward under the more superficial ones; some fibres, not shown, also run across without any definite connection with the

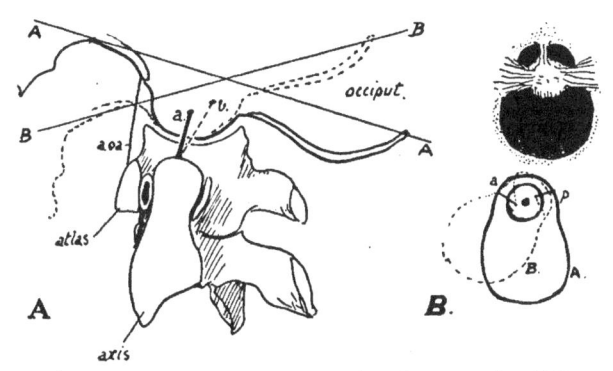

FIG. 14.—Schemes to illustrate the action of the lateral occipito-odontoid ligaments in (A) flexion and (B) rotation of the head.

process. The area of insertion into the odontoid is on its postero-lateral aspect, so that the deeper fibres reach the bone in front of the position of the centre of rotation, while the "anterior" and superficial fibres, crossing these, reach the process behind this centre. Rotation takes place between the Atlas and Axis, but for this movement the Atlas and Occipital may be looked on as one mass, so that the ligaments check the rotation as seen in Fig. 14, B.

Here it is shown that when rotation occurs, as for instance from the position A to B, that is to the right, the right posterior fibres p are carried back on the margin of the foramen magnum and so stretched, and the left anterior fibres are rendered taut by being carried forward, both of these sets of fibres being attached to the odontoid in a relation to the centre of rotation opposite to that of the direction in which they are carried by the occipital bone.

A certain amount of "screw" also occurs. A glance at the surfaces shows that, as the atlas turns, one of its surfaces must descend while the other ascends. The descent leads to slackening of the ligament on that side, so giving more range. Any decrease in range on the other side is compensated by slight tilting of the head to the side of the ascent, a movement possible to some degree at the occipital joint.

Anatomy of Skeleton

The cruciate and occipito-odontoid ligaments are covered behind by the *membrana tectoria* or *occipito-axoid ligament*, a broad thin band that extends to the basilar groove on the occipital bone from the back of the body of the Axis: it has the spinal dura mater in contact with it and attached to it, and it becomes continuous with the posterior common ligament on the Axis. There are some variable lateral fibres of this sheet, attached to the occipital behind the anterior condylar foramina.

The capsules of the joints on the two first vertebræ are loose, and do not interfere with the movements, save in the case of the occipito-atloid capsule and extension of the head. There is, however, on the postero-internal aspect of the atlo-axoid joint a strong *accessory* band that might conceivably help to limit rotation toward the opposite side. The band is sometimes considered to be part of the lateral sheet of the membrana tectoria.

Detailed Consideration of Cervical Vertebræ.

The anterior common ligament covers the bodies centrally, and on each side of this are the prevertebral muscles. Take one of the middle members of the series and observe that the central *ligamentous* area is raised and flattened, while the lateral *muscular* areas are hollow and extend on to the fronts of the transverse (costal) processes. Also notice that the size relation between the areas is not constant : as we ascend the column the ligamentous area decreases in width, with a corresponding increase in the muscular hollows. This indicates the increasing breadth of the ligament from above downwards and shows plainly how it is only a narrow cord when it reaches the anterior tubercle of the Atlas : because this tubercle is made by the ligament, it has its tip directed downwards.

The observant eye can easily read these facts on the column, to which the ligament is attached strongly in the neighbourhood of the discs, but loosely on the remaining front of the body.

The muscular area, partly on body and mostly on the costal processes, presents roughnesses for the attachment of the vertical portion of Longus colli : the oblique portions of the muscle are inserted into the costal processes, and these usually show ridges for the tendons of the lower oblique part on the front surfaces of the fifth and sixth vertebræ (Fig. 15A). The large upper hollows are occupied by the Rectus anticus major in addition to the Longus colli, and the origin of this muscle by tendon is indicated on the anterior tubercles of 3, 4, 5, and 6 C. V. The Atlas, viewed from the front, presents the largest muscular hollow, altogether occupied by R. anticus major lying on the bone, while the Longus colli comes to its attachment on each side of the base of the anterior tubercle : it has already been seen that the direction of the anterior tubercle is determined by the attachment of the anterior common ligament, but a roughened area on its upper part shows where the anterior occipito-atlantoid ligament continues the line of the common ligament to the skull.

The third, fourth, fifth, and sixth vertebræ are the only ones possessing noticeable anterior tubercles as terminations of their costal processes : these are the only segments that have muscles attached here, and it seems not unlikely that not only the presence of the tubercles, but also their direction, is determined by the muscular attachment. Thus the lower oblique fibres of Longus colli, and the lower strong fibres of Scalenus anticus, are inserted into the anterior tubercles of 5 and 6 C. V., and their downward pull, only antagonised by the lower fibres of R. anticus major, may account for the fact that these tubercles are at a lower level with regard to the transverse element than are those of 3 and 4 C. V. In the latter vertebræ the upward pull is by stronger fibres of the anterior Rectus and the upper oblique fibres of Longus colli, only

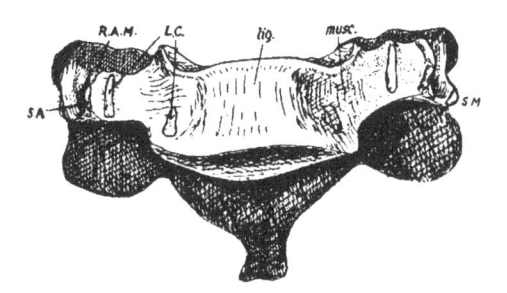

Fig. 15A.—Sixth cervical vertebra from the front, showing the raised ligamentous area (*lig.*) and depressed muscular area (*musc*). Muscle attachments: *SA*, Scalenus anticus; *RAM*, Rect. ant. major; *LC*, Longus colli; *SM*, Scalenus medius.

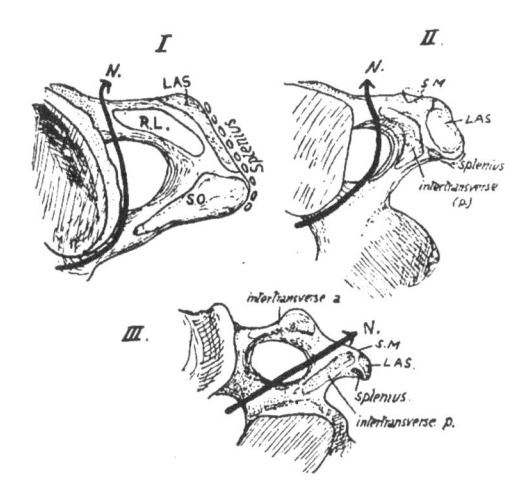

Fig. 15B.—Upper aspects of transverse processes of first three cervical vertebræ, showing their muscular attachments and the course of the anterior primary nerve-division in relation to these (*N.*). *RL*, Rectus lateralis; *LAS*, Lev. ang. scapulæ; *SO*, Sup. oblique; *SM*, Scalenus medius. Observe that in *III.* there are two intertransverse muscles with the nerve between them, but in the first two vertebræ only one (*post.*) is present (*RL* in *I.*).

antagonised by weak upper fibres of Scalenus anticus. The downward direction of the brachial nerves may also affect the fifth and sixth processes. The Rectus and the upper oblique fibres are frequently fused at their attachments, so that no separate markings are found on the bone; but occasionally separate markings are seen for the oblique fibres, internal to the tubercle. Observe that the posterior tubercles project beyond the level of the articular masses, whereas the anterior ones do not do so.

The lateral view of the cervical column discovers differences in the transverse processes below the Axis. They are sufficiently striking from the surface (Fig. 16) and can be summarised as follows. Third vertebra has its costal process small and on an altogether higher level than the true transverse element, a very oblique costo-transverse bar, and a very narrow and shallow neural groove.

In the fourth the obliquity is not so marked and the groove is deeper, with a higher anterior tubercle.

In the fifth the costo-transverse bar is more nearly horizontal, and the groove is wider and deeper in correspondence with the larger size of the anterior division of the fifth nerve.

In the sixth the groove is extremely wide, a character that distinguishes this from the other vertebræ, especially when taken in conjunction with its large anterior tubercle, ridged in front. This tubercle is occasionally much increased in size, forming the *carotid tubercle* or *Chassaignac's tubercle*.

In the seventh the anterior element is very small and may be incomplete, and has no muscle attachment: the posterior element, on the other hand, is comparatively massive, and roughened at its extremity for the lower scalene and levator costæ fibres. It can be appreciated that the small size of the costal element affords room for the vertebral artery to run up to the sixth transverse process: the articulated bones will show that the artery must lie just in front of the seventh costal element and the seventh nerve that runs behind it, and therefore immediately lateral to the Longus colli.

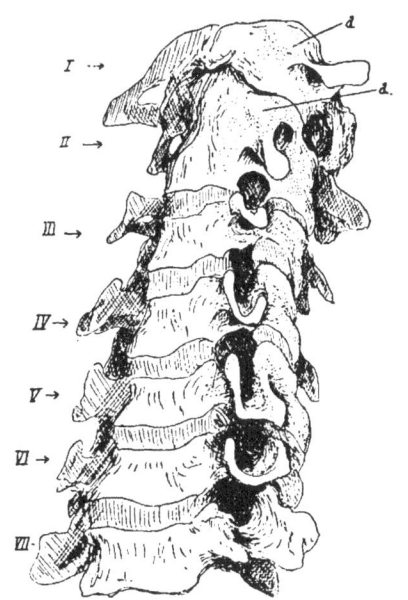

FIG. 16.—Cervical vertebræ articulated to show the transverse processes for comparison.

Examine the neural groove in any of the middle or lower segments, and notice that the issuing nerve must pass behind the vertebral artery in its course. This is the anterior primary division of the nerve corresponding in number * with the vertebra: the posterior primary divisions arise in the intervertebral foramen and turn back over the transverse element and round the articular process, making a slight groove on the third segment (see Fig. 17).

The first two cervical nerves differ from all other spinal nerves in issuing *behind* the articular masses. The explanation is that the articular masses concerned are

* Each cervical nerve is above its vertebra: the eighth is therefore above 1 D.V. and below 7 C.V. So the first dorsal nerve is below its vertebra, as are the remaining spinal nerves.

morphologically different from those lower down. If the upper surfaces of the bodies of the middle and lower cervical vertebræ are inspected, a prominent *neuro-central* lip is seen standing up outside the junction of the costal element and neural arch with the centrum : in the recent state a small synovial cavity lies between this lip and the bone above, outside and behind the intervertebral disc.

From its position this cavity is in front of the issuing nerve : the cavities increase in size in the upper part of the column, and their bases become the functional articular processes on the Atlas and top of Axis, so that the issuing nerves lie behind them. These small cavities are evidently in series with the costo-vertebral joints in the dorsal region ; inspection of the direction of the bony fibres in the surface texture at the junction of costal element with body suggests that they are ossified representatives of the stellate ligaments of the rib heads.

The Vertebrarterial (costo-vertebral, costo-transverse) foramen transmits the artery with its accompanying veins and a sympathetic plexus from the lower cervical ganglion ; it is subdivided in the fifth and sixth segments, the small posterior part being for a vein, and the subdivision may be present in the seventh and is occasionally suggested by bony spicules in the third and fourth. The seventh foramen only transmits a vein and a filament (grey ramus) from the ganglion running up to join the seventh nerve. The ganglion is situated beside the vertebral artery below the level of the seventh costal process, so that its filament to the eighth nerve reaches it directly, but that to the seventh must ascend through the foramen.

At the top of the column the line of the arterial channel turns sharply outwards in the Axis and is thus able to reach the foramen in the Atlas : observe that the transverse process of the Atlas extends further out than any other in the series, the next in range being that of the seventh.

The laminæ increase in depth from the third downwards. They tend to overlap each other, so that the Ligamenta subflava, which are attached to the upper borders of each lamina, pass up deep to the lower borders of the laminæ above and reach their anterior surface : their attachment here is marked by a transverse ridge on the neural surface of the laminæ. The superficial or posterior surface is roughened by marks of tendinous insertions of Multifidus spinæ : these are near the lower edges and the spines, on to which they extend. Outside the laminæ are tuberculated areas on the backs of the inferior articular processes of the lower four segments : the upper fibres of Multifidus arise here, and outside them is the origin of the more superficial next layers, Semispinalis, Complexus, and Trachelo-mastoid. The compound nature of these markings on the articular processes is sometimes apparent, particularly on the fifth vertebra, where a strong fasciculus of Multifidus frequently makes the inner portion of the marking into a more or less prominent tubercle (Fig. 17).

The insertions of the still more superficial Cervicalis ascendens and Transversus cervicis are on to the backs of the transverse processes, further out and further forward.

The divergent points of the spinous processes afford insertion to Semispinalis colli, and the small prominences on their inner sides, that give them an angled appearance, mark the attachment of the bilaminar ligamentum nuchæ. A median ridge on the under aspect of the spine shows the attachment of the weak interspinous ligament, corresponding with the median upper ridge on the bone below.

The seventh spine is non-bifid, and this is frequently the case also in the sixth—and above this—a condition normal in lower races ; but even when this occurs there is a distinct suggestion of its double nature, which is not found in the last cervical spine.

The three-sided spinal canal increases in size from the Axis to the fifth and then decreases, so corresponding with the position of the thickest part of the cervical enlargement of the cord. The front wall is the back of the body, but this is covered

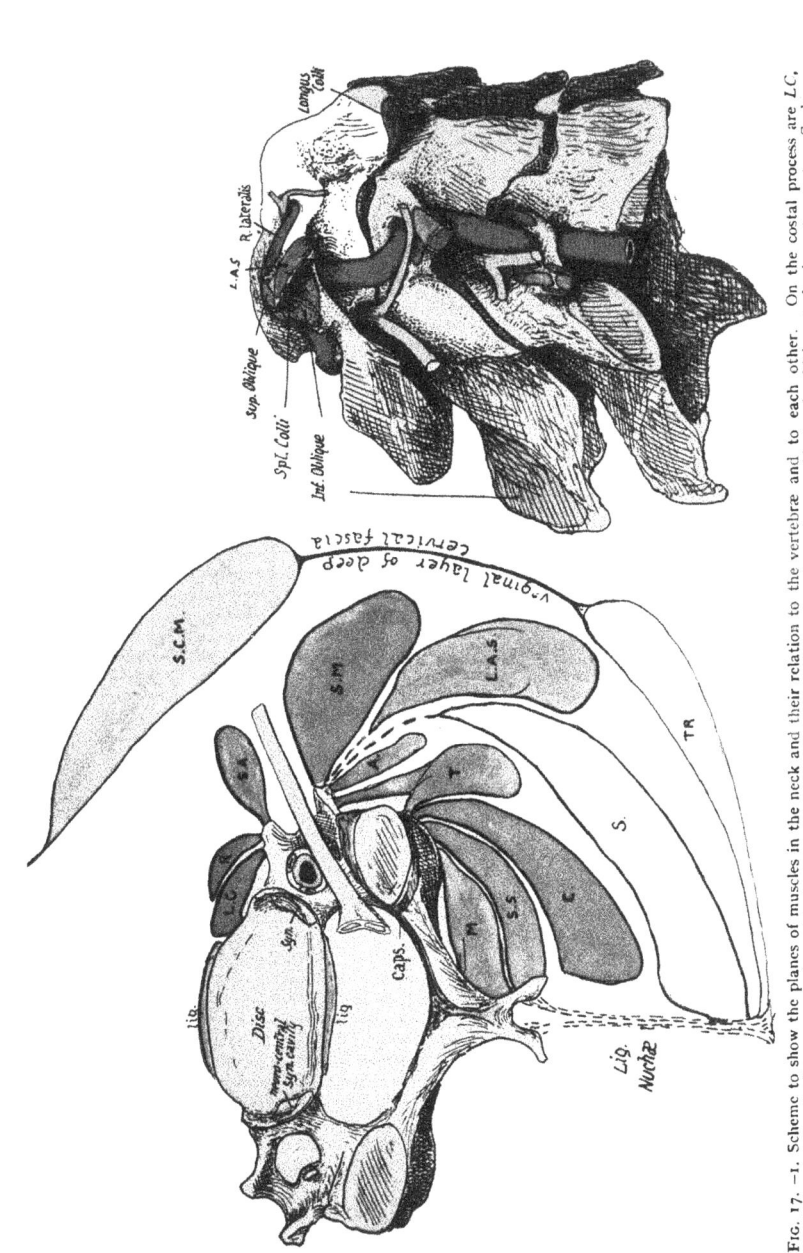

FIG. 17.—1. Scheme to show the planes of muscles in the neck and their relation to the vertebræ and to each other. On the costal process are *LC*, Longus colli; *R*, Rect. ant. major; *SA*, Scal. anticus; these separate the process from the carotid sheath which rests by its outer part on Scalenus anticus or, above the level of this muscle, on the issuing nerves and Scal. medius. Scalenus medius (*SM*) and Lev. ang. scapulæ (*LAS*) arise most externally from the posterior tubercle, while the post-vertebral muscles are attached further back and further in. These are attached in order, the superficial outside the deep; *A* is the plane of Cerv. ascendens; *T* of Transversus and Trachelomastoid; *S*, Splenius; *C*, Complexus; *SS*, Semispinalis; and *M*, Multifidus. Deep to the plane of Trachelomastoid the muscles are attached to the articular processes. *SCM*, Sternomastoid, and *T*, Trapezius, are in the plane of the deep enclosing or vaginal layer of fascia o of the neck. 2. The first three cervical vertebræ, showing the relations to them o the nerves, muscles and vertebral artery. The muscles are named on the atlas but the drawing is for comparison with Figs. 15, 18 and 20.

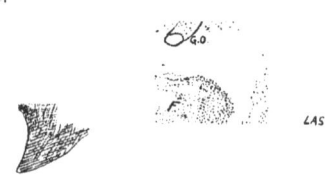

Fig. 18.—Postero-lateral view of axis. *OI*, Inferior oblique; *SS*, Semispinalis; the area for Multifidus is on the under surface below Semispinalis; *RM*, Rectus post. major. *F* is a roughened area, not always very evident, which is not due to the attachment of a muscle but to the presence of a fibro-fatty pad which lies between the diverging Semispinalis and Inferior oblique to allow the latter to follow the movements of the head; the post. primary div. of the second cerv. nerve (great occipital, *GO*) runs above this pad below the Inferior oblique to turn up round this muscle, and thus rests on the lamina. On the transverse process are: *Tr*, Intertransverse; *LAS*, Lev. ang. scapulæ; *Spl*, Splenius colli. Compare Figs. 15 and 17.

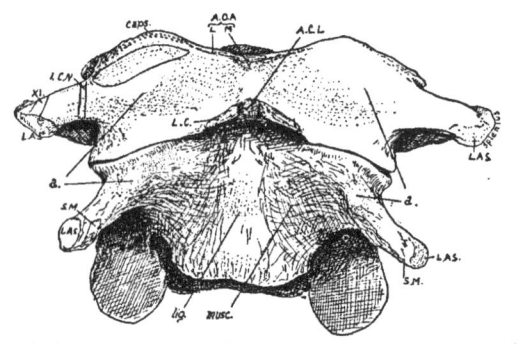

Fig. 19.—Front view of atlas and axis. The rudiments of the anterior tubercles are shown at *a*. On the axis the narrow ligamentous area (*lig.*) indicates the narrowing of the ant. common lig. to its attachment on the atlas at *ACL*; the concave areas on each side of it are filled by Rect. ant. major, Longus colli also passing up deep to this to insertion (*LC*) on atlas. On the atlas the spinal accessory (*XI*) and suboccipital (*1CN*) nerves cross the transverse process; internal to the last the Rectus anticus minor arises. The tip of the odontoid is seen above the anterior arch. *AOA*, anterior occipito-atloid lig., lateral and middle. The hollow between *a* and the anterior tubercle is filled by Rect. anticus major.

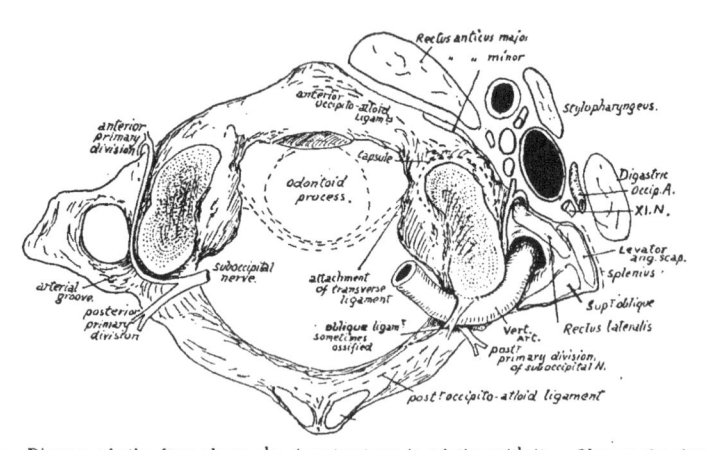

Fig. 20.—Diagram of atlas from above, showing structures in relation with it. Observe the situation of the suboccipital nerve, the relation of its two divisions to the vertebral artery, and the emergence of its anterior part between R. ant. minor and R. lat. The posterior wall of pharynx is in close contact with R. ant. major and, between the two muscles, is related to the atlas. R. post. minor is on the posterior tubercle.

in the recent state by the posterior common ligament : deep to the ligament, veins issue from the back of the body (Venæ basis vertebræ) by foramina which tend to be paired but may be multiple.

It remains to consider the upper two vertebræ.

The **Axis**, on its front surface, shows a narrow ligamentous area which indicates the cord-like nature of the upper part of the anterior common ligament : outside this the muscular hollow is overhung above and externally by the front edge of the upper articular process, roughened by the capsular ligament. On the base of the transverse process, where it projects from under the articular edge, is an ill-defined tubercle, to which some oblique fibres of the capsule are attached.

This tubercle, seen in Figs. 16 and 19, is the proper " anterior tubercle " of the transverse process, and the *quasi*-process projecting from it is really the costotransverse bar connecting this tubercle with the true transverse element behind ; so the tip is only the " posterior tubercle." The correspondence of the parts of the " transverse processes " in this vertebra and the next are shown in Fig. 15, B, and it is evident that the intertransverse muscle on the top of the second process is only the representative of the *posterior* intertransverse on the third, while the anterior intertransverse is wanting * in the former case ; thus the anterior primary division of the second nerve, descriptively issuing " internal to the intertransverse muscles," is really in the same plane as the lower nerves that come out between them, and lies, as they do, on the costo-transverse bar. The nerves are shown, with their relations to the artery, in this figure and in Fig. 17.

The oblique ligamentous fibres correspond with the upper bands of the stellate ligaments of the rib-heads in the thorax, and lie in front of the articulation. which we have seen to be of the nature of a costo-central joint.

The tip of the transverse process has a sharp posterior lip : the aponeurosis of the Splenius colli is attached to this. In front of this and rather above it, on the tubercle, is a rounded small surface for Lev. anguli scapulæ, and internal to this a slight ridge marks the origin of the top fibres of Scalenus medius. The intertransverse muscle is attached to the top of the tubercle and true transverse limb. extending slightly on to the costo-transverse bar.

The groove on the neural arch behind and outside the upper articular area is for the anterior primary division of the second cervical nerve : the posterior division, or great occipital, crosses the arch a little further back, lying on it under cover of the inferior oblique (see Fig. 18). Two areas can be seen and felt on the surface of the spine and lamina : one concave, and looking upwards and outwards on the spine and extending forward about half-way to the articular surface, the other a rough and narrower area looking downwards and outwards and extending from the lower margin of the spine toward the lower articular eminence. The upper one gives origin to the Inferior Oblique, and the lower affords insertion to Semispinalis colli and Multifidus spinæ, the last named also reaching the lower aspect of the lamina : this makes the lamina thick below, giving it a three-sided section, with a sharp upper border for attachment of lig. subflavum. The crest on the upper and back aspect of the spine gives origin to the Rectus posticus major.

The neural aspect of the laminæ shows a transverse line above its lower border for deep fibres of the lig. subflavum.

The back of the body joins the back of the odontoid process at the level of a ridge

---

* There are muscle fibres of this group to be found sometimes attached to the " anterior tubercle," internal to the issuing nerve. Usually these form an oblique ligamentous band.

connecting the margins of the articular surfaces; a corresponding ridge may be visible in front.

The back of the body exhibits two venous foramina with a median ridge, and is tuberculated for the attachment of the Membrana tectoria, and accessory alto-axoid ligaments: the crus inferior of the cruciate ligament is attached to the transverse line and the bone below this.

The odontoid process articulates in front with the arch of the Atlas and has a cartilage-covered facet upon it. Its sides are pierced by small foramina, and its posterior surface is grooved by the transverse band of the lig. cruciatum: a bursa on cartilage intervenes, which may extend out on the front of the ligament and communicate with the occipito-atloid joint. The top of the process is divided by a triradiate line into three areas, of which the front is for the median, and the two posterior for the lateral occipito-odontoid ligaments.

The process, being the separated body of the atlas, is ossified from a bilobed or double centre distinct from that of the true body of the axis, and a plate of cartilage separates the two elements at birth. A few years later the spreading ossification involves the peripheral parts of the cartilage, but the central portion remains as cartilage until past middle life. The transverse line mentioned as showing the limits of the process behind is below the level of the inner part of the articular surface, and many bones indicate on inspection what is suggested by this fact, that the inner corner of the articular surface is ossified from the odontoid centres. Possibly there is an occipital body included in the apex of the process.

At birth the bone is in four pieces joined by cartilage—the dens, the true body, and the two halves of the neural arch. These are consolidated by the sixth or seventh year. The lines of junction of the neural centres with those of the true body lie below the levels of the pedicles and transverse processes.

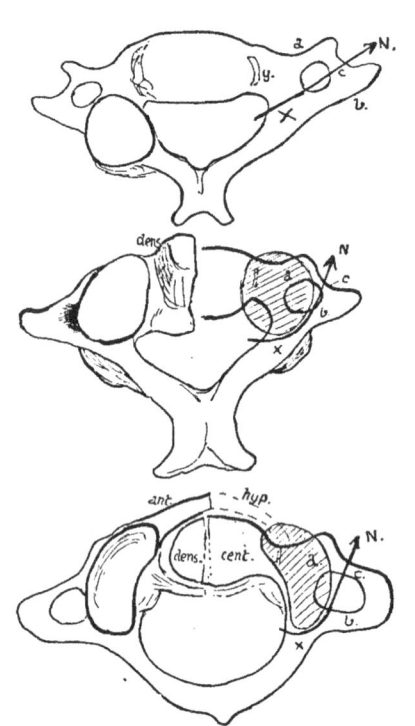

FIG. 21.—Schemes to show the modifications of the costal processes in the two upper vertebræ consequent on the adoption of an enlarged neuro-central articulation as a main joint.

Atlas (Fig. 20).—The tip of the anterior tubercle points down, and has the anterior common ligament attached to it: on each side of this (Fig. 19) the Longus colli is inserted on a rough area. The anterior occipito-atloid ligament has a strong median band on the upper part of the tubercle (so that a non-ligamentous area of bone separates this from the lower ligament), and at each side its lateral part makes a ridge on the bone running into the markings of the upper capsule.

Rectus anticus minor arises from the bone in front of the capsule: this muscle does not as a rule make much definite marking on the bone, but it does not

Vertebral Column

extend out on to the transverse process, and is limited below by an ill-defined prominence continuous with the front limb of the transverse process. Below and internal to the prominence is a hollow which is filled by Rectus anticus major without affording origin to it.

The prominence just mentioned, shown in Figs. 16 and 19, answers to the anterior tubercle of the transverse processes lower down. The modifications consequent on the growth of the neuro-central articulations in the upper two vertebræ are shown schematically in Fig. 21. In the drawing the right-hand side indicates the fundamental principle that is common to the cervical segments, with the modifying factor, while the result is shown on the left-hand in each figure. The first figure represents the typical state, with a costal process (a), a tranverse process (b), and a costo-transverse bar (c). Articular processes can be neural (x) or neuro-central (y). In the typical cervical region they are neural, and lie, therefore, on the neural arch at the base of the transverse element, leaving the costal process clear. The second figure shows how, in the Axis, the upper *neural* articulation fails, and the *neuro-central* one enlarges, spreading outwards as shown by the shaded area, so that it covers the costal process and causes the course of the arterial canal to be turned outwards; thus the condition already noted in the Axis is attained. The last drawing shows how a comparable condition is reached in the Atlas without much change in the direction of the foramen, because the articular surface is not so broad. Thus the costal process can be looked on as buried by the articular masses in the upper bones, and the partly-suppressed tubercle is all that remains to indicate its position, while the anterior boundary of the arterial foramen in these bones is really the costo-transverse bar. So the nerves (N) are seen to issue in corresponding and similar relations on all the bones.

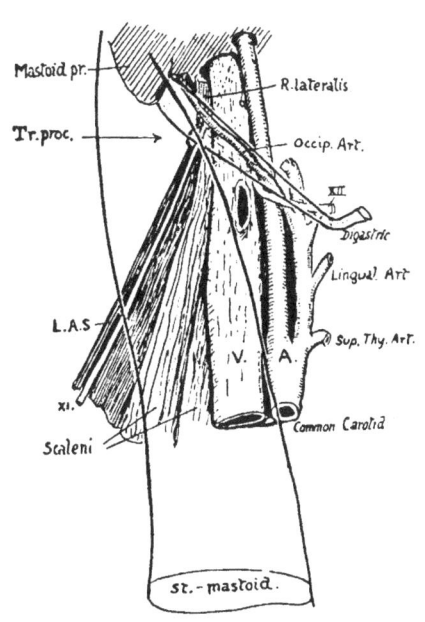

FIG. 22.—To show the anterior relations of the transverse process (*tr. proc.*) of the atlas. The posterior belly of Digastric and Sternomastoid are represented as transparent, thus exhibiting the Rectus lateralis and eleventh nerve related to the process. This is the only transverse process which projects beyond the vein.

The transverse process of the atlas (Fig. 15) has an external border that shows a sharp lower ridge: this ridge affords insertion to the aponeurosis of Splenius colli (S), and the concave surface internal to it, on the lower surface of the process, is roughened for the insertion of the Inferior Oblique. Above the ridge, on the upper aspect, are two impressions for Superior Oblique (S.O.) and Levator anguli scapulæ (L.A.S.), the latter reaching the front corner of the process and the former extending from the posterior corner to the arterial groove. The Rectus lateralis has a faint marking behind and internal to the Levator, extending along the front limb of the process to the slight groove that lodges the anterior primary division of the suboccipital nerve. The nerve runs forward round the outer side of the articular mass (Fig. 20) lying deep to the inner border of the artery, and emerges between the Rectus lateralis and anticus minor, thus lying immediately behind the carotid sheath: after giving branches it turns down in front of the transverse process to join the next nerve. The carotid sheath is altogether in front of the process, and does not extend laterally beyond it as in the lower segments, partly because the vein here is smaller, but mainly owing to the breadth of the Atlas : the tip of the process projects beyond the vein, and is crossed by the spinal accessory nerve as this leaves the sheath after passing behind or in front of the vein, while the posterior belly of the Digastric covers all the structures and separates them from the Sterno-mastoid. In this way the spinal accessory nerve passes on to the Levator anguli scapulæ under cover of the Digastric, and when it emerges from under the lower border of that muscle it is lying on the Levator directly under cover of the Sterno-mastoid (Fig. 22).

The transverse process can be felt deeply on pressure through the Sterno-mastoid, below the mastoid process and a little behind the mandible.

The relation of the vertebral artery, as it turns round the articular process, to the nerve differs from that obtaining in the other segments : this is probably owing to the different value of the artery, which should be looked on here as an enlarged spinal branch, not comparable with the direct artery below this level. As the vessel runs from the transverse process to the posterior arch, it overhangs the outer border of the bone, which is notched here by a vein passing to the vertebral system. This notch is occasionally converted into a foramen by ossification in the fibrous tissue that bridges it in and supports the artery.

The same conversion may occur in the notch that marks the end of the arterial groove, behind the articular mass : this is usually bridged by a ligament, the " oblique ligament of the atlas" attached in front to the overhanging edge of the mass and extending back to the outer extremity of the ridge on the posterior arch that marks the line of the posterior occipito-atloid ligament.

The upper articular facet shows a tendency to division by a constriction of its area. Opposite this constriction there is a vascular pit on the upper surface of the internal protuberance caused by the lower articular surface : the vascular spot has the tubercle for the transverse ligament in front of it, while a small projection for the accessory atlo-axoid ligament may be seen below and internal and somewhat behind it.

On the under surface of the bone may be seen the line for the thin Lig. subflavum, evidently passing into the fibrous tissue that bridges over the venous notch in the bone. On the under side of the transverse process is the concavity for the Inferior Oblique, bounded by the ridge for Splenius colli, and a roughness on the anterior limb of the process marks the attachment of the intertransverse muscle. The markings of Longus colli and anterior common ligament are well seen from below.

The anterior surface of the posterior arch is smooth and is separated from the dura mater by a layer of soft vascular tissue : the front wall of the neural canal is formed by the ligaments that cover the back surface of the odontoid process.

DORSAL VERTEBRÆ.

The distinguishing mark of this series is the presence of costal facets on the bodies. Other general characters, as seen in a middle vertebra in the series, are as follows :—

Body somewhat triangular or heart-shaped viewed from above, and slightly deeper behind than in front, carrying an upper and a lower costal demi-facet on each lateral aspect, which are covered with fibro-cartilage in the recent state.

Upper *intervertebral notches* very shallow or even absent, lower notches correspondingly deep.

Vertebral Column

Transverse processes strong, directed backwards and outwards, with somewhat clubbed extremities carrying a costal facet, for the tubercle of the rib, on their front aspects.

Articular processes have joint surfaces in a nearly vertical plane, but looking *very slightly* outwards as well as backwards in the upper pair, and in the opposite direction in the lower pair.

Spinous processes long and directed obliquely downwards, ending in a slightly marked tubercle.

Spinal foramen more or less circular and small, the canal being well covered behind by the overlapping of the broad laminæ and spines.

At the ends of the series the vertebræ become modified, and the ribs, which typically articulate by their heads with the bodies of two vertebræ and the intervening disc, show a tendency to confine their articulation to one vertebra. These changes enable certain of the segments to be recognised fairly easily and certainly.

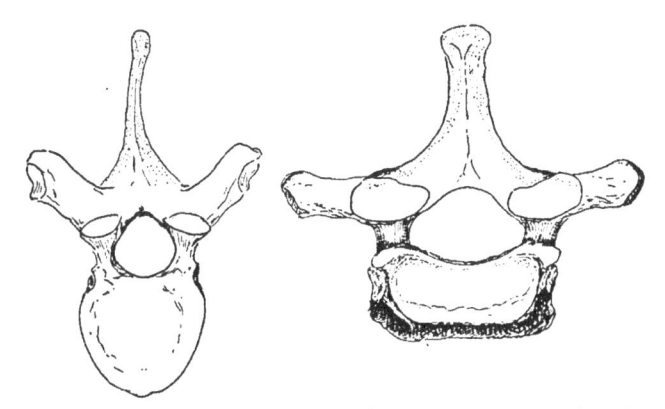

FIG. 23.—First dorsal (right) contrasted with typical dorsal (left) to show their characters, which have been somewhat exaggerated.

The **FIRST DORSAL VERTEBRA** is like the last cervical in its general shape, but possesses no foramen in its transverse process, and this, with the presence of the costal facets, marks it at once as dorsal. The facets are in the form of one more or less complete on the upper part of each side of the body, with a small part of a second facet below, and one on the transverse process (Fig. 23).

The upper surface of the body is concave from side to side like the cervical bodies.

The upper intervertebral notches are deep, and the articular surfaces behind are slightly more oblique than in the usual dorsal type and do not face at all outwards.

The thick spine is longer than that of the last cervical, but otherwise rather like it, and is directed almost horizontally.

The spinal foramen is more of the cervical shape, and is somewhat larger than the typical dorsal foramen, because it contains the terminal part of the cervical enlargement of the cord.

In the lower end of the series the modification of the costal articulations is marked, in conjunction with other changes associated mostly with the decrease in the development of the ribs carried by the vertebræ.

Thus the tenth rib frequently articulates with one vertebra only, the tenth dorsal, so that the **ninth dorsal vertebra** may only present on its body a large " demi-facet " above, and none below : usually, however, there is at least an indication of the former presence of the lower rib, if not an actual articular surface, and there is always a facet on the transverse process.

The **TENTH VERTEBRA** has a whole facet, or nearly a whole facet, on its body above, and none below, with one on the transverse process : this last, however, may not show a recognisable articular surface.

The **ELEVENTH AND TWELFTH DORSAL VERTEBRÆ** are alike in having complete single facets on each side of the bodies with none on the transverse processes, which are reduced to thick, stunted, and tuberculated projections. The spines are also reduced to short and somewhat pointed projections extending nearly horizontally backwards.

Though alike in these characters, the two vertebræ may usually be distinguished by their articular processes, for whereas those of the eleventh are dorsal in character, those of the twelfth are only dorsal above, the lower ones being turned out like those in the lumbar series : it must be added, however, that occasionally, though rarely, this double character is present in the eleventh segment, in which case both upper and lower processes in the twelfth vertebra exhibit the lumbar type of conformation.

In these last two dorsal vertebræ the spinal foramen widens owing to the presence of the lumbar enlargement of the cord.

The **centrum** of a typical dorsal vertebra is covered in front and laterally by the fibres of the wide anterior common ligament which separates it from the intercostal arteries that pass out on it to their destinations, and from the other structures that are described as being in relation with the column in the posterior mediastinum. But one of these, the Aorta, usually leaves its impress on the middle vertebræ notwithstanding, in the shape of a raising of the concave side aspect of the body, so that it presents a flat surface for the support of the artery through the ligament. This flattening of the left side may affect the upper and lower margins as well, giving an asymmetrical appearance to the bone examined, and is accompanied by a difference in the aspect of the texture of the bone as seen on the surface. The " aortic impression " produced in this way is best marked on the fifth body, but can be followed down from this for the succeeding two or three vertebræ, finally fading towards the front of the bones at about the ninth segment ; in some columns the impression cannot be demonstrated certainly. Large vascular foramina can be seen, mainly on the sides of the bodies, which transmit veins through the ligament, or round its edges, to the azygos veins : in the bodies of the vertebræ these lateral channels are connected with the posterior one, which makes its exit by means of one or two foramina on the neural surface of the body, through which veins pass beside the posterior common ligament to join the spinal venous plexus (anterior longitudinal spinal veins), which extends behind the vertebral bodies for the length of the column.

Indications of the diverging nature of the fibres of the stellate ligaments of the rib-heads can be seen in the markings that lie round the facets, and the attachment of the anterior common ligament to the upper and lower parts of the bodies can usually be recognised by the markings on the bones.

The dorsal bodies vary in their shape and in the relations between their measurements. This can be illustrated by the outlines of the upper surfaces of the bodies taken from a normal column (Fig. 24). Comparing the sagittal measurement with the transverse between the margins of the upper facets, the first body gives 9 : 18, the

Vertebral Column

second gives 8 : 13, in the third it is 8 : 12·5, while in the fourth it suddenly changes to 10 13. After this the measurements give :—

5, 11 : 13 +	9, 13 : 14·5
6, 11 : 13 +	10, 13 : 14·5
7, 12 : 13 +	11, 14·5
8, 12·5 : 13·5	12, 15·5

In the last two bodies the facets do not reach the margin of the body.

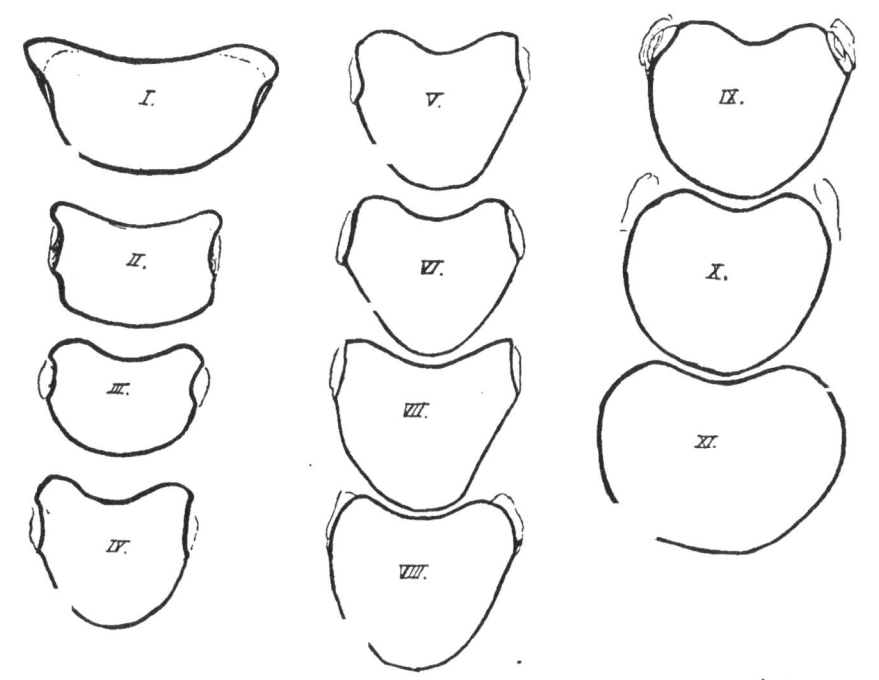

FIG. 24.—Tracings to scale of the upper surfaces of the dorsal bodies. Notice that the third is the smallest body.

Considering these figures it may be said that the fourth is the first body in which the typical dorsal shape is undoubtedly apparent. Moreover, the body of this vertebra is, owing to this fact, more wedge-shaped than that of the third segment, so that it gives the impression of being smaller when seen from the front : it is evident, however, that it is not the smallest body, which is really the body of the third. From this down the bodies increase in size, and in the last two the increase is considerable, especially in the transverse diameter, thus showing a tendency to approach the lumbar type. In the same way the bodies, ascending from the fourth, show an increasing approach to the cervical type, and the upper surface of the first body is strongly suggestive of the appearance of a cervical vertebra : the first vertebra also presents the two antero-lateral prominences on the lower border of its body that are seen in the cervical region.

The dorsal region is naturally the most fixed portion of the true vertebral column, and its bodies afford evidence of the nature of the dorsal curve in their slightly lesser vertical depth in front : this agrees with the primary nature of the curve, and the comparative immobility of the region is related to the shallowness of the intervertebral discs.

Observe that the upper facets on the tenth are largely on the free pedicles, and those of the ninth to a lesser extent ; the facets on eleventh and twelfth are also largely on the pedicles, but do not reach the upper margins at all. This simply indicates the reduction in size of the rib, with consequent contraction, so to speak, of its area on the neural arch. It has been pointed out already that the rib is purely neural in its articulation, and the line of the neuro-central synchondrosis, though not apparent in the adult bone, lies just in front of the facet on the body, so that the facet is not on the centrum, but on the extreme front end of the neural portion. But while the fully-developed rib extends its articulation to the limit of the arch that carries it, the reduced bone withdraws from this extremity, and thus comes on to the pedicle.

The thick **pedicles** are attached to the backs of the bodies high up, so that the upper intervertebral notches are almost non-existent : in the first segment, however, the prominence of the neuro-central region in front, as in the cervical series, makes the notch well marked. In the case of the second vertebra the pedicles are placed so far up that the plane of the upper surface of the body passes through them obliquely : it must also be observed that the pedicles of the upper segments are directed slightly upwards as well as backwards.

The **transverse processes** are strong for the support of the ribs : in the last two vertebræ, where the ribs do not articulate with the processes, they are much modified. Their facets depend for their direction and appearance on the position of their ribs in the series : thus the true or sternal ribs are the upper seven, the remainder being " false " ribs, and the upper facets are well-marked, large, concave surfaces, situated on the front aspects of the transverse processes and in a vertical plane, whereas the facets for the lower ribs are smaller and flatter and tending to lie on the upper aspect of the process and to look upwards.

The seventh facet is usually a transition form in these characters, but occasionally, when the eighth rib reaches the sternum, the eighth facet becomes transitional, and the seventh has the characters of those higher up. The tenth facet is small and badly marked, or may apparently even be absent.

The differences in the facets may have to do more with the attachment of the diaphragm to the false ribs than with any inherent result of their asternal position, but, however this may be, the variations are sufficient in themselves to indicate that the direction and method of movement of the ribs differ in the upper and lower series ; but of the nature of the movements divergent opinions are held by observers.

On the first transverse process the facet is at its extremity, and may look almost directly outwards : in some short processes it may even look outwards and backwards, but this may be the result of disease. The process in the first D.V. is directed most transversely, those below this showing increasing obliquity backwards.

In the eleventh vertebra the process may be a mere shapeless mass, or it may show an arrangement of tubercles on the same plan as in the last dorsal vertebra (Fig. 25).

In the twelfth vertebra there are three tubercles to be seen, the *upper*, *lower*, and *external*, of which the first two at least are for muscular attachments : these will

Vertebral Column 31

be considered when dealing with the region of the transverse process in the lumbar series, and it is only necessary to recognise them here (see p. 35).

Certain roughened areas for fibrous attachments can be made out on the transverse process of a typical dorsal vertebra (Fig. 26). On the front, a somewhat hollow and triangular area marks the attachment of the middle costo-transverse ligament; below this a convex roughness is for the superior costo-transverse ligament of the rib below. On the top of the process is the impression of the inter-transverse muscular and tendinous bands, and the corresponding mark on the lower aspect is nearer the outer

FIG. 25.—The two last dorsal and first two lumbar vertebræ, to show the tuberculated modification of these dorsal transverse processes, best marked in the twelfth.

end. At the outer end is a thick lip which is behind the outer border of the facet and gives attachment to the external or posterior costo-transverse ligament: it runs below into a tuberculated impression for the origin of the Levator costæ muscle, and the tubercle is better marked in the lower members of the series in conformity with the larger size of the muscle in them (see Fig. 42).

On the posterior surfaces are slight markings for tendinous fibres of Multifidus and Semispinalis, with Rotatores spinæ above these: the markings are more noticeable in the upper half of the region, owing to the additional origins of Complexus, Transversus cervicis, and Trachelo-mastoid, but the different attachments cannot be distinguished with any certainty on the bones. The Longissimus Dorsi is inserted into the processes.

These post-vertebral muscles also make markings on the sides of the spines, but separate indications are not to be found : possibly the more tuberous state of the spinal surface in the upper three or four dorsal vertebræ is associated with the insertion here of Semispinalis dorsi.

Notice how the **laminæ and spines** overlap the succeeding bones and cover in the canal : such an arrangement makes it probable that the ligamenta subflava have their upper attachments higher up on the neural surface of the broad laminæ than is the case in the cervical region, and inspection of this surface shows that it is so, the upper line made by the ligament crossing at the mid-level of the lower articular surfaces and indicating that the ligament becomes continuous externally with the capsule of the joint. The presence of strong ligamenta subflava in this region shows that these are not primarily concerned in covering the canal : they are to resist flexion forwards of the column, and we therefore find that they increase in strength from above downwards, reaching their maximum in the lumbar region. The line on the neural surface of the laminæ enables us to divide this surface into an upper, separated from the dura mater by vascular fatty tissue, and a lower for the attachment of the interlaminar ligaments and fibrous tissue continuous with that of the interspinous ligaments.

The obliquity of the spinous processes varies in the different parts of the series, being greatest in the middle, where the spine of the seventh dorsal vertebra is usually the longest and most obliquely directed. With the exception of the upper three or four, the spines of the series are somewhat tapering and thin, but enlarging at their extremities to form tubercles : these are covered by supra-spinous ligaments and give origin to Trapezius, Rhomboids, and Latissimus dorsi muscles, according to their position.

Below the seventh, the obliquity of the spines decreases, being practically not apparent in the last two vertebræ of the region. Individual spines frequently deviate to one side or the other, without any change in the central position of the bodies.

It is clear that, as a result of their directions, the levels of the tips of the spines as felt on the surface of the body do not correspond with those of the bodies to which they belong, and it is important when considering levels or surface markings of internal structures to bear clearly in mind this distinction between bodies and spines. Each spine differs from the others in its relation to the levels of lower bodies, but a rough rule may be remembered for practical purposes, that in the middle vertebræ, fourth to ninth, the tip of a spine is about opposite the disc between the two bodies next below its own body; the other levels can be obtained without much trouble on an articulated column.

The rounded shape of the **spinal foramen** is first reached in the fourth segment : it persists throughout the series until the lumbar enlargement of the cord, which commences just above the level of the eleventh dorsal vertebra, widens out the canal laterally in that bone and in the twelfth, so that the foramen in this last is the widest and largest in the whole series.

LUMBAR VERTEBRÆ.

Distinguished by their large size and massive structure, with absence of transverse foramen and costal facets. They present large kidney-shaped *bodies*, slightly deeper in front than behind : *pedicles* thick, with very shallow upper notches : *transverse processes* comparatively slender, with the small *accessory tubercles* behind them. *Articular processes* vertical and curved transversely so that the lower pair look

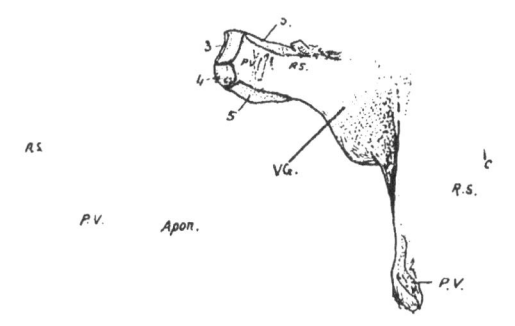

6.—A mid-dorsal vertebra, showing markings on processes of the neural arch. C, capsule of costo-transverse joint; 1, superior costo-transverse ligament; 2, 3, middle and posterior (external) costo-transverse ligaments; 4, origin of Levator costæ; 5, intertransverse musculo-tendinous bands. The post-vertebral muscles occupy the vertebral groove, *VG*; the deepest of these are Rotatores spinæ, *RS*, while the attachments of the more superficial muscles (Multifidus, Semi-spinalis, and Spinalis dorsi on the spine) are indicated at *PV*. All these are covered by the aponeurotic sheets (*apon.*) of origin of Trapezius, Latissimus, and vertebral fascia.

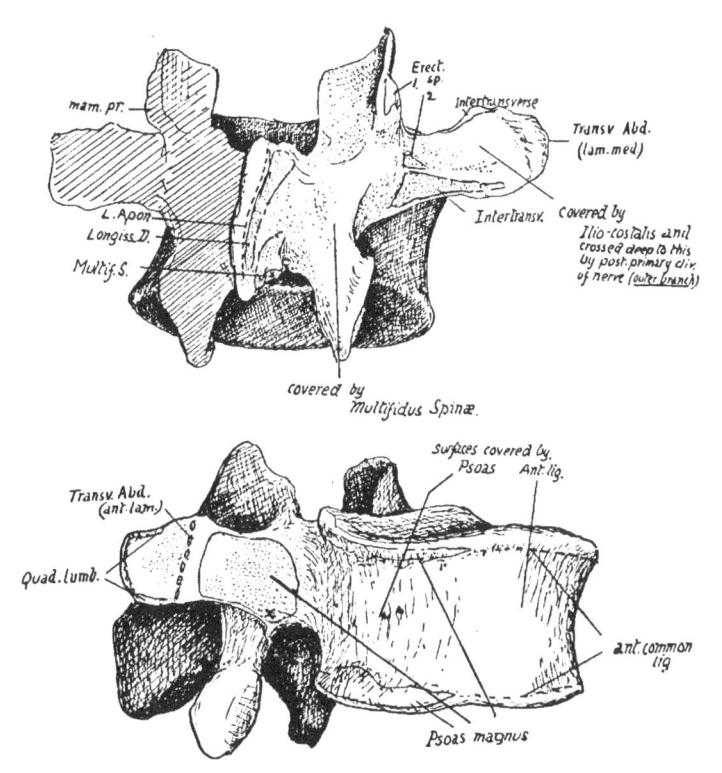

7.—A mid-lumbar vertebra. In the upper figure observe that the mammillary process gives origin to Multifidus (1), while the accessory process, further out, has Longissimus (2) attached to it. The post-vertebral muscles lie against the side of the spine, cover the lamina, and external to this lie on the transverse process and intertransverse structures and middle lamella of Transversalis. In the lower figure notice the division of the transverse process into two areas by insertion of anterior lamella of Transversalis, with Quadratus lumborum, deep to this lamella, covering the process external to the line, and Psoas, superficial to the lamella, on the bone internal to the line. The tendinous origin of Psoas is marked at *x*. The lumbar vessels lie on the body between the origins of Psoas from it.

Vertebral Column

forwards and outwards and are nearer together than the wider upper pair, which are curved to look inwards and backwards to receive them. *Mammillary tubercles* are present on the back aspect of the *upper* processes, to afford origin to post-vertebral muscles. *Laminæ* deep but short, and the *spine* is horizontally directed, thick and square-cut, when viewed from the side.

The lumbar column rests on the top of the sacrum, with which it makes a prominent *lumbo-sacral angle*, owing to the sudden change in direction as the sacrum turns backwards and downwards : the sharpness of the angle, however, is modified by the construction of the last lumbar body to form a transitional bone. Thus it occurs that this body is wedge-shaped, markedly deeper in front, and its spine is much smaller owing to the limitation of space resulting from the backward turn of the sacrum : this last character is more marked in the female, owing to the more pronounced sacral bend. The spinal foramen is triangular and large, containing the nerves that constitute the cauda equina with their coverings.

FIFTH LUMBAR VERTEBRA.—In addition to the modifications already noticed in the body and spine, the transverse processes are not very long, but are thick and strong, arising not only from the arch, but spreading their bases forward on to the body : it thus happens that the issuing nerve must pass over this base, so that the upper intervertebral notch is described as grooving it.* From this broad base the process narrows to a blunt tip, to which the ilio-lumbar ligament is attached. This strong band, helping to support the column between the two innominate bones, accounts for the strength of the process, and also for its direction, which is upwards as well as outwards. The lower articular processes articulate with the sacral processes, which are far apart, so that on the last lumbar vertebra the lower processes are practically as far apart as the upper. The upper and lower processes together form a mass which projects somewhat into the neural canal, sharpening its outer angles, and in front of this the pedicles show inner margins thinner than in the higher bones, below which a sloping surface leads to the lower intervertebral notch.

Examination of the upper and lower borders of the lumbar **bodies** enables one to recognise the position of certain attached structures (Figs. 27 and 35).

Each border, along the lateral aspect of the body, is rendered heavy and thick by the connection with it of a fibrous sheet that is thrown across the concavity of the body and affords origin to the Psoas : the sheet and the muscle may encroach on the bone to a greater extent than in the figure, as may be seen on some bones by observing the extent and depth of the roughened area.

The concave surface between these ridges lodges the lumbar vessels and twigs from the sympathetic.

The anterior common ligament covers the front of the body between these lateral areas, and the ridges to which it is fixed are less prominent than at the sides ; it separates the aorta and origins of the lumbar arteries from the bone. The crura of the diaphragm are attached to this ligament and through it to the upper two or three lumbar bodies, but no distinct marking can be assigned to them on the bones.

The Psoas also obtains origin from the front of all the lumbar **transverse processes.** The extent of this origin appears to be somewhat variable, but, being by muscular fibre, it leaves no mark on the bone ; there is, however, a tendinous slip generally in the lower and inner part of each of these origins, and a corresponding slight roughening is to be found at the inner end of the lower border of the process.

* The transverse process of the fourth exhibits a tendency to forward extension like the fifth.

34   Anatomy of Skeleton

These origins from the fronts of the transverse processes account for the presence of the lumbar nerves in the substance of the Psoas, for it is apparent that each nerve as it turns down over the front of the transverse process next below (see Fig. 35), must lie on a part of the muscle and so enter its substance; the ilio-inguinal and ilio-hypogastric nerves do not cross a transverse process, so they lie behind the muscle and emerge from under its outer border, and the fifth lumbar nerve also does not lie in the muscle, whereas the lower part of the fourth, passing down to join it across the fifth transverse process, must run in the muscle for a short part of its course.

There are tendinous and muscular fibres between the transverse processes, and markings for these may be found on the upper and lower edges of the processes and adjacent parts of the posterior surface.

The middle or main lamella of the aponeurosis of origin of the Transversalis is

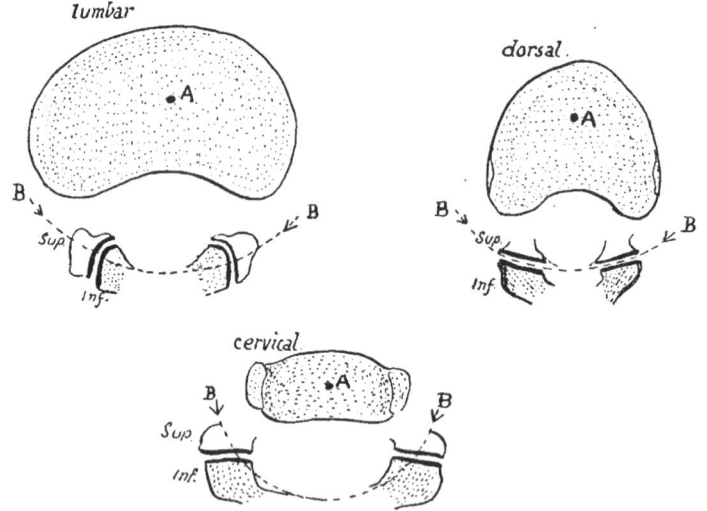

FIG. 28.—Schemes to show the effect of the shape of the articular processes on the movement of rotation between the bodies. *A* is approximately the centre round which rotation takes place; *B* is part of the circumference of a circle drawn round *A* and passing through the articular processes. It is clear that the shape and plane of these processes in the lumbar and cervical regions inhibits any rotation between the vertebral bodies.

attached to the outer end of the transverse process, extending on to its back surface, where a roughened area near the tip marks its connection with the bone. Fibres radiate out from each process into this aponeurotic sheet, as shown in Fig. 35, and these are modified above into the costo-lumbar ligamentous attachments and below into the ilio-lumbar ligament; between these they open out into the middle lamella. The lamella is thin between the processes, and has some inter-transverse fibres in it which are probably part of the inter-transverse musculo-tendinous system, with the fasciæ of which it otherwise blends.

The Quadratus lumborum lies in front of this middle lamella and also covers the outer end of the transverse processes, to the upper and lower parts of which it is attached: the muscle is covered by the anterior lamella, and this reaches the transverse processes at the inner margin of the Quadratus, under cover of Psoas. Thus a

Vertebral Column

secondary line can be found running across the transverse process for this lamella, dividing the front surface into an inner part for the Psoas and an outer covered by Quadratus lumborum.

The accessory tubercle or true transverse element gives origin to a slip of Longissimus dorsi, and the Multifidus spinæ lies further in, on the back of the **laminæ and articular masses,** to which it is attached : the mammillary process is on the back margin of the upper articular process, and affords origin to some fibres of Multifidus.

Examine the line made by the strong ligamenta subflava on the neural surfaces of the laminæ. Evidently the ligament does not run into the capsule, as this is turned

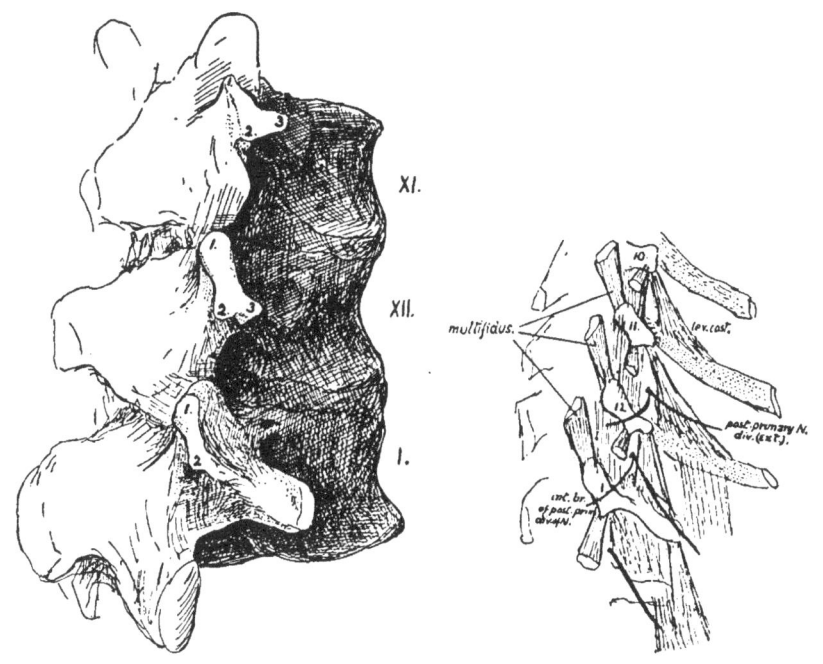

FIG. 29.—First lumbar and last two dorsal vertebræ, to show the transverse processes. 1, upper (mammillary) ; 2, lower ; and 3, external (accessory) tubercles. The smaller drawing is a scheme of a dissection, including the tenth dorsal vertebra.

out and away from it : as a result the edge of the ligament is free in the hinder part of the intervertebral foramen just behind the issuing nerve.

The lower articular processes of the fourth segment are further apart than in those higher up, but not so widely separated as in the last lumbar vertebra, in which they are practically as far apart as the upper pair. There is a slight change in the curvature and direction of the articular surfaces from above downwards, but in all the vertical plane in which they lie, directed mainly in an antero-posterior direction, effectually inhibits the power of rotation in the lumbar column. This is shown in the accompanying scheme (Fig. 28), and compared with the conditions in the cervical and dorsal regions.

There is occasionally a distinct costal element in the ossification of the first lumbar transverse process, but this does not form the whole of the process, and the correspondence of this with a dorsal rib is not quite certain : it is interesting to consider this region in the lumbar series and compare it with the transverse processes of the lower dorsal vertebræ.

In Fig. 29 the last two dorsal and first lumbar vertebræ from a well-marked set of bones are represented and the tubercles on the transverse region are shown and numbered ; at the same time a scheme of a dissection to show the muscle-attachments is given, which includes the process of the tenth dorsal vertebra. The mammillary tubercles (1, 1) are plainly enlarged "muscular" tubercles for Multifidus spinæ, while the accessory tubercles (2, 2) evidently serve the same function for the Longissimus dorsi ; the different relation to the articular processes only has to do with the different positions of these processes in the dorsal and lumbar regions. But there is another tubercle on the last two dorsal vertebræ, numbered 3 in the figure, and this appears to be in series with the costal support of the tenth transverse process and, like it, to give origin to Levator costæ fibres in each segment. In these two vertebræ, therefore, we may perhaps look on the third tubercle as in some measure comparable with the others, being accentuated as a point of muscular attachment, or we may consider it a representative of the " costal tubercle " or rib-carrying process ; probably its value might be expressed by describing it as a costal tubercle specialised as a muscular process. But in the first lumbar transverse process there is great difficulty in recognising this third tubercle. Is it elongated to form the " transverse process," or is it lost in the base of this process, which is a true rib in value and supported on the third tubercle fused with it ? The first of these alternatives is suggested to the eye by the schematic figure and the second by the conditions seen in the larger drawing, but in each case, the opinion would only be founded on appearances. Examination of a first lumbar gives one the decided impression that the proximal parts of a rib are fused with the bone, and the rare occurrence of a costo-transverse foramen here supports the view. Lumbar ribs articulating with the body also support it, but other ribs occur that are only at the ends of the transverse processes and these seem to favour the first of the alternatives. Possibly the extra costal centre in the transverse process represents a fused head end of a rib and the occasional lumbar rib at the extremity may represent the shaft belonging to it.

Ossification of Vertebræ.—The vertebræ are preformed in cartilage, the cartilaginous structure showing a centrum and two separate halves of a neural arch, with the various processes represented at first only in concentrated mesenchyme into which the chondrifying process extends later.

Ossification begins during the seventh week by the formation of three *primary* centres, one for the centrum and one for each half of the neural arch. The centre for the centrum is probably double in origin, fusing rapidly into one.

The primary centres appear at different times in different regions : thus those for the arches appear first in the cervical region (Axis) and succeed one another from above downwards, while the centrum ossifies earliest in the lower dorsal region and " spreads " from thence in both directions. So the cervical arches can be said to ossify before their centra, while the reverse is the case in the lumbar and lower dorsal regions : there appears to be considerable individual variation in the rapidity of the successive appearances of centres, but it may be said that by the end of the third month there are primary centres in all the true vertebræ.

At birth the corresponding primary bony parts are distinct, but joined by cartilage. The neural halves join dorsally during the first year or so, fusion commencing in the lumbar region, and a few years later the neural arches effect junction with the bodies, the process commencing in the dorsal vertebræ.

After puberty *secondary* centres appear and fuse with the primary bone by the age of twenty-one. These centres occur in the cartilage covering the upper and lower surfaces of the body (epiphysial plates) and the tips of the transverse and spinous

Vertebral Column

processes. The epiphyses on the body only ossify the *periphery* of the cartilaginous plate, the central part remaining cartilaginous, but in advanced life this also may become ossified.

The account just given applies to a typical vertebra such as the human dorsal segment. Highly-modified vertebræ, as the first two cervical, show corresponding modifications in ossifications, but even the slighter differences in other bones introduce certain extra centres. Thus in the *cervical* region the bifid spines have double epiphyses, while the costal processes of the seventh and occasionally of the sixth (and even the fourth) have separate centres of ossification, instead of becoming ossified by extension from the neural arch and transverse process, as in the higher vertebræ. In the *lumbar* region the mammillary processes have separate centres (like the twelfth dorsal), and there is said to be an occasional "costal" centre in the transverse process of the first lumbar, and occasional double centres for each half of the neural arch in the fifth lumbar (and very rarely in others).

Of such double centres one forms the pedicle, upper articular and transverse process, and the other forms the lower articular process and lamina : occasionally a suture persists in the adult, passing obliquely between these two portions.

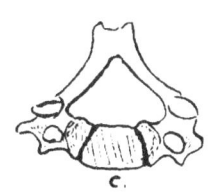

FIG. 30.—Cervical vertebra from a child to show how much the neural arch is concerned in this region in the formation of the "body." C. is the proper centrum.

The **Axis** has the ordinary primary centres plus those of the body of the Atlas (odontoid process). It also has the corresponding secondary centres, with the exception, perhaps, of those for intervening discs between its body and odontoid : the disc on the upper surface of the odontoid is represented by an apical epiphysis. At birth the bony parts are four, with wide tracts of cartilage intervening, and it does not become consolidated before six years of age : the four pieces are—(1) centrum ; (2) odontoid and top of body ; (3) and (4) two halves of neural arch. The odontoid portion unites first with the others, about three, and the remainder join each other between the ages of four and six.

The odontoid ossification includes the top of the body and the inner part of the upper articular processes.

The **Atlas** has the usual two centres for the neural arch, from which it and the greater part of the lateral masses are formed bilaterally. At birth these halves are separate, and are only joined in front by fibrous and cartilaginous tissue of the anterior arch. This begins to ossify in the first year from one central or two lateral centres, there appearing to be great individual variation in this matter : the front part of the lateral mass and upper articular surface is formed in this ossification.

The neural halves join dorsally about the fourth year, and with the anterior arch usually between the fifth and ninth years, but the consolidation of the bone may be incomplete in some cases even at puberty.

Secondary epiphyses have been described for the tip of the transverse process, and occasionally for the posterior and even for the anterior tubercle.

The vertebræ are *intersegmental* in position—that is, they are to be considered as situated morphologically opposite the "septum" between two neighbouring segments * of the body wall. This is brought about by a secondary division into anterior and posterior parts of the mesoderm

---

* The term *segment* is now being used, not in the descriptive sense in which it has hitherto been employed but in its stricter morphological sense, implying the theoretically fundamental and serially homologous foundation of the body wall, indicated in the embryo by the primitive segmentations of paraxial mesoderm and its extension into the wall, with its "segmental" nerve and vessels.

Anatomy of Skeleton

of each primitive segment as it begins to spread inwards round the notochord. The posterior half of one segment joins with the anterior part of the next segment on its caudal side, and the chondrification that makes the rudiment of the vertebra takes place in the composite mass thus produced; it contains, therefore, mesodermal tissue from two segments, so is situated opposite the intersegmental line, and its neural arch and processes, costal, transverse, and spinous, are all situated in the intersegmental plane.

The ingrowing cells (*sclerotome*) of the primitive segments form a sheath for the notochord, representing the membranous stage in the phylogenetic development of the column; later chondrification takes place here, and later still ossification. The intervertebral discs are remains of the cartilaginous structure, and the anterior arch of the Atlas is looked on as a persisting and ossified *hypochordal bar*, which is the cartilaginous portion ventral to the sheath of the notochord in the intervertebral regions, disappearing in the other parts of the column.

The notochord is destroyed in the vertebræ by the formation of the bones, but persists in the nucleus pulposus of the intervertebral discs.

SACRAL VERTEBRÆ.

Five (or sometimes six) vertebræ below the lumbar region are fused in the adult into a single bone, the *Sacrum*, which forms the back wall of the bony pelvis, being

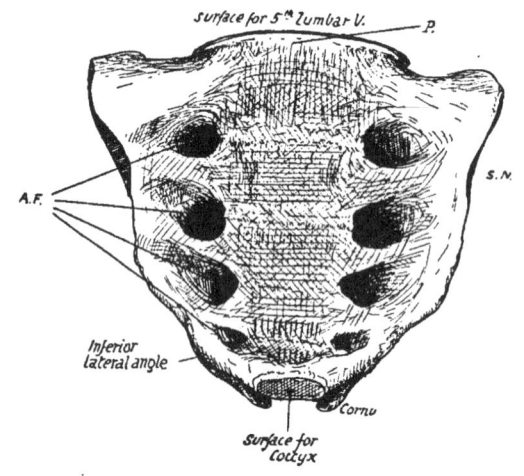

FIG. 31.—Front aspect of sacrum. *A.F.* anterior sacral foramina; *S.N.* lateral sacral notch; *P.* sacral promontory. Apex at lower end.

situated between the two innominate bones and articulating with the upper and back parts of their inner surfaces.

The Sacrum is a wedge-shaped bone, with its base uppermost. Looked at from the front it presents an *anterior surface* (Fig. 31) concave from above downwards and slightly from side to side, of which the central part is evidently composed of the fused bodies of the sacral vertebræ. Outside these are four *anterior sacral foramina* on each side, in nearly parallel rows, for the exit of nerves, and again outside these are the *lateral masses* of the bone, marked by *neural grooves* leading out of the foramina.

The surface is bounded externally by the *lateral margin*, the edge of the lateral mass: this margin has a depression, the *sacral notch*, opposite the second segment,

Vertebral Column

and ends just below the fourth foramen in the *inferior lateral angle*, while the last segment projects below this as a blunt *apex*.

The convex *posterior surface* (Fig. 32) has in the middle the fused neural arches that cover in the *sacral canal*: the canal is open above and below, and at the lower end one or two of the neural arches may be partly deficient. Various bony projections

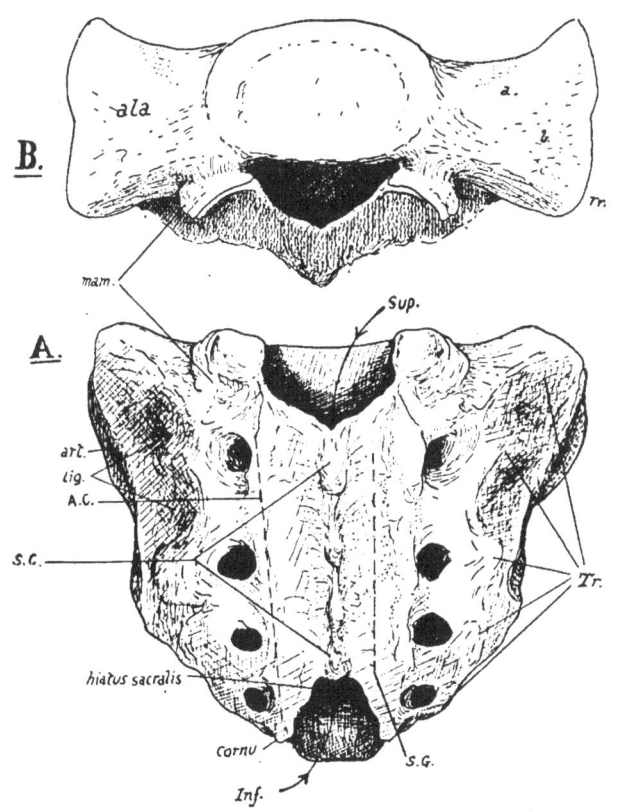

FIG. 32.—A. posterior aspect of sacrum. The arrows pass through the upper and lower openings into the sacral canal. *Tr.* transverse processes; *S.C.* crest made by spinous processes; *A.C.* articular crest; *S.G.* sacral groove. B. view from above, showing the upper aspect of the lateral masses, forming the alæ. *a.* impression for nerve; *b.* roughened markings for ligamentous fibres.

may be recognised as lying in series with the processes of the true vertebræ. In the middle line is a row of connected spines, the *spinal crest* (S.C.), outside this is a narrow *sacral groove* (S.G.), floored by fused laminæ and bounded laterally by a row of low tubercles in series with the articular processes, the *articular crest* (A.C.); outside these are the four *posterior sacral foramina*, and then the back aspect of the lateral mass. The sacral groove and its neighbouring processes give origin to post-vertebral muscles, and the two articular crests end below in the projecting *cornua* which represent the processes of the last sacral vertebra for connection with the Coccyx. As the last laminar arch is wanting—and sometimes the fourth also—the articular crests are longer than the spinal, and the interval between them, opening into the sacral canal, is termed the *hiatus sacralis*.

The lateral mass has a rough double hollow on its postero-lateral aspect, opposite the first two segments, making a *ligamentous* area (*lig.*) for the strong posterior sacro-iliac ligaments—and in front of and below this, on its lateral aspect, a slightly concave *articular* or *auricular surface* (*art.*) for the innominate bone : this corresponds in extent with three sacral bodies.

The upper aspect of the sacrum, or *base* (Fig. 32), shows the first sacral body centrally, and the *alæ* on each side of it, expanding outwards ; each ala is the upper surface of the lateral mass, and has an inner and anterior smooth part slightly grooved for the lumbo-sacral cord, and an outer and posterior rougher ligamentous part. Behind the first body is the upper opening of the sacral canal, somewhat triangular in shape, flanked by a large *superior articular process* on each side, and between this and the body is an *intervertebral notch*. A *mammillary tubercle* is below the hinder edge of the articular surface on the back of the bone : it is not always present.

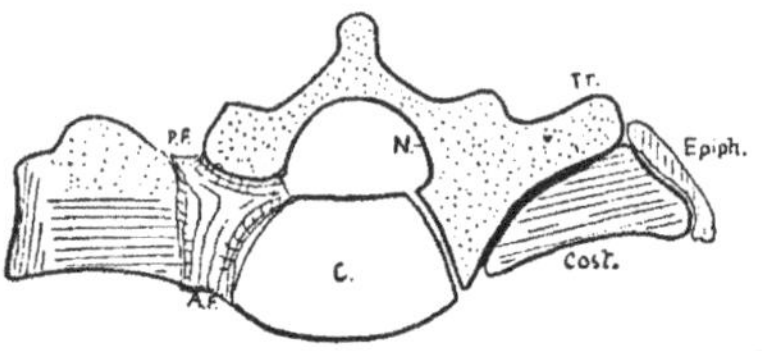

The front surface of the bone looks more downwards than forwards in its upper part, when the bone is in position, as may be seen in the articulated skeleton, and thus the front part of the upper border of the first body becomes the most prominent forward part of the bone and constitutes the *sacral promontory*. When the lumbar column is in position, the *lumbo-sacral angle* of about 120 degrees is formed here between the first sacral and last lumbar segments, the angle being rounded by the intervening disc, which is much thicker in front than behind.

FIG. 33.—To show the constitution of the lateral piece of the sacrum. *Tr.* transverse element ; *Cost.* costal element ; *Epiph.* marginal epiphysis ; *N.* neural arch ; *C.* centrum ; *A.F.*, *P.F.* anterior and posterior sacral foramina.

The female sacrum is broader than the male bone in proportion to its length. The anterior curve of the bone from above down is a continuous one in the male, but in the female there is usually a more sudden curve in the lower part, the upper being nearly flat. This is not a constant feature.

Examine the sacrum in conjunction with the rest of the column, and it becomes evident that the lateral masses can be analysed into transverse and costal elements. The *transverse* processes (Fig. 32, *Tr.*) of the third and fourth elements can be recognised as prominent tubercles on the dorsal surface of the bone, just outside the foramina : again, it is easy to see the transverse element of the first segment in the thick prominence that forms the upper lateral angle of the bone, and the second transverse point can be recognised without difficulty in the middle of the ligamentous sacro-iliac area. Now look at the bone from the front, and there is no difficulty in seeing the *costal* elements in the bars of bone that separate the anterior foramina : all this is simply an expression of the fundamental structure of the column, and the anterior primary divisions of the nerves come forward between the ribs while the posterior divisions pass back between the transverse processes.

Looking at the bone from above, it is easy to see that the lateral mass is built up by the fusion of these costal elements and transverse processes, and the same

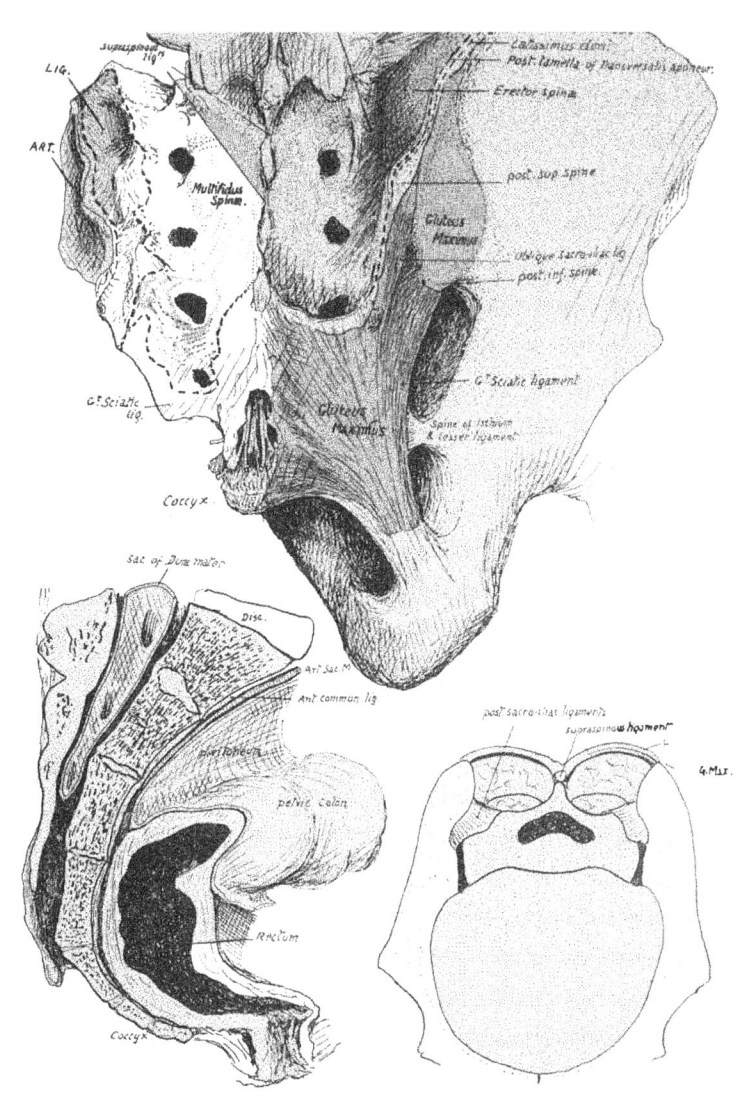

FIG. 34.—The upper figure shows the sacro-coccygeal region from behind. On the left side are the areas for Multifidus, sciatic ligament; posterior sacro-iliac ligs. (*LIG*), and articular surface (*ART*). On the right the Multifidus ha altogether removed; it is covered in by the mass of Erector spinæ, which is arising from the posterior sacro-iliac ments and the adjacent part of the ilium. These post-vertebral muscles are shut in by a thick aponeurosis wh fastened to the spinous ridge and to the iliac crest externally: it is a compound of post-vertebral aponeuros aponeurotic origin of Latissimus dorsi, and its iliac attachment is shown by interrupted lines. It is fused below with the sciatic ligament and is shown cut here; this continuity with the ligament explains how the Gluteus maximus ar ligament obtain origin from the lower part of the sacrum. Multifidus does not extend down to the lower part of the bon here the ligamentous fibres are attached to the sacrum and cover in the hiatus sacralis centrally, with the nerves, etc., ly it. In the drawing the hiatus and coccyx have been exposed by removal of this covering of fibrous tissue. It sho noted that Multifidus really arises from the surface of a fibrous sheet which is attached to points on the bone, and not d from the bone itself; the junction between the nerves takes place between this sheet and the bone. The lower left shows a section through the middle line of the sacrum and coccyx, illustrating their relations with the rectum peritoneum. The remaining figure is a scheme of a section across the pelvis to show how the sacrum is "susper between the innominate bones by the posterior sacro-iliac ligaments, which must, therefore, be exceedingly thick and s Behind these the post-vertebral muscles fill up the deep fossa between the posterior parts of the iliac crests and the spines. Compare this with the first figure *LD*, plane of Latiss. dorsi fused with post-vertebral aponeurosis.

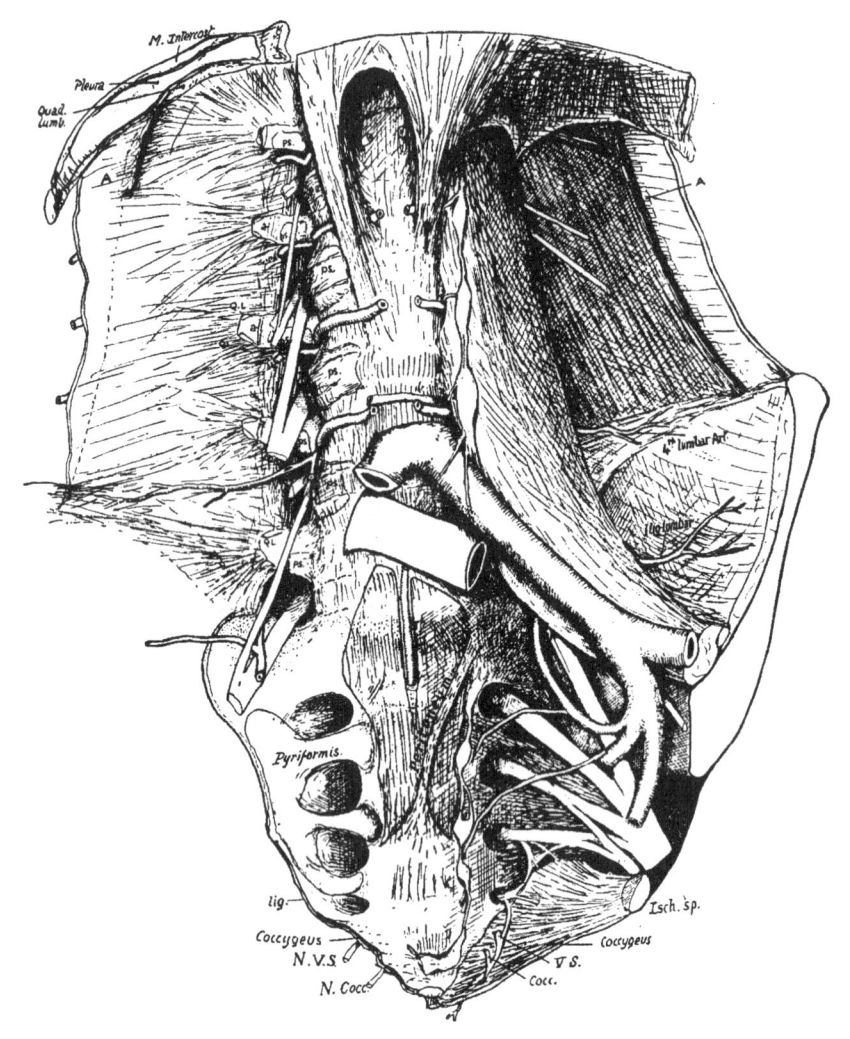

35.—To show the anterior relations of the lumbar column and sacrum. Semidiagrammatic, the lower portion being repre ented as somewhat flattened out in order to exhibit the structures. For the lumbar region the figure should be compared wit Fig. 27, and for the sacrum, with Fig. 34 (2). *A*, cut anterior lamella of Transversalis aponeurosis: *a*, its attachment o transverse process: *ps*, origin of Psoas: *QL*, Quadratus lumborum; *lig.*, attachment of small sacro-sciatic ligament continuous with Coccygeus: *N. VS, N. Cocc*, anterior divisions of last sacral and coccygeal nerves. The aorta is cut awa down to the bifurcation, leaving the lumbar arteries in place; these are exposed on one side by removal of the Psoas The first two arteries run under the crus of the Diaphragm on each side as well as the Psoas. They come into contact wit the bones as soon as they come off the anterior common ligament. Observe that the main (middle) lamella of Transversali aponeurosis contains fibres radiating from the ends of the transverse processes, and these are continuous with the costo lumbar set above and with the ilio-lumbar bands below: all these are behind the plane of Quadratus lumborum. The ban seen in front of the lower end of the muscle is in the anterior lamella and does not really belong to the proper ilio-lumba ligamentous system. The central part of the sacrum is covered for its upper half or so by peritoneum, forming the pelvi mesocolon along an oblique line, and supports the rectum below. The peritoneum has been cut to expose the middle sacra artery. Pyriformis origin extends on to the bodies, and this is usually the only part of the origin that is tendinous, makin definite markings on the bone. The sympathetic chain and lateral sacral arteries reach the bone under the peritoneun internal to these origins. The small sacro-sciatic ligament is attached to the edge of the bone as far up as the lower level o the articular surface; it is continuous below with the Coccygeus, the fibres of which muscle also extend on to the front of the bon and join the fibrous tissue which covers this aspect of the bone and the coccyx and separates them from the rectum. The las sacral and coccygeal nerves must pierce the Coccygeus to reach a plane anterior to the bone. Other details are considered i the text.

structure can be traced down through it until it reaches its end in the badly developed elements of the fifth segment. In addition to these fundamental parts of the lateral mass, there are epiphyses that cover its lateral aspect (Fig. 33). There are also ossifications of inter-transverse, costo-transverse, and other ligaments.

The articular surface for the sacrum on the innominate bone extends back on to the posterior inferior spine of that bone, so that this bony point is opposite the third sacral segment : the great sciatic ligament is attached above to the inferior spine, and from this extends on to the lower part of the transverse tubercle of the third segment. The upper part of the tubercle has the superficial oblique fascicles of the sacro-iliac ligaments attached to it. Below this level the great sacro-sciatic ligament is fastened to the posterior aspect of the lateral margin of the bone and to the transverse tubercles, extending down to the coccyx, and more superficial fibres extend further in to obtain attachment to the articular masses and cover some outer fibres of the Multifidus spinæ arising from the bone. Gluteus maximus spreads from the ilium down this ligament and thus gets indirect origin from the bony points of the lower part of the sacrum (see Fig. 34).

Fibres of Multifidus spinæ can be found arising in the sacral groove as far down as the fourth segment, and extending outwards between the third and fourth foramina : above this the fibres reach the sacro-iliac ligaments externally, so that the more superficial Erector spinæ is separated by it from the sacrum and only reaches bone (ilium) outside the ligamentous area, though it has some slight attachment through its covering aponeurosis to the median spinous ridge.

It thus comes about that the posterior primary divisions of nerves, coming through the first three foramina, are in the substance of Multifidus spinæ and are connected by loops here, on the bone : their outer branches run down and out to get on to the great sciatic ligament deep to Gluteus maximus. Lower down the loops between the third, fourth and fifth posterior divisions lie on the bone.

At the extreme edge below the articular surface is the attachment of the Coccygeus fibres, extending down to the coccyx (see Fig. 35). The back and upper fibres of this sheet are ligamentous and constitute the small sacro-sciatic ligament, so that this ligament separates the muscle from the great sciatic ligament, being in contact to some extent and fused with the latter and being continuous with the former.

The fifth sacral nerve emerges at the side of the lower sacral aperture, and its anterior primary division turns outwards and forwards, below the cornu and behind the last segment of the bone. At the outer border of this it reaches the plane of the Coccygeus and pierces this to appear on its pelvic aspect, where it is joined by the descending filament from the fourth sacral and turns down close to the coccygeal attachment of the muscle. The anterior division of the coccygeal nerve runs a similar course at a lower level, behind the first coccygeal body, piercing the muscle lower down to join the fifth sacral nerve in front of it : the nerve thus composed pierces the Coccygeus again near the tip of the coccyx to supply the skin in that region.

The front surface of the bodies—particularly in the middle segments—may show longitudinal markings, which indicate that some at least of the fibres of the anterior common ligament are involved in the general ossification of cartilage and ligament that consolidates the bone ; but in any case the greater part of this ligament is continued down on the front of the bone, forming the anterior sacro-coccygeal ligament at its lower end : the middle sacral artery runs down on the ligament.

At the side of the bodies the parietal pelvic fascia is attached to the margin of the foramina, so that the issuing nerves lie deep to the fascia : they lie on the Pyriformis, which arises from the lateral mass opposite the middle three pieces and outside the first foramen, and extends in on the costal bars between the foramina, as far as the limiting markings on the sides of the bodies. Lateral sacral vessels lie on the fascia and reach the sacrum internal to the foramina, and here also the sacral sympathetic chain touches the bone. Superficial to the structures mentioned, the peritoneum lies in front of the bone in its upper half or more, while in the lower part the rectum rests directly on the bone. All these relations are shown in Fig. 35.

The ala, or upper surface of the lateral mass, lies deeply under cover of the Psoas for the most part, but its inner and front part carries the lumbo-sacral cord and ilio-lumbar vessels, and these appear from under cover of the inner border of the muscle. The last lumbar nerve emerges between the upper sacral articular processes and the back of the intervertebral disc, and its posterior primary division turns back round the process to reach the post-vertebral muscles, while its anterior part runs forward, outward and downward on the ala, being joined here by the branch from the fourth. Thus the nerve appears to emerge through an osseo-fibrous foramen completed externally by arching fibres of the ilio-lumbar ligament. All these structures are placed deep to the Psoas. The fibres of the ligament pass to the transverse tubercle of the first sacral vertebra and the bone in front of this and account for some of the roughness generally found here: the remaining markings, extending round the margin of the articular area, are caused by the weak anterior sacro-iliac ligaments.

The sacral canal narrows from above down; smaller canals leave it at intervals and run down and out, dividing into two terminal parts which at once open at the anterior and posterior foramina. It thus happens that each foramen in front is directly opposite its corresponding posterior opening, and arteries (from lateral sacrals) entering the anterior foramina to supply the structures in the canal can give branches that in a very short course reach the back openings and emerge there: these correspond with the posterior branches of the intercostal and lumbar vessels, and are the proper continuations of the branches of the lateral sacral.

Passing through the upper aperture are the sacral and coccygeal nerves (part of cauda equina), the filum terminale, and spinal membranes. In the bone the upper four nerves pass out through their proper foramina, and the loose sac of dura mater, with its enclosed arachnoid, only extends about half-way down the sacrum: here it is pierced by the filum terminale, but a very thin layer from the dura is continued on this structure.

Therefore are only found, passing through the lower aperture, the diverging last sacral and coccygeal nerves, and the centrally placed filum terminale, running on to end in the ligamentous tissue covering the back of the coccyx.

This aperture is continuous with the "hiatus": the hiatus is present over the fifth segment and frequently over the fourth. Where the bone is deficient the canal is covered in by ligamentous bands attached to the bony margin.

The condition of complete consolidation is attained comparatively late in the sacrum, not being found before the age of twenty-five or later. Possibly the persistence of non-fused segments and parts is associated with the pelvic growth that precedes sexual maturity, and consolidation commences in the lower part of the bone a year or two after the age of puberty; but it must also be remembered that, while growth continues, the lengthening limbs and changed proportions of the trunk lead to alterations in the position of the centre of gravity and in the lines of transmission of weight, so that the later consolidation of the upper part of the bone, delayed till growth has ceased, may be connected with the fact that this is the part of the bone through which weight is transmitted to the innominate bones.

Even when the bone is firmly joined there are still remnants of the intervertebral discs to be found on section (Fig. 34): these can be discovered up to past middle life.

Vertebral Column

The sacrum presents sexual distinctions. The female bone is broader than the male in proportion to its length, and shows a different anterior curve: in the male the curve is more or less uniform from above downward, but in the female it is sharply marked at the lower part of the bone, the upper portion being almost flat.

The difference in breadth between the bones is well expressed by the "sacral index," in which the breadth (at the base) is multiplied by 100 and divided by the anterior length; this gives an average index for British males of about 112 and for females 116. Indices above 100 class the sacra as *platyhieric*, those with indices below 100 being *dolichohieric*. The lower races tend to be dolichohieric, although their females are usually platyhieric. The proportionate breadth of the sacrum is a human character; the gorilla gives an index of 72.

The sacrum is occasionally found to contain six segments instead of the normal five. This may result from inclusion of the proximal coccygeal element or, less commonly, from the inclusion of the last lumbar; in the latter case the added bone may present lumbar characters on one side and sacral on the other. In other rarer cases the sacrum may possess only four segments: if this condition is due to failure of fusion with the highest segment there will be twenty-five presacral vertebræ.

Opposite views are held as to the possibility of the human sacrum being in process of shifting its position along the column: certain observers hold that the sacral consolidation is tending to progress cranially, while others deny this or even maintain that it may be shifting caudally. The first and oldest of these views does not appear to receive support from embryology.

Ossification.

The centres from which the bone is formed are those of its constituent vertebræ *plus* costal and epiphysial centres in the lateral mass, but the number of these last is variously estimated by different observers: possibly this bone, like others, is not absolutely constant in the number of its small secondary centres.

Each segment is laid down in cartilage, including the body and the neural half-arches with their transverse and articular processes. Costal chondrification forms the costal bars, which quickly fuse with the other cartilaginous parts proximally, and with each other distally to make the basis of the outer part of the lateral mass carrying the articular surface. Thus the bone as a whole is preformed in cartilage. *Primary* centres appear in the cartilage at very variable times, those for the proximal segments generally much earlier than the distal centres. The centres for the bodies appear between the third and seventh or eighth month, those for the neural arches after the mid-fœtal period, and the costal centres shortly after the neural centres: within these limits there seems to be much individual variation.

The centres extend very slowly, so that at birth the sacrum consists mainly of cartilage. Secondary centres begin to show about puberty and continue to appear until after twenty-one: they include epiphysial plates for the bodies, epiphysial plates for the lateral masses, and a variable number of centres for the different bony prominences.

The centres may be classed as follows:—

I. PRIMARY.

1 for each body	5
1 for each half neural arch	10
1 for upper three costal bars	6
	21

(Costal centres may be 2 or 4 on each side.)

Anatomy of Skeleton

II. Secondary.

(a) (*Constant*)
Upper and lower epiphysis for each body	10
Epiphysial plate for auricular surface	2
Epiphysial plate for margin below this	2
	14

(b) (*Variable*).
Epiphysial tips for spinous, transverse, and mammillary processes.
Additional centres for the margins.

The fusion of all these component parts commences in the lateral region, where the costal elements join the neural arches before the sixth year. A little later the arches and costal bars join the centre, the more distal ones fusing first. As soon as this has occurred, the laminæ commence to unite from before backward, completing the process at puberty, and then the costal processes unite with each other distally.

After this the epiphysial plates of the bodies commence their junction, from below upwards, the process not being completed until twenty-five.

Lastly, about this time, the lateral epiphysial plates and the remaining secondary centres consolidate with the bone.

COCCYX.

This small bone is composed of the fused remains of four or more vertebral bodies, of which the first alone, as a rule, carries rudiments of transverse processes and upper articular processes: the latter are termed *cornua*, and are attached by ligaments to the sacral cornua. The upper surface of the first coccygeal body articulates with the lower end of the sacrum with intervening fibro-cartilage. The number of segments in the bone is variable, and may said to be four or less in women and four or more in men, speaking generally: ventral processes have been described as occasionally present, representing either costal elements or the remains of hæmal arches.

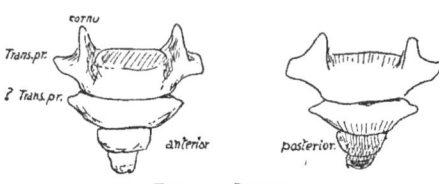

FIG. 35 —Coccyx.

The first segment is often separate, the others more rarely: fusion is said to take place later in women, no doubt in accordance with the functional requirements of the pelvic outlet.

The last segment represents the fusion of three or more rudiments that are present in the embryo: it is the extent of this fusion that determines the number of segments present in the adult bone: the fourth segment is frequently larger than the third.

The bone is directed forwards and downwards from the sacrum, so that its ventral surface looks upwards and forwards. The muscular floor of the pelvis reaches its margin, and ligamentous fibres from this and the lesser sciatic ligament run on to its front surface and cover it, with the fibres of the anterior ligament: these separate it from the coccygeal ganglion and the rectum (Fig. 35). The great sciatic ligament and Gluteus maximus reach the margins behind the structures already mentioned, and fibres from these spread over the dorsal surface of the bone, covering in the filum

Vertebral Column

terminale, which runs to the back of the bone from the lower sacral opening, and the coccygeal nerve, which turns out behind the first segment. The " ano-coccygeal ligament " and external sphincter are attached to the end of the bone.

Thus the coccyx could be described as embedded in fibrous tissue and affected by the contractions of the muscular pelvic floor and the Gluteus maximus: the fibrous tissue on its dorsum contains filaments of the posterior primary divisions of the coccygeal and last sacral nerves.

The rudiments of at least six coccygeal vertebræ are to be found in embryonic life. The distal ones fuse, and there are usually four at birth, still cartilaginous. **Ossification** begins usually in the first year in the first segment, but this is very variable in individuals. The ossification of the succeeding segments goes on with wide limits until puberty, or even after this. Centres are also described for the cornua and for epiphysial plates in each segment: also a fifth segmental ossification has been described, fusing later with the fourth. Possibly some, at any rate, of these can be looked on as individual variations

Ossification appears to be later in women than in men.

CHAPTER III

THORAX

THE thoracic skeleton is attached dorsally to the vertebral column. It contains a series of twelve *ribs* on each side, connected with a central *sternum* in front by means of *costal cartilages*.

Of each series of twelve ribs only the upper seven have their cartilages directly connected with the sternum : the next three cartilages each join the one immediately above, so only reach the sternum indirectly, and the two last ribs have their short cartilaginous ends terminating without such junction, lying in the muscles of the body wall.

The upper seven pairs are termed *true* or *sternal* ribs : the remainder are *false* or *asternal*, the last two ribs on each side being sometimes referred to as " floating ribs." There are sometimes eight sternal ribs.

The wall of the thorax is completed by intercostal muscles and membranes in the *intercostal spaces* between the ribs, and pleura lines it on its inner side.

The ribs and sternum constitute a firm but movable thoracic cage protecting the viscera. The firmness due to the bony elements obviates collapse from atmospheric pressure, the lungs being expanded. Hence the intrathoracic pressure can be below that of the atmosphere. The power of movement of the walls is associated, of course, with respiration.

RIBS AND COSTAL CARTILAGES.

A bone from the middle of the series will show the general characters and various parts of a rib. Such a specimen (Fig. 37) exhibits at its vertebral end an expanded *head* (A), joined to the remainder of the bone by a somewhat narrower *neck* which has a rough ridge, the *crista colli superior* (C), for the upper costo-transverse ligament. The head has two *articular facets*, which are covered with fibro-cartilage, for the bodies of the two vertebræ that carry it, separated by an *interarticular crest* (B) for the attachment of the ligament that fastens it to the intervertebral disc. The neck has a smooth pleural surface in front, but is ridged behind by the middle costo-transverse ligament : there is often an inferior crest on the neck.

The rest of the bone, outside the neck, is the *shaft* or *body* (D), and where this joins the neck is the *tubercle* on the back and lower aspect of the bone : the tubercle presents an inner *articular* part (*art.*), with a facet for the transverse process of its vertebra, and an outer *non-articular* part (*n. a.*) for the external or posterior costo-transverse ligament.

The shaft passes outwards from the tubercle for a little distance, and then turns rather sharply forwards and outwards : this turn makes the *angle* of the rib. Observe that the angle is properly a sharp curve in the bone, but carries on its surface a secondary marking made by the attachment of the Ilio-costalis and vertebral aponeurosis.

Beyond the angle the curving shaft is continued in a forward, inward and downward direction : it has an *upper* border, thick and rounded behind and sharper in front, a *lower* border sharp behind and rather rounded in front, an *outer* and an *inner* surface.

The inner surface, covered by pleura in its greater extent, presents below a *costal groove* (S) for intercostal vessels, best marked behind : the intercostal muscles are attached to the margins of the grove.

The anterior end of the shaft has an oval hollow (CC) which receives the rounded end of the costal cartilage without any intervening synovial cavity.

A slight bend in the shaft may be noticed a little distance from its anterior extremity : this is the *anterior angle*. The rib is bent in three directions : (1) a general curve forwards and then inwards ; (2) a curve upwards of the head and neck and

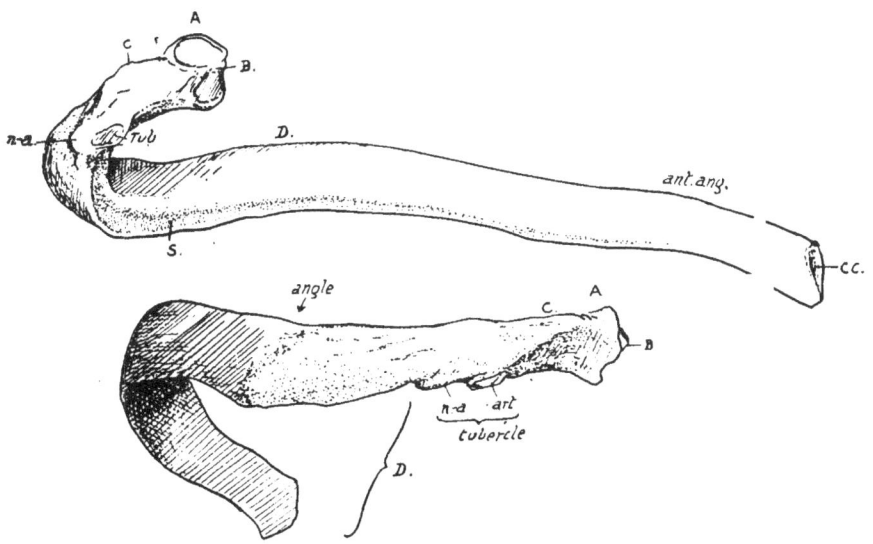

FIG. 37.—A middle left rib. *A*. head ; *B*. ridge for interarticular ligament ; C. crista superior ; *D*. shaft. Upper figure shows the inner or p eu surface of shaft, and the subcostal groove S. Lower drawing shows the posterior part of the rib, between head and angle, which has intercostal structures attached to it and is completely covered by post-vertebral muscles over these ; the most superficial of these last reach the bone at the angle and make the secondary lines here. The head articulates with two vertebræ and the tubercle rests on the transverse process of the lower one.

adjoining portion of shaft, connected with a downward turn at the front end in many ribs ; (3) a twist on its longitudinal axis that begins at the angle, so that the outer surface of the bone shows a tendency to look upwards as well as outwards as it is followed forward.

COSTAL CARTILAGES.

These are bars of hyaline cartilage rather more rounded than the ends of the ribs with which they are joined. Each is covered by a thick *perichondrium* continuous with the periosteum of the rib, and this is sometimes partly ossified in old age.* The cartilages that reach the sternum, with the exception of the first, articulate with that bone by diarthrodial joints. There are also synovial cavities at the junction of the eighth with the seventh, the ninth with the eighth and the tenth with the ninth, and

---

* The perichondrium of the first rib becomes calcified much earlier.

usually there are additional interchondral articulations between the fifth, sixth and seventh cartilages. The cartilages run at first in the direction of the front ends of their ribs, so that, from the fourth down, they go at first downwards and then come up with an increasing curve, towards the sternum: the third and second are nearly horizontal, and the first is directed downwards as well as inwards.

The second is most prominent anteriorly, and can be recognised at once in the living subject: the succeeding cartilages can be counted down from this.

Having become familiar with the names and positions of the various parts and with the general shape of the ribs, it is as well to consider them as a whole in the articulated skeleton before proceeding to study the details of the individual bones.

Looking at them in position, it at once becomes evident that the ribs are placed obliquely in the wall of the thorax, and there is a slight increase in their obliquity from above downwards, so that the intercostal spaces are somewhat wider in front than behind: the upper spaces are also rather wider than the lower ones. The bony thorax, the greater part of the walls of which is made by the ribs, is somewhat barrel-shaped, flattened from before backwards, and broader below than above. There is an evident convexity from above downwards in the line of its side wall, and this line turns in markedly at the top. Fig. 38 shows the effect of this form of the thorax on the direction in which corresponding surfaces of the ribs are turned: it is seen that the highest part of the curve is about the ninth rib, and that the line, turning in below this, causes the pleural surface of the lower ribs to look upwards as well as inwards, while above the ninth they look down. This downward aspect of the pleural surface reaches its acme in the upper two ribs, owing to the increased curve in the line, and on the first rib the pleural surface looks almost directly downwards. But it is clear, notwithstanding, that the lower surface of the first rib corresponds with the inner (and upper) surface of the last rib.

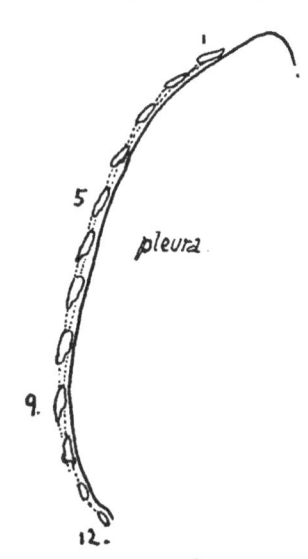

FIG. 38.—Scheme to show the effect of the shape of the thorax on the direction of the surfaces of the ribs.

The antero-posterior compression of the thorax is a human character, a greater comparative depth being found in lower animals and in the fœtus.* The adult measurement between the lower end of the sternum and the vertebral column is about 8 inches, and at the top of the sternum about 2—2½ inches, while the broadest part of the cavity, about the level of the ninth rib, is about 11 inches. Owing to the anterior deficiency that forms the *costal arch* or *subcostal angle*, the length of the posterior part of the cavity (12—13 inches) is about twice as long as its vertebral measurement in front.

These last measurements, with the obliquity of the ribs, account for the *upper aperture*, bounded by the first ribs and dorsal vertebra and top of manubrium, sloping downward, while the large *lower*

---

* Like other recently acquired characters, this shallowness of the thorax is variable. It can be expressed by an index $\frac{\text{depth} \times 100}{\text{breadth}}$ (at the nipple level). The normal average is about 71, but varies upwards to 80, or even 90 or more. These high indices are frequently if not usually in tuberculous chests, and may represent an arrest in evolution of the human thorax: the evolution goes on during ontogeny, the index being considerably over 100 before birth, decreasing to about 80 at puberty, and then reaching the normal level in the next eight to ten years.

ч

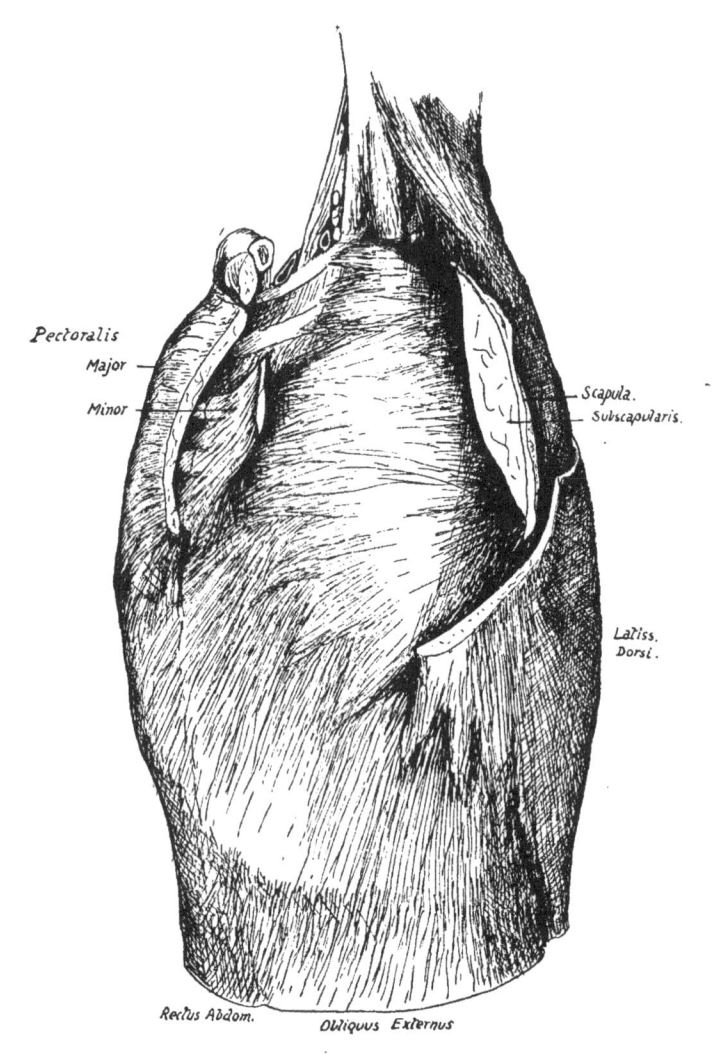

FIG. 39.—The antero-external relations of the ribs and costal cartilages. The front and back walls of the axilla have been in great part cut away, exposing the Serratus magnus as the inner wall. Observe the front parts of the two upper ribs not covered by the muscle, and Scalenus posticus passing deep to it.

Thorax

aperture, bounded by the cartilages of the false ribs, the last ribs, and the twelfth dorsal vertebra, is in a plane sloping in the opposite direction—at least, in front of the tenth rib.

Looking into the cavity, the bodies of the vertebræ are seen to make a prominent median dorsal projection into it, so that a transverse section would give a reniform shape to the cavity. This vertebral ridge, together with the backward direction of the ribs in the region of their necks, leads to the existence on each side of it of a deep and broad *pulmonary groove* that is filled by the thick posterior border of the lungs.

On their outer side the ribs are covered for the most part by large flat muscles which lie on them. Fig. 39 shows these relations: the pectorals have been in great part removed on the one hand, and on the other the scapula and Subscapularis, with the exception of a strip of their inner parts, have been cut away: thus the Serratus magnus is exposed, covering the greater part of the upper eight ribs, with the External Oblique below this, covering the lower eight and the cartilages. The Latissimus dorsi lies on those ribs that are not covered by Serratus magnus above the External Oblique; its origin is by digitations in series with those of the Serratus, but being more vertically directed it passes into a more superficial plane, first covering the ribs below the Serratus. The cartilages of the false ribs are covered by External Oblique except in front, where the Rectus lies on them and separates them from the aponeurosis of the oblique muscle; here the Pectoralis major obtains some origin from the aponeurosis.

We thus see that, so far as the ribs are concerned, three large muscles cover them nearly altogether, and arise from them. Examine the line of the "anterior angles" in the articulated skeleton, and it is evident that this corresponds with the line of interdigitation of External Oblique with Serratus magnus and Latissimus dorsi: in fact the bend here is apparently due to the pressure of these flat muscles, and if a well-marked "anterior angle" is examined three surfaces can be found meeting on it, as shown in Fig. 40. The lower surface is for the origin of a digitation of the External Oblique, the area in front of this is covered by the digitation of the same muscle arising from the rib next above, while the posterior area is covered by and gives origin to Serratus magnus.

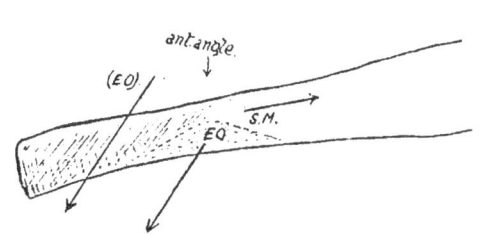

Fig. 40.—Anterior angle of a middle rib; the arrows show the direction of fibres of External Oblique, some arising from the rib and making the lower surface, others coming from the bone above and thus pressing in the front part of the rib. Serratus magnus arises from the posterior area, runs in the direction of the arrow S.M. and moulds the rib lying under it.

Above the fifth rib the definite "anterior angle" is not so clear, but the ribs show signs, in some cases, of the origin of Pectoralis minor in front of Serratus magnus.

The digitations of Serratus magnus from the two upper ribs, on the most in-turned part of the wall, support the axillary sheath in the top of the axilla. Immediately in front of these are the pectoral structures of the anterior wall of the axilla, lying on these ribs.

Further back, the first rib is above the level of the Serratus, and is crossed here by the Scalenus posticus, which goes to the second rib under cover of the top of Serratus magnus.

Though covered by these large muscles, most of the ribs can be felt on the side of the body. By pushing the fingers up in the axilla the third rib can be reached. The second rib can only be found in front, where its cartilage is very apparent.

Now look at the thorax from behind. The large flat muscles only cover the outer aspect of the ribs, and the surface pertaining to them only extends to the line of the angles of the bones: internal to this line the ribs are covered by the muscles of the Erector spinæ group, and even outside the line the upper and lower ribs are covered by the Serrati postici passing to their attachments on them. The line of the angles is not vertical: it is furthest from the middle line usually on the eighth rib, and turns in above and below this, so that it practically coincides with the tubercle on the first rib and thus brings the outer border of the muscle group up to their insertions on the cervical transverse processes. In the same way, below the eighth rib, the width of the muscle mass decreases toward that of the back of the sacrum.

Now observe the change in the relations of the ribs that results from movement of the scapula. That bone covers the second to the seventh rib, but is separated from

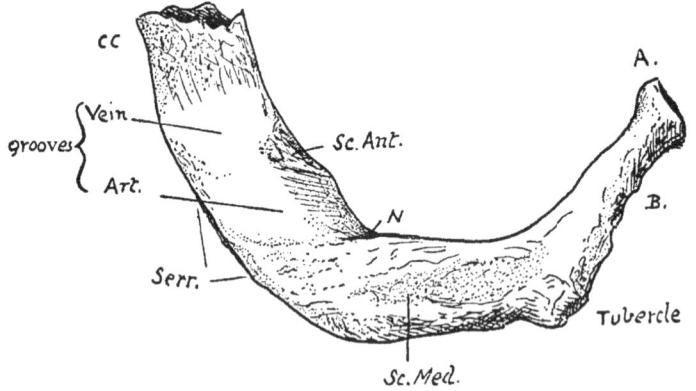

FIG. 41.—Upper surface of left first rib. *A.* head; *B.* neck; *CC.* calcified costal cartilage; *Serr.* origin of Serratus magnus from outer border; *N.* groove made by first dorsal nerve.

them by Subscapularis and Serratus magnus. When the shoulders are "brought together" the vertebral margin of the scapula and its attached structures may ride over the post-vertebral fascia, so that Serratus magnus covers the bones as far as the angle. But when the angle of the scapula is brought outwards and the whole bone rotated, as in raising the arm above the shoulder, the Serratus magnus and vertebral border come away from the angles, and the Rhomboid muscles get into relation with that portion of the bones that was formerly covered by Serratus magnus external to the angles of the ribs: this can be appreciated by rotating the scapula on its clavicular attachments in the skeleton. Owing to the presence of the scapula, the upper ribs cannot be directly examined from behind. Below the seventh, however, the ribs can be felt, though the twelfth is often impalpable in women.

A brief examination shows that the upper and lower members of the series differ decidedly from the majority of ribs: the chief points of distinction are as follows:

First Rib.— Articulates with one vertebra. Short, flat, and comparatively broad, presenting *upper* and *lower surfaces* with *inner* and *outer margins*, instead of the typical arrangement. Head rounded, with a *single* facet for the first dorsal body, neck elongated, tubercle pronounced, no angle. Upper surface marked by scalene muscles and structures passing between them to axillary sheath, showing from behind forwards a rough area for Scalenus medius in front of the tubercle, a groove for subclavian artery (or first dorsal nerve, or both), a *scalene tubercle* near inner margin for Scalenus anticus, and a shallow groove at anterior end for subclavian vein. The partly calcified shell of the costal cartilage is frequently attached to the bone, causing it to appear longer than usual. There is no subcostal groove. The whole bone is more or less in one plane, though the head and neck sometimes show a little downward curve— not upward, as in the lower bones.

Second Rib.—Surfaces look more upward and downward than outward and inward : in fact, the bone has somewhat the appearance of a much elongated edition of a first rib. Head has *two* facets, the upper one very small. Said sometimes to have no angle, but this is not a true description of most bones : angle a short distance from tubercle. A prominent boss of bone about half-way round the outer side of shaft for origin of Serratus magnus. Subcostal groove very broad on hinder part of shaft. This rib also, like the first, lies more or less in one plane.

Tenth Rib.—Head may have one or two facets, according to the individual articulation, with the tenth dorsal vertebra only, or with the ninth and tenth. Tubercle may have a definite though small articular facet, or this may be absent.

Eleventh and Twelfth Ribs.—These do not articulate with transverse processes so they do not present tubercles : they are only carried by their corresponding vertebral bodies, so their heads have single facets. The eleventh rib is longer than the last, shows a definite angle, and has a trace of subcostal groove : the last rib has neither tubercle, groove, nor angle, and varies in length from 1 to 7 or 8 inches. Each of these ribs carries a free pointed short cartilage, embedded in the deeper layers of the abdominal wall.

For the *detailed examination of ribs and cartilages* it will be well to take a middle rib and work out on it the various parts and markings, and then to follow the modifications of these that occur as one passes towards the ends of the series.

Examine first the vertebral end of such a rib (Fig. 42). Roughnesses are found round its articular surfaces for ligamentous bands connecting it with the vertebral bodies. These bands are best developed in front, where they form the "stellate" ligament : this is described as having upper, middle, and lower parts diverging to the two vertebral bodies and intervertebral disc which carry the rib-head, but the division is usually indistinct. The interarticular band is thin, separating the two articular cavities, and is probably derived from the upper fibres of the capsule, left *in situ* when the articular surface extends to the upper vertebra from the lower, to which it properly belongs.

Looking at the inner end of the bone from the front, it is found to present two areas at its vertebral end, the lower of which (P) is pleural and directly continuous with the pleural surface of the shaft : the upper area (X) is confined to the neck and shaft between the neck and angle, and is separated by a soft fatty tissue from the membrane which supports the pleura. The intercostal vessels and nerve run through this tissue in front of the upper costo-transverse ligament (the nerve is shown in the figure), but do not come into contact with the bone until they get near the angle. The depth of X depends on the growth of the Crista colli superior. The "endothoracic fascia," the membrane supporting the pleura, is attached along the line A separating the two areas and to the lower border of the rib above (AA in Fig. 42, C).

These two areas are usually apparent on the bone, but the finger will at any rate detect them at once: the line between them can be traced out to become continuous with the line of attachment of the Internal Intercostal muscle.

The membrane can therefore be looked on as continuing the plane of the Internal Intercostal, and it contains on its deep aspect some aponeurotic fibres continuous with those of the muscle so that in this sense it may be termed the posterior intercostal membrane *—just as the anterior membrane continues the plane of the External Intercostal in front.

The lower border of this part of the bone is a prominent bar that strengthens the curve of the neck, and is continued into the inner lip of the groove on the shaft, to which the Internal Intercostal is attached.

Behind this prominent rim, the under surface of the neck and neighbouring shaft has a broad groove (XX) which corresponds with the surface X on the upper part of the rib, and is in relation with the same soft fatty tissue: it is bounded behind by the plane of the middle costo-transverse ligament (MCT) and capsule of the costo-transverse articulation. Like the upper area X, this lower groove depends for its breadth and depth on the development of the Crista—in this case the *Crista inferior*—that bounds it behind. The Crista *superior* affords attachment to the upper costo-transverse ligament, of which the more oblique anterior fibres reach the top of the crest and the less oblique posterior ones extend somewhat on to the posterior surface. The External Intercostal is continued as an aponeurotic layer along the dotted line B to the ligament. The markings for all these structures are apparent on the bone.

On the back of the neck there is a rough ligamentous area in the lower part (MCT) for the middle costo-transverse ligament, extending down to the lower border and out to the costo-transverse capsule. Between this marking and that for the upper ligament the bone is in its " natural state," and lies in front of the inter-transverse tissues without affording attachment to them.

On the posterior surface of the shaft, between the angle and the tubercle, it is possible to distinguish three areas of markings on the bone. The upper E is for the External Intercostal, as is also the lower and larger one EE, while the intermediate district LC is for the Levator costæ; these muscles are shown *in situ* in the lowest drawing, which exhibits their relations to each other and neighbouring structures. E, frequently an even more narrowed area than in the bone figured, is continued out along the top margin of the shaft and inwards into the line B, while EE reaches the tubercle internally and the lower margin of the bone externally. LC is a well-marked part that extends in above the non-articular tubercle, but externally it does not reach the angle, and the smooth piece of bone left in this way is covered by fibres of Accessorius, which, with Ilio-costalis and the vertebral aponeurosis, is attached to the bone at the angle and makes the secondary lines here. Tendinous slips of Longissimus reach the bone between LC and EE, usually only in the middle six or seven ribs.

E and LC are continuous, owing to continuity of the muscular fibres, and some of the ridges of the levator area can be traced occasionally into EE just in front of the tubercle. This last indicates that the fibres of the Levator may occasionally be continued across the bone into the next External Intercostal.

These possibilities of continuity between the Levator costæ and the External Intercostals can be understood by reference to the drawing of the dissected part:

* I am aware that this is not the usual description, but it appears to me to accord more truly with what is found on dissection.

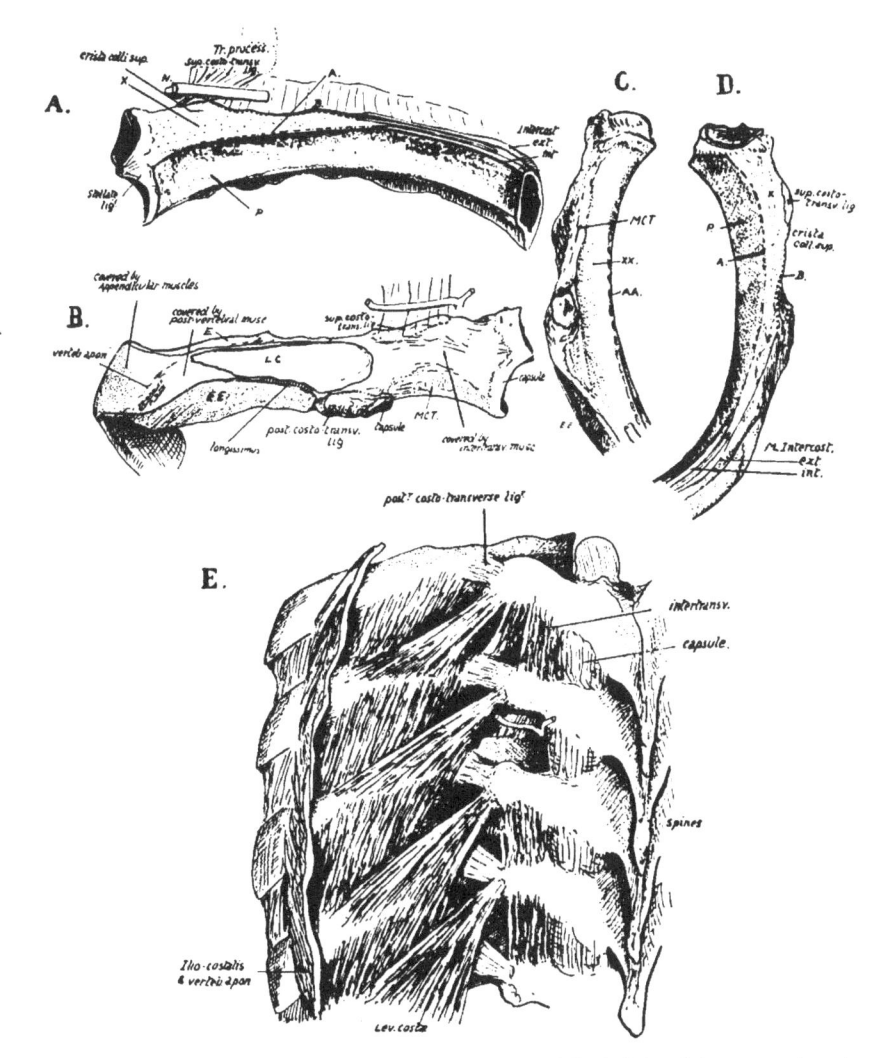

42.—The proximal end of a middle left rib. A from the front, B from behind, C from below, and D from above. In the line A marks the attachment of a fascia continuous with the inner intercostal muscle and therefore the "posterior intercostal fascia": it separates the pleural surface (p) from the area x which supports fatty tissue. Above this th nerve (N) runs through the tissue in front of the ligament. The External Intercostal is prolonged back as an aponeuroti sheet along B to the ligament. In Fig. B observe the different direction of the fibres of the upper ligament seen fron behind: LC, Levator costæ, E and EE are External Intercostals of upper and lower spaces respectively, MCT is middle costo-transverse ligament. Iliocostalis and accessorius are attached just internal to the vertebral aponeurosis, and Longissimus between LC and EE. In C and D the references are as in the other figures. XX and AA in C are area corresponding with X and A on the upper aspect of the rib below. E is from a dissection showing the structures in c ion with the dorsal parts of the ribs. In one space the inter-transverse muscles have been removed, and the posterior division of the nerve is seen dividing into outer and inner branches: the former turns out between the superior costo ns e se ligament and the Inter-transverse muscles to appear on their outer side, while the inner branch runs back round the articular processes. Compare the arrangement of the muscles with Fig. B; also observe that the lower Levators are partly continuous with the outer Intercostal muscle in the next space.

in the lowest spaces in this there is marked prolongation of fibres from the Levator to the next Intercostal, and these fibres, found in the lower spaces, are distinguished as forming " Levatores longiores."

The non-articular portion of the tubercle is for the posterior costo-transverse ligament : observe that the marking for this ligament is partly on the back and partly on the under surface of the bone, where the line of External Intercostal runs into it.

Now compare the posterior parts of the remaining ribs, leaving the first rib for separate consideration. The length of the neck is greatest in the middle ribs and decreases above and below. The same is true of the degree of development of the Crista superior, which is largest in the middle and next succeeding members, so that as one ascends or descends the series the area X decreases considerably in actual size and in fact hardly exists in the terminal ribs. There is a corresponding diminution in the size and distinction of the lower area XX, and the ligamentous surface MCT is smaller and comes to look more downward.

These differences are associated with the degree of development of the superior costo-transverse ligament on the one hand and of the middle ligament on the other. In the last two ribs there is no middle ligament, and the roughened areas lying on the postero-inferior aspects of these bones outside their heads are for the scattered fibres of the posterior ligament further in than in the tenth rib because of the shortness of the last two transverse processes.

The posterior costo-transverse ligament is stronger on the upper ribs than the lower, and so we find the non-articular part of the tubercle better developed in the former and forming rounded masses. Probably in association with this is the occurrence of epiphyses for the non-articular portion in the upper five or six ribs, and not in the lower ones : the strongest of the ligaments is on the middle ribs. These ligaments are strengthened outer bands of the capsule of the tubercle, with which they are continuous, and their development does not appear to be so necessary in the joints that are not in a vertical plane : on examining the articular surfaces of the tubercles we find that those of the upper six are in a vertical plane, while below this they tend to come into a horizontal position, the seventh being transitional : but when the eighth rib is sternal, the seventh tubercular facet is vertical, the eighth is transitional, and the ninth and tenth are less completely horizontal.

The length of the back of the shaft between the tubercle and angle decreases as we pass up from the eighth rib ; it also lessens as we descend from the eighth, but the decrease is of a different sort. On the eighth rib the length of the part under consideration is about a quarter to a fifth of the whole length of the shaft of the bone, and this proportion still holds, or even shows a very slight increase, in the ribs below this ; but as we ascend we find that it bears an ever-lessening ratio to the length of the shaft, being about one-eighth at the fifth rib, and one ninth at the second. Thus we might say that the decrease below the eighth level is a proportionate one, whereas that above this rib is disproportionate as well as actual.

The lessening breadth of this area (covered by post-vertebral muscles) upwards is of course dependent on the narrowing in of the muscle mass to bring its outer attachments on to the cervical transverse processes, but with it is necessarily associated decrease in size of the muscular areas on the bones, and this is particularly noticeable in the case of those for the Levatores costarum : these muscles become smaller and more blended with the underlying External Intercostals as we ascend, being hard to distinguish on the second rib and absent on the first.

The secondary markings made by the outer column of the Erector spinæ and the aponeurosis can be seen on all the bones except the first : even on the last rib, although

Anatomy of Skeleton

there is no true angle in the form of a bend of the shaft, yet there is frequently a secondary marking in series with that on the eleventh.

There is another thing to be noticed in connection with the angle: the *pleural surface* of the rib at this part of the bone shows a tendency to narrow somewhat in the upper ribs, and this reaches such a degree in the second that the pleural surface for a little distance is nearly linear on this bone, widening on the neck and on the under surface of the shaft. The meaning is not clear, but the result appears to be attained by the encroachment of the internal intercostal above, and this muscle with the subcostal groove below, and it is one of the minor points which, taken together, can be used to distinguish upper from lower ribs. In connection with the hinder parts of the ribs it remains to point out that the upward curve formed by these with their shafts is most marked in the seventh rib and decreases from this. This is best shown by putting the bones in a row, resting each bone on the lower border of the shaft as far as possible, when the vertebral ends will be seen to be raised from the table and

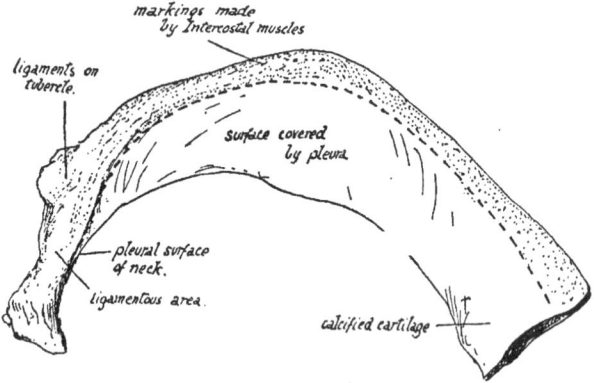

FIG. 44.—Lower surface of first rib.

a line joining them forms a curve with its highest point made by the seventh head and the lowest by the second and tenth.

There is not much to be added to what has already been said about the shafts beyond the angles (pp. 48—50).

There is a tendency to the formation of a *supra-costal groove* between the attachments of the intercostal muscles on the upper ribs. The subcostal groove is perhaps better marked in the lower ribs—except 10, 11, and 12—than in the higher: the floor of the groove is pierced by several small foramina directed towards the vertebral end of the bone. The lower margin of the shaft is sometimes drawn down to form a prominent flange, at and outside the angle, especially in the eighth and ninth ribs, a condition well marked in the foetus: it is interesting to note that downgrowths from the lower margins of the middle ribs, forming *uncinate processes*, are found among the Sauropsida, in crocodiles, some lizards, and many birds.

The general relations of the upper surface of the **first rib** have been already pointed out (p. 50). The extent of insertion of Scalenus medius is shown in Fig. 43. There is no distinct Levator for this rib, the muscle being continuous with the mass of Scalene fibres, but the area that receives what are apparently corresponding fibres is indicated in the figure: it is seen to extend in beyond the level of the tubercle.

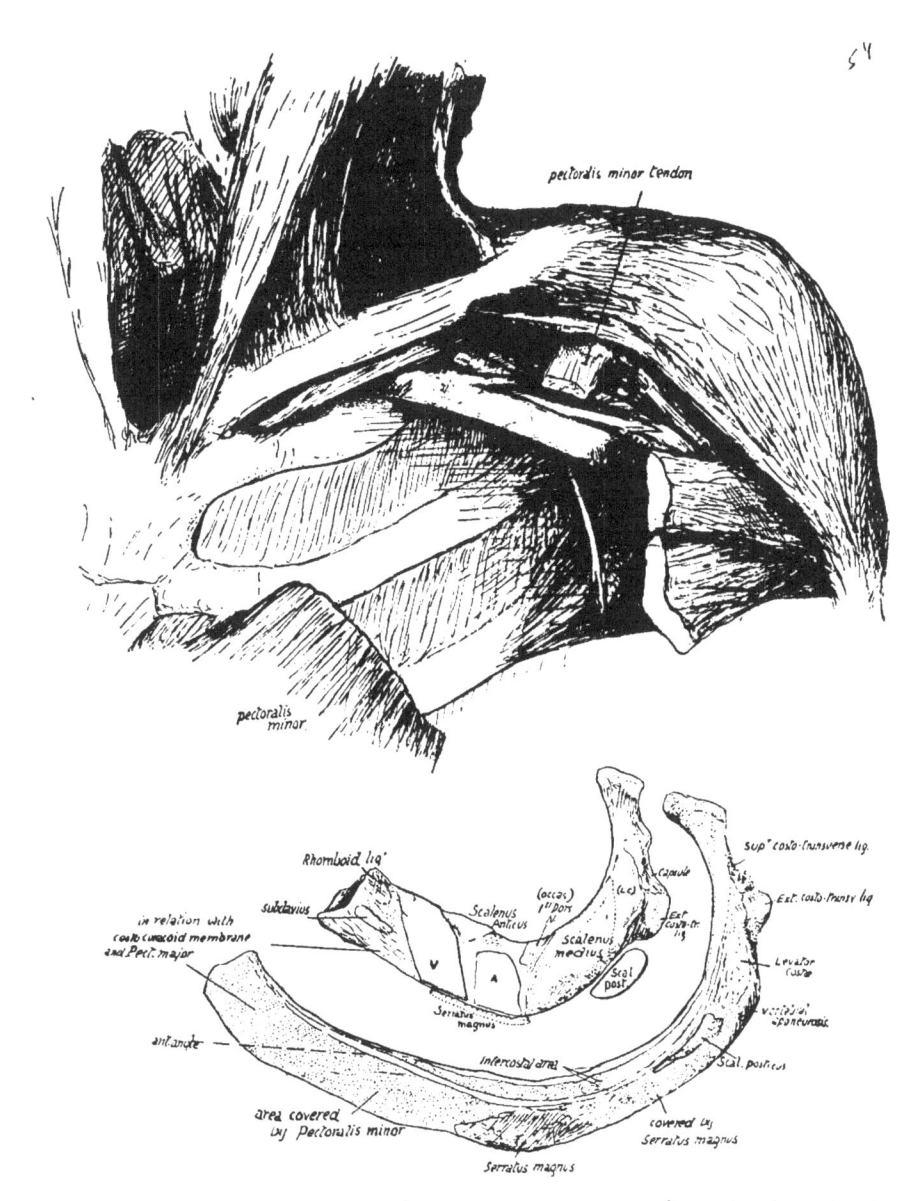

43.—The lower figures are the first two ribs seen from above. The upper figure is from a dissection to show the course of the contents of the axillary sheath. Observe that these go backwards as well as outwards from the first rib, so that the come to lie on the Serratus magnus at once, even though this muscle arises comparatively far back on the two upper ri This leaves the front portions of the ribs uncovered by either the muscle or the vessels, and here the pectoral structu must come into relation with the ribs: the first rib is here in relation with the costo-coracoid membrane and the second with Pectoralis minor. Areas for both these structures may be seen on the second rib, as is shown in the lowest figure, and t anterior angle marks the continuity of the two areas. The lower figures require little comment, but it may be pointed o that the Subclavian and rhomboid attachments on the first rib appear there largely because there is some calcification of t cartilage, that the sharp groove for the 1st dorsal nerve is not always evident, and the posterior part of Scalenus medi insertion, marked off by a dotted line and labelled *LC*, probably represents the Levator costae found as a separate musc lower down.

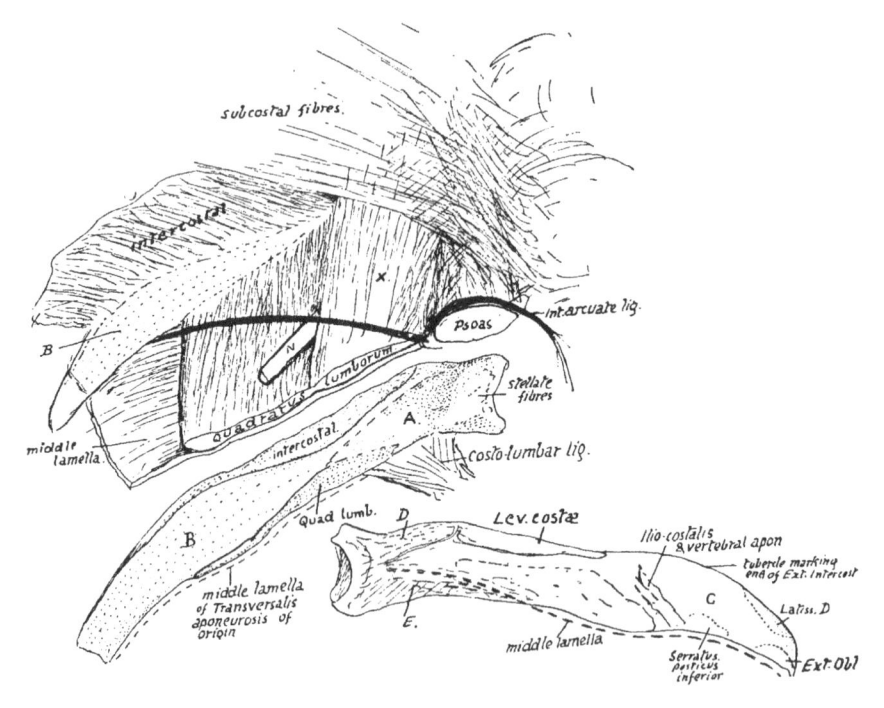

Fig. 45.—Last rib. Right side. See text.

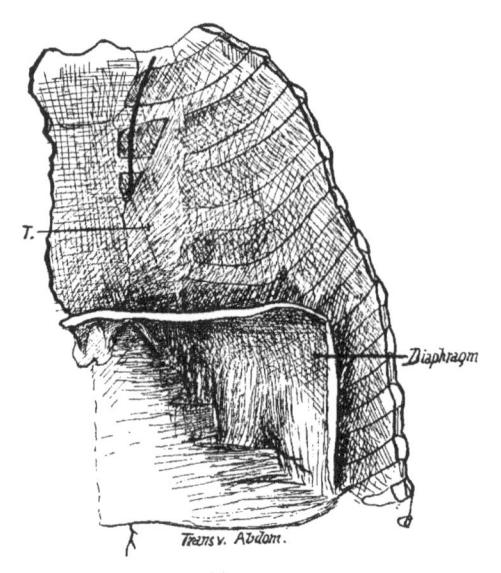

Fig. 46.—Deep View of front wall. *T*, Triangularis sterni.

Thorax

Its inner margin affords attachment to Sibson's fascia (and the Scalenus pleuralis if present), and its outer margin has a marking about half-way round for the origin of Serratus magnus : behind this the Scalenus posticus crosses the margin, just in front of the tubercle, and passes deep to the digitation of the Serratus to reach the second rib. In front of the origin of Serratus magnus the margin is in relation to the costocoracoid membrane and Pectoralis major.

The lower surface of the first rib is nearly altogether pleural in its front half, the intercostal muscles being only attached along the outer margin. The various areas are shown in Fig. 44.

The narrowing of the pleural surface to a line opposite and beyond the angle in the **second rib** has been noticed, as has also the smallness of the insertion of Levator costæ and its continuity with the upper external intercostal. Pass the finger along the most prominent convexity of the bone just outside the angle and an irregularity will be detected which marks the attachment of Serratus posticus superior to the rib. Above and in front of this, near the upper margin, is another impression for Scalenus posticus, and these two muscles lie deep to Serratus magnus. The origin of Serratus magnus is from the prominent boss on the shaft, and immediately in front of this is a flatness, or even a shallow concavity, about an inch and a half broad, on which the Pectoralis minor rests when the arm is hanging by the side. These relations are shown in Fig. 43.

The inner surface of the **last rib** looks in an upward and inward direction in its whole extent, particularly at its vertebral end : if this is borne in mind the common mistake of placing the bone on the wrong side of the body will not be made. Its lower margin is sharper than the upper border and has the costo-lumbar ligamentous bands inserted into it : Quadratus lumborum is in front of this plane, so its insertion into the bone is along the lower part of its front or inner surface, where it has a marking on the bone. The marking is due to the presence of aponeurotic fibres in the muscle where it lies behind the Diaphragm, and to the anterior lamella of aponeurotic fascia covering it in front. The Internal Intercostal also encroaches somewhat on this surface near the top border, but with these exceptions the surface is pleural (see Fig. 35).

But this simple arrangement of the structures about the last rib does not always hold, and bones are frequently met with which show by their markings that additional relations were present with them. Such a bone is shown in Fig. 45, and a sketch of a dissection is given above it which represents a condition that was also probably present in the body from which the bone was taken.

In the dissection the Quadratus lumborum is seen to give a stout aponeurotic sheet (X) (in front of the last dorsal nerve (N)), which passes up in front of the bone to reach the interrupted line above A in the middle figure. This sheet is covered in the dissection by thin aponeurotic sub-costal fibres running inwards and downwards to the side of the vertebral column, so that the inner end of the bone is covered by these two layers of aponeurosis in front. Thus we can recognise, on the front of the bone, the muscular areas for Intercostal and Quadratus lumborum muscles, and an area A covered by aponeurosis passing to the ridge that bounds it above, with a pleural area B.

On the back of the rib (lowest figure) the rough area D is for the scattered costo-transverse fibres, and the lower, E, for the costo-lumbar ligaments connecting it with the first lumbar transverse process. Outside these is a marking above for Levator costæ, and, below this, irregular markings showing the insertion of mixed fibres of the Erector spinæ mass. This area is bounded externally by secondary markings made by fibres of the definite Ilio-costalis and covering aponeurosis. Outside this is the area C, covered by Latissimus dorsi and giving origin to some of its fibres as well as to External Oblique and Serratus posticus inferior.

The middle or main lamella of the aponeurotic origin of Transversalis is fastened to the lower border of the rib and joins the deep aspect of the costo-lumbar ligaments.

COSTAL CARTILAGES.—The *costal arch* or *sub-costal angle* is made by the lower six cartilages on each side. Occasionally the eighth, however, reaches the sternum on one or both sides, more frequently on the right. The increasing obliquity of the cartilages downwards only extends to the tenth : the last two not having indirect connection with the sternum are not turned up but are sharp-pointed tips embedded in the muscles of the body wall.

Interchondral and chondro-sternal articulations are secured by anterior and posterior accessory bands on the capsules of the joints : in the sternal joints these radiate on to the corresponding bony surfaces. The anterior bands are stronger. Synovial cavities are in all the interchondral and in most of the chondro-sternal joints except the first : double in the second and occasionally in the fourth and third. The connection with the ribs is by continuity of covering membrane and some amount of fixation between the cartilage and bone : a synovial cavity has been occasionally found in the first costo-chondral joint, but not in any of the others.

The cartilages increase in length from the first to the seventh and then decrease : their thickness is in general greater as they are followed upwards ; they become somewhat smaller as they approach the sternum.

The cartilages afford attachment above and below to the Internal Intercostal muscles and anterior intercostal membranes, which fill up the interchondral intervals.

The deep aspect of the true cartilages, from the second down, is covered by Triangularis sterni, which arises from them, and crossed by the internal mammary vessels : that of the first cartilage (and part of the second and third) is covered by pleura and crossed by the vessels. In the case of the seventh cartilage the vessels are the superior epigastric, and the outer part of the cartilage, outside the area of Triangularis, is, like the remaining cartilages, taken up (see Fig. 46) by the alternate origin of digitations of Diaphragm and Transversalis : these cartilages that form the costal arch are crossed on their inner side, between the digitations, by intercostal nerves and vessels from musculo-phrenic running into the abdominal wall.

The lower edges of the false cartilages, where these edges form the costal arch, receive the insertion of the upper fibres of the Internal Oblique, and where the cartilages are not joined, in the last two spaces, the fibres of this muscle become directly continuous with those of the Internal Intercostals.

The front surfaces of the true cartilages are covered by Pectoralis major, which also arises from them, with the occasional exception of the first and seventh. Outside and below the pectoral area, the Rectus abdominis covers and is attached to the fifth, sixth and seventh cartilages, and may even extend higher : thus its upper part may come under cover of the pectoral muscle, which gains some origin from its sheath.

In this position the Rectus lies immediately on the cartilage, but when it comes down over the costal arch it lies on the aponeurosis of Internal Oblique inserted into the border of the arch, and this aponeurosis separates it from *muscular* fibres of Transversalis arising from the deep aspects of the cartilages.

Outside the part covered by Rectus the cartilages of the false ribs are covered by External Oblique, but do not give origin to any of its fibres (see Fig. 39).

The first costal cartilage requires some further notice. Its upper surface has an articular surface for the clavicle, on the inner margin of which the inter-articular fibro-cartilage is attached. Outside and rather behind the joint is the rhomboid ligament : in front of this the cartilage is covered by Pectoralis major and, deep to this muscle, gives part origin to Subclavius. The subclavian vein lies behind the ligament and is an upper and back relation of the cartilage to a small extent, and it is below the vein that the pleura and internal mammary artery come against the cartilage. The Sterno-thyroid muscle arises from the inner end of the cartilage (see Fig. 47), so that this part has no pleura in relation with it, and the deep layer of the omohyoid

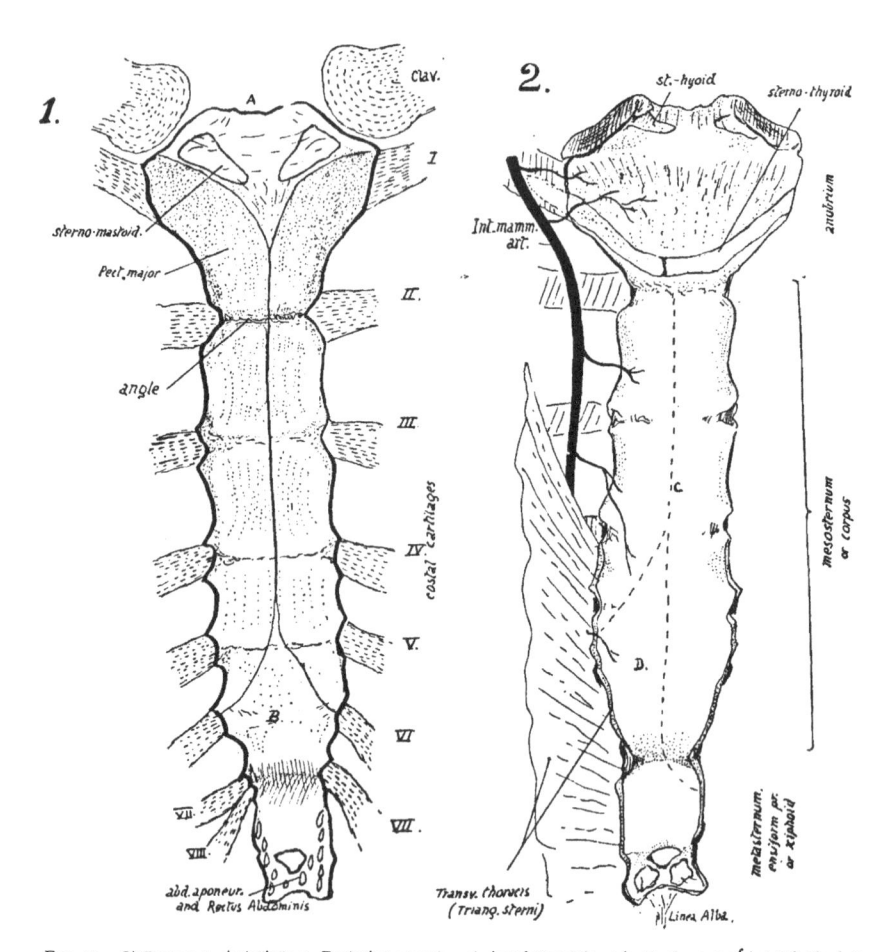

FIG. 47.—Sternum : 1, Anterior ; 2, Posterior aspect. *A*, jugular notch and attachment of interclavicular ligament. *B*, area covered by interlacing fibres (tendinous) continuous with Pect. major, and thus somewhat roughened. Posterior surface below manubrium is covered by thick fibrous tissue, on which *C* indicates lines of pleural reflection, and *D* the area in relation with pericardium, where the left pleura passes away from the middle line. The costal cartilages are shown in their places ; on the right the eighth reaches the sternum, an occasional occurrence.

Thorax

fascia passes down the rhomboid ligament to enclose the muscle, carrying down the layer of scalene fascia that covers the vein, so that both these are attached to the inner part of the cartilage below the Sterno-thyroid.

Development.

Ribs are laid down in cartilage : chondrification appears in the blastemal costal processes of the vertebræ (that extend ventrally into the body wall) and is not continuous with the cartilage of the vertebræ, but connected with it by the continuous blastema.

The ends of the cartilaginous ribs are connected by the sternal plate of that side and the ribs only extend at first partly round the body : the sternal plates have partly met and fused in the middle line when ossification begins in the ribs.

Bony centres appear near the angles : first in the middle ribs, in the eighth week, and other ribs rapidly following, that for the first usually appearing before that for the twelfth. Ossification extends in both directions in each rib. At birth the head, tubercle, and part of neck are cartilaginous (see Fig. 8). At puberty epiphysial centres appear for (1) the head and crista, (2) the articular, and (3) the non-articular tubercle : these all join the main bone after twenty, the head epiphysis joining last— about twenty-four.

The ribs below the sixth do not have a non-articular tubercular epiphysis, and the last two bones have no tubercular epiphysis at all.

STERNUM.

The sternum is a long flat bone, evidently composed of several fused segments, that lies in the front wall of the chest, having the true costal cartilages attached to its sides (Fig. 47).

The bone consists, as already mentioned, of several parts, of which the first piece is the largest and is termed the *manubrium, presternum*, or *episternum*. The succeeding pieces that carry the remaining cartilages are taken together as constituting the *body* of the bone, *mesosternum* or *gladiolus*.

The last piece is an irregular prolongation of the bone, usually bifid at its extremity and lying below and between the seventh cartilages, and is termed the *xiphisternum, ensiform process, xiphoid cartilage*, or *metasternum*.*

The manubrium is the thickest and strongest portion. It articulates with the next piece by a layer of fibro-cartilage that does not ossify until late in life, and the two pieces do not lie quite in the same plane, but form an angle, so that the line of their junction is prominent, and is known as the *angle* of the sternum, or *angle of Louis*. Below this the body is composed of four segments, frequently termed *sternebræ*, and, although these segments are firmly united in the full-grown bone, the lines of junction are usually apparent as three transverse ridges on the front aspect of the body. They are not usually visible behind, though as a rule some indication of the division is to be found near the margins. Each sternebra is flat, or very slightly concave, on its front aspect, and flattened behind. The sternum as a whole, however, shows a slight forward convexity. The metasternum, partly bone and partly cartilaginous, is set in a plane deeper than that of the body, but its lower end may project forward : its upper part may be partly hidden by the cartilages of the seventh ribs.

---

* The older names, manubrium, gladiolus and xiphoid, were given with reference to the fancied resemblance to a gladiator's sword, viz., handle, blade, and point. The other terms are better, in that they refer to the situation of the parts, although the term " episternum " is open to some objection on grounds of comparative anatomy.

The bone is broadest across the manubrium, and then opposite the junction of the fifth cartilages with it. It is narrow at the angle and still more so at the junction with the xiphoid.

The top border of the manubrium presents a central notch, the *suprasternal, interclavicular*, or *jugular notch ;* outside this is an articular facet for the clavicle on the upper lateral angle of the bone : the facet is concave from within out, slightly convex from before backward, and looks upwards and outwards and slightly backwards. The first rib cartilage reaches the bone, just below this facet, on the upper part of the side margin, and the second cartilage is connected with it opposite the angle, partly attached to the manubrium and partly to the first segment of the body. Thus the manubrium has one and a half cartilages connected with it on each side. On the body the third, fourth, and fifth cartilages are attached respectively opposite the lines between the segments as in the figure (Fig. 47), the sixth on the side of the last segment, and the seventh partly on this and partly on the xiphoid. In the figure the eighth cartilage on one side also reaches the xiphoid, an occurrence occasionally met with.

Thus each sternebra bears two half facets on each side for the cartilages : in the case of the second cartilage the articular cavity on the sternebra is separated from that on the manubrium by an interarticular ligament attached to the cartilage in the angle, making a double chondro-sternal articulation, and a corresponding condition may obtain in the third and fourth joints, but frequently in these, and usually in the remainder, the interarticular ligament is increased so that it obliterates one or both of the small cavities.

It may be noticed in this connection that there is usually some asymmetry of the bone, most apparent in the levels of its lateral articulations.

The manubrium, on its front surface, shows two ill-defined lateral hollows below the clavicular level, from which the Pectoralis major muscles arise. The vertical slightly raised area between these hollows forms a T-shaped ridge with a transverse thickening that extends between the clavicular facets. Examination of the back of the bone shows a corresponding transverse elevation, and thus the thickest part of the manubrium, and indeed of the whole bone, is situated between the inner ends of the clavicles, evidently to withstand the pressure of these bones and shocks transmitted through them.

On the front ridge, running down and in from the facet, is a rough area on each side for the attachment of the sternal tendon of the Sterno-mastoid. Above and between these the bone is hollowed, covered by subcutaneous tissues and forming the floor of the supra-sternal hollow, and on the top margin is a roughness for the attachment of the interclavicular ligament.

The Pectoralis major meets its fellow in the middle line a variable distance above the angle ; its area of origin then takes the front surface of the mesosternum to the middle line, as far as the level of the fifth cartilages, where the area narrows towards the sixth cartilage : this narrowing may commence one segment higher. All this origin of the muscle is by fleshy fibres, so that no corresponding rough marking occurs on the bone ; but the bare surface on the bone seen in the drawing below the areas for the muscles is really covered by a thick layer of interlacing tendinous fibres directly continuous with the Pectorals (see also Fig. 39), and these fibres, decussating in the centre, stream down over the end of the bone on to the surface of the xiphoid, where they are blended with the aponeurotic abdominal layers of the linea alba : these aponeuroses are the sheath of the Rectus, and some fibres of this muscle also reach the xiphoid.

Probably these fibrous portions of the pectoral masses account for faint markings that may often be felt on the last segment of the body, frequently suggesting to the touch that there is another dividing ridge extending between the two sixth cartilages. The pseudo-ridge cannot be of the same nature as the other transverse ridges, for these mark the line of junction of the sternebræ, of which each one is developed from distinct centres. The paired centres for the last segment but one appear shortly before birth, and those for the last segment within a year after birth, and no further centres occur in the body, so that there cannot be an additional true transverse ridge. The metasternum commences to ossify some years later than the last segment.

The posterior surface of the body is covered by a very thick layer of vertically arranged aponeurotic fibres, which extend down on to the ensiform process. The layer is not shown in the figure, but it separates the bone from the pleural sacs, whose anterior edges are shown by the interrupted lines. At the *margins* of the body and xiphoid, from the level of the fourth cartilages downward, the Triangularis sterni is inserted. This muscle arises from the cartilages as far up as the second ; its upper fibres, as indicated in the drawing, are very oblique, becoming less so as they get lower, until it merges into the Transversalis abdominis (Fig. 46).

The Diaphragm arises by two muscular slips from the back of the xiphisternum.

On the manubrium there is no thick membrane. The Sterno-thyroids arise by a curved line of continuous origin along the lower part, from one first cartilage to the other, and they lie directly against the bone. Observe that the right muscle as a rule transgresses the middle line in its area of origin.

The Sterno-hyoids arise by a much smaller origin near the top of the bone. None of these infra-hyoid muscles arise by tendon, so no ridges on the bone mark their situation. The deep layer of the " omohyoid fascia " reaches the bone below the Sterno-thyroids, with the pretracheal fascia, just at or above the angle : the superficial layer of the omohyoid fascia is attached to the upper and back part, above the Sterno-hyoids.

The internal mammary artery is shown in the figure, running down immediately outside the Sterno-thyroids, and giving branches between the muscles and the bone : these branches give vessels into the numerous foramina on the back of the manubrium. Lower down the artery gives branches to the back of the body, and even when it lies in front of Triangularis sterni some smaller twigs may come through that structure. Some of these vessels may pass at once deep to the outer edge of the posterior membrane, but the larger branches appear to pass through foramina of some size in the membrane to get to the bone.

The disposition of the membrane suggests that it is the degenerated remnant of a muscular sheet, of which the upper part only remains now in the infra-hyoid muscles attached to the manubrium.

The sternum consists of a thin shell of compact bone enclosing a coarse cancellous tissue, the meshes of which are filled with red marrow : the other bones of the thorax, ribs and vertebræ, also contain red marrow.

The male sternum is longer in proportion to its breadth than the female bone : the difference in length is mainly in the body of the bone, the female body being *less* than twice the length of the manubrium, while in the male it is *more* than twice the length, but part of this difference depends on the fact that the male manubrium is on the whole somewhat smaller proportionately than in women.

The bone slopes slightly forwards as well as downwards in the body, forming with the vertebral column an angle of 20 to 25 degrees, so that the antero-posterior depth of the chest is considerably greater at the lower end of the sternum than at the level of the manubrium, in accordance with the position and shape of the heart and pericardium. Although it is covered by the great pectoral muscle in the greater part of its extent, yet, as this muscle is thin on the bone, the sternum is practically palpable from the surface over its whole area except on the depressed xiphoid : this lies in the floor of the infrasternal depression. The angle is particularly prominent and marks the level of the second costal cartilages, from which the succeeding cartilages may be counted down ; but it should be pointed out that in some uncommon cases the prominent angle is opposite the third cartilages, the manubrium including what would otherwise be the first piece of the body. It is therefore advisable if, when examining a chest, the angle gives the impression of being lower than usual, to take care to ascertain that the angle is really of the normal type and has the second cartilages opposite it. The abnormality is well known, and is usually considered a reversion to the arrangement normally present

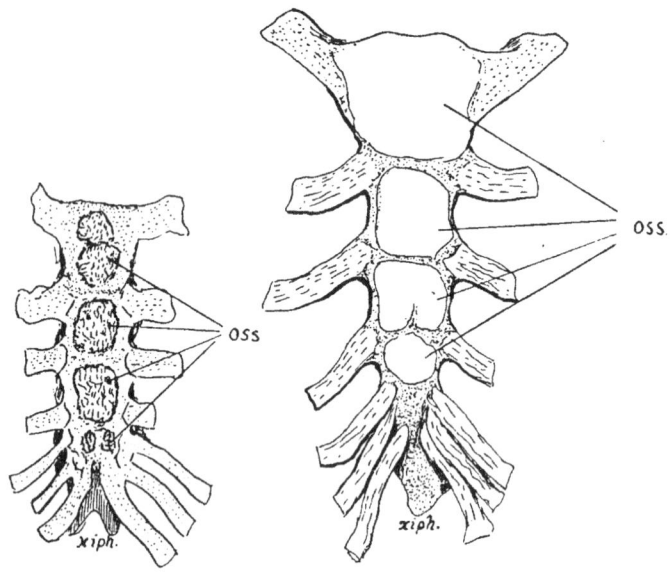

FIG. 48.—Showing variations which may occur in ossification. 1 is a sternum at birth : observe the double centre for manubrium and the single one (perhaps fused) for second piece of body. This is practically a normal specimen. 2, from a young child, although there is considerable growth of the whole sternum, does not yet seem to have any centres in the last piece and thus is hardly normal. Observe that the bone is shaped still in cartilage with separated centres—a condition which remains for some years as the centres extend slowly. When there is more than one centre for the manubrium the added centre appears later than the main one ; this applies also to extra centres found sometimes in or between the sternebræ.

in some of the higher monkeys, but we cannot recognise the conditions that are responsible for its occurrence in certain individuals.

The morphological position of the sternum, the value of its various parts, and their correspondence with the different portions of the sternal apparatus in the lower vertebrates, are controversial questions on which little light is thrown by the ontogenetic development of the human bone. Its comparative shortness in man is correlated with other changes in the thorax resulting from the assumption of the erect position.

In its earliest state the sternum is represented by two bars of condensed tissue, the *sternal plates*, which are continuous with the ventral ends of the upper eight ribs.

At this time the ribs are very short, so that the sternal plates are on the ventro-lateral aspects of the thorax and separated by a wide interval; the clavicles reach the top of each plate and become connected by a dense "episternal bar," which can therefore be looked on as connecting the two sternal plates as well as the clavicles. The sternal plates come together at their cranial ends about the sixth week, and fuse in a caudal direction. The episternal bar is probably partly included between the cranial ends of the plates, and partly remains as the interclavicular ligament.

As the plates come together chondrification extends into them from the ribs, so that the single sternum is at one stage represented by paired cartilages connected by dense mesenchyme. It is necessary to state that some observers maintain that the sternum arises independently of the ribs.

The mode of origin just described suggests that a presternal structure has been included in the manubrium, but the nature of such a structure remains doubtful: it is interesting, however, to observe that the ossification of this part of the bone is very irregular as to its number of centres, supporting the view that the manubrium is of a compound nature.

The two plates, joined mesially, form the cartilaginous sternum, but their caudal ends do not fuse, remaining as the cartilaginous costal arch. The xiphoid process is a secondary growth backwards from the end of the fused portion. Junction may be incomplete in places, especially in the lower part, so that when ossification occurs there is a corresponding failure in bone formation and a foramen remains in the bone, a condition not uncommon: the xiphoid is frequently perforated.

Ossification begins in the sixth month in the upper end of the cartilaginous sternum and progresses from before backwards by the formation of successive centres in the various segments.

Centre or centres for the manubrium appear sixth month.
,,   ,,   first piece of body seventh month.
,,   ,,   second   ,,   eighth month.
,,   ,,   third   ,,   ninth month.
,,   ,,   fourth   ,,   during first year.

Thus *at birth* there are centres present (Fig. 48) above the last segment of the body, but not in that segment.

The centre for the xiphoid appears several years later.

The number of centres for the segments is variable—as a rule one (but often more) for the manubrium, one for the first sternebra, and two for each of the others placed side by side. A single one for the xiphoid.

The course of ossification just described is liable to frequent changes. Beyond the variation in number of the centres for the manubrium, perhaps the most common modification is an earlier appearance of the body centres, so that the last segment has centres at birth, the occurrence of that for the first segment, perhaps, even preceding the centres for the manubrium.

Fusion *between the sternebræ* occurs irregularly, generally progressing from below upwards, occasionally asymmetrically, commencing some years before puberty and being completed about twenty-five. Before it is completed a thin epiphysial plate usually appears for the clavicular facet, and fuses quickly.

The bony ensiform usually *unites with the body* in middle life, whereas the cartilage of the *angle* does not ossify, or only does so in advanced years.

CHAPTER IV

LIMBS: UPPER LIMB

1. THE LIMB GIRDLES AND SKELETON OF THE FREE LIMBS.

The term " limb " is popularly applied to one of the appendages that are freely movable on the " body " and distinct from it, but it must not be forgotten that these are only what one may call " free " limbs, and are carried by a more fixed part related to them and embedded in the walls of the trunk. So in dealing with the skeleton of the limbs we can talk of that of the free limb, and that of the embedded portion with which the skeleton of the free limb articulates.

These embedded parts, really applied to the surface of the proper body-wall, constitute the *pectoral* and *pelvic* girdles: in the former the skeleton is capable of movement and not firmly articulated with the trunk skeleton, but in the latter, where stability and strength are necessary for carrying the weight of the body, the girdle is firmly fixed to the sacrum and forms with it the bony *pelvis*.

The pectoral girdle consists of *Scapula* and *Clavicle* on each side: the clavicle is the only bony connection between the scapula and the trunk skeleton, articulating with the sternum. The upper arm, the proximal segment of the free limb, has the *humerus* as its bony skeleton, articulating with the Scapula and carrying the bones of the forearm (*radius* and *ulna*) at its distal end.

The pelvic girdle is formed by two *innominate* bones which articulate with each other ventrally and with the sacrum dorsally. Each os innominatum is composed of three bones joined together, the *Ilium*, *Ischium*, and *Pubis*, and has a deep articular surface (*acetabulum*) which carries the *femur* that supports the thigh. The Ilium is on the dorsal and upper side of the acetabulum, the Ischium and Pubis on its ventral side, and the junction of one innominate bone with the other is effected by the meeting of their pubic portions in the *symphysis*. All the three parts of the bone are concerned in forming the acetabulum.

The resemblance in details between the upper and lower limbs are matters of common observation, and the general resemblance between their bony parts is very apparent. The meaning of such likeness and the nature of the relationship and morphological identities of the limbs, if any, has been the subject of much controversy and many theories, and appears likely to remain in that position for an indefinite time: those interested should consult the special works that deal with the matter, but some of these views will be shortly spoken of when considering the individual bones of the limbs.

The mode of development of the vertebrate limbs is hidden in the æons that have passed since the earliest chordate came into existence, but we may, perhaps, assume that they were evolved from lateral appendages used by the hypothetical early vertebrate as a means of moving through the water in which he is supposed to have dwelt.

As the bulk of the animal increased, or the size of the appendages, it became necessary to render them firmer by the growth of a skeletal frame-work in them, and, as a result of the continued operation of such causes, to provide a more resistant basis in the wall of the body from which the enlarged and strengthened appendage could act with greater force: thus we have a skeletal basis of a free limb movable on a skeletal support embedded in the body-wall.

Since such limbs were serving the same purpose (with very slight differences as the result of position), the limbs and their skeleton would develop on more or less parallel lines, and their increasing strength would be correlated with parallel growth of their embedded support. We are

Limbs : Upper Limb

only concerned with a theoretical extension dorsally and ventrally on the wall of the body, as shown in Fig. 49, so that we would obtain the rudiments of pelvic and pectoral arches similar in structure and consisting of *dorsal* and *ventral* parts supporting at their junction the skeleton of the free limb. The ventral portion was double, consisting of two bars articulating perhaps with a central ventral part of the trunk skeleton (? sternum) or with their fellows of the opposite side : thus greater stability was secured for the action of the free limb. The dorsal part was broad and single and could reach the dorsal skeleton of the trunk by extension. In this we have the foreshadowing of the human pelvic arch at least, the dorsal ilium and the two ventral bars, pubis and ischium ; the pectoral girdle had a dorsal " scapula," and ventral " coracoid " and " pre-coracoid " bars.

If we now imagine that these strengthened limbs are being used for purposes of support on land as well as propulsion in the water, we can understand that the slight differences in the con

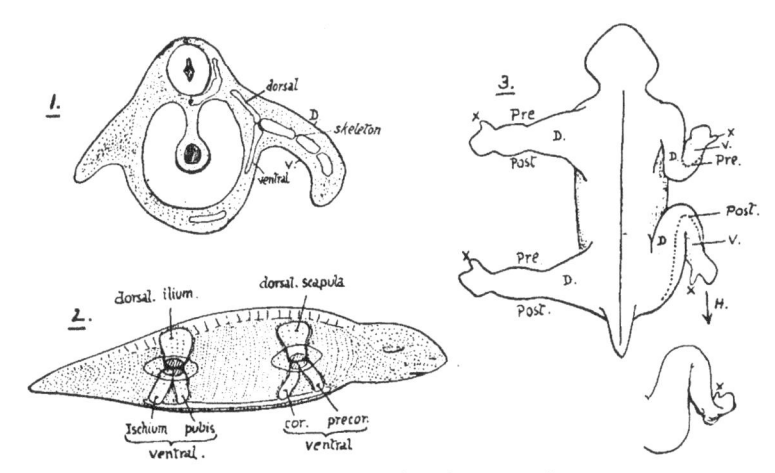

FIG. 49.—1 and 2 are intended to illustrate the hypothetical growth of the bony supports of the limb in the vertebrates. In 1 a short lateral appendage is shown on the left side, made more useful by thickened and resistant tissues in its centre ; on the right side the limb is lengthened and in consequence its supports are strengthened and made bony, these are segmented to allow of movement in the limb, and a *point d'appui* for the action of the whole limb is provided in the form of a skeletal support in the body-wall. Such supports extend dorsally and ventrally. 2 shows the animal from the side with the dorsal and ventral extensions in the body-wall in the pectoral and pelvic regions ; the limits of the limb attachments are seen and the shaded areas at the dorso-ventral junctions mark articular surfaces for the skeleton of the free limbs. 3 contrasts the corresponding parts of the two limbs. Dorsal View. *Pre.*, *Post.* preaxial and postaxial borders ; *D.* dorsal aspect ; *X.* thumb and big toe. The common type is altered on the right as the limbs come into position, and the ventral surface, *V.*, looks forward in the fore limb and back in e hind limb. *H.* points to the form of the human (or plantigrade) foot derived from this

ditions of action of the limbs will begin to graft differences of detail on the fundamentally similar structure of these parts, and these will be accentuated as the specialisation in function of the fore limbs progresses. Thus we might consider the fixation of the pelvic girdle and great mobility of the pectoral girdle as being characters secondarily acquired, in common with lesser differences in structural detail, and grafted on the fundamentally similar construction of the limbs.

Looked at in this way, it becomes doubtful whether such a thing as true *homology* can be said to exist between the upper and lower limbs ; the likeness between them is the result of similarity of relations and not of genetic similarity—the dorsal scapula and the dorsal ilium, for instance, have been called into being by the same necessities of function, and not because they both come from the same stock, so to speak, and similar views might be held about the other corresponding structures.

In other words, there is an analogous resemblance rather than an homologous one between the corresponding parts of the limbs, and, where the functions have become most divergent, there the likeness becomes least traceable, but always can be found the essential correspondence that marks the early similarity in function.

So it does not seem advisable to attempt to establish resemblances more than the most general; but such attempts have been made with great ingenuity—to homologise soft parts as well as the skeleton, and to establish the correspondence in detail. Reference will be made later to some of these views: it is sufficient at present to say that the " body " of the scapula and the Ilium " correspond " because they represent the dorsal moiety in their respective girdles, while the Ischium and pubis correspond in the same way with the coracoid process—which is the only ventral part left in the human shoulder girdle, and may be a true coracoid or a pre-coracoid. Similarly the femur and humerus are the bony supports of corresponding proximal segments of the free limbs.

If we now consider the development of the human limbs, we find that they go through stages that could be said to represent their evolution as just described so very shortly and generally. Each limb begins as a bud jutting out from the body-wall; this has its mesenchyme applied to the surface of the body segments. It elongates and shows signs of dividing into its various parts, and at the same time its mesenchyme commences to lay down the condensations that will form its skeleton.

The skeletal basis is mapped out first in the proximal part of the limb and in the tissue immediately carrying it: the distal part of the bud grows rapidly, and its skeletal basis is laid down rapidly, but the extension of the condensation of the limb girdles follows it a little later, spreading into the wall of the body. In the same way, chondrification of the girdles is secondary to the change in the free limb. As development proceeds the girdle, which is at first very small compared with the free limb, comes nearer its proper size, and later still it joins the sacral region and its fellow in the case of the pelvis and in that of the shoulder settles into position on the thorax. Although we must not look on ontogenetic development as necessarily a recapitulation of phylogeny, yet it is interesting to observe that this short sketch of embryonic formation is in complete accord with the sketch of assumed evolution.

It is in the later stages of development that one must look for the importation of secondary modifications, and it is after the mapping out of their skeletal bases that differences begin to be apparent between the limbs, which are at first very much alike. It is not necessary to deal with these differences here, save in one or two particulars.

Fig. 49 (3) is intended to show how the limbs may be considered to have come into their present positions. They are to be thought of as at first lying in the same direction, at right angles to the long axis of the body: under these conditions each limb has a dorsal and a ventral surface and anterior or cephalic and posterior or caudal borders. From their relations to the axis of the limb these borders are termed *preaxial* and *postaxial*.

The conditions in the four-footed animal are attained by bringing the thigh forward so that its preaxial border comes against the body, and flexing the knee, while at the same time the upper arm is brought back with its postaxial border against the body and the elbow bent: thus it becomes evident that in the hind limb the preaxial border is now mesial or internal, but in the fore limb the postaxial border is internal. The orthograde type is easily attained from this by bringing the legs down and thus making a rotation through 180 degrees at the hip-joint. The preaxial digit is the thumb in the hand and the big toe in the foot, and in this way these digits are found on descriptively different sides of the limb, and the same may be said for the bones of the leg: the outer bone of the leg (fibula) corresponds with the inner (ulna) in the forearm. Moreover, the dorsal or extensor side of the original lower limb comes to look ventrally as a result of this rotation.

In this way the two limbs may be compared, but it is questionable if the comparison is worth anything, or if the terms " pre-axial " and " post-axial " have any value beyond that of any other merely descriptive term. If, however, it is thought that true homology exists between the limbs, the appreciation of these points becomes of great importance, and in any case the " correspondences " just indicated can be at once admitted without defining the nature or value of the relations between the structures.

Limbs: Upper Limb

It has been pointed out more recently, that there is a mirror-relation between the limbs—one limb corresponding with the other " the other way round "—and this is a suggestive conception of the limbs that does not depend on homology : but the question is rather outside the province of this work, and it is not possible to devote further space to its consideration.

The primary elements of the girdles and of the free limb skeletons are laid down in cartilage, and the process of ossification follows in a general way the evolution of the limb as sketched above—that is, the centres for the proximal segments of the free limbs appear first, followed shortly by those for the next segment, and at about the same time ossification begins in the girdles : in the extreme segments the process is variable in its onset and delayed in parts, as it is in some portions of the girdles.

2. UPPER LIMB.

The buried pectoral girdle consists of two bones on each side—the *Scapula* or shoulder blade, on the back and outer side of the upper part of the thorax, supported here by muscles and by the *clavicle* or collar bone, which articulates at one end with the scapula and by the other with the top of the sternum : so the girdle is completed in front by the manubrium.

The *humerus* is the bone of the upper arm, articulating with the scapula and carrying at its lower end the *radius* and *ulna*, that constitute the bony skeleton of the forearm. The Radius is the preaxial bone, the Ulna postaxial. The radius carries the skeleton of the hand, and is capable of pronation and supination by rotating round the ulna.

The bones of the upper limb in men are frequently a very little longer on the right than on the left side. In the adult the limb, hanging by the side, extends to about half-way down the thigh.

CLAVICLE.

A long bone connecting the acromion process of the scapula with the upper end of the sternum and directed outwards and backwards : there is frequently a slight downward direction also to be found in female skeletons.

The bone has a double curve—whence its name *—in the horizontal plane. The inner curve takes about the inner two-thirds of the bone and has a posterior concavity that arches over the axillary sheath (see Fig. 51). The outer curve is shorter and sharper, with an anterior concavity, and takes the outer third of the bone. The inner two-thirds of the bone forms a somewhat rounded bar, but the outer third is flattened, so that it can be described as possessing *upper* and *lower surfaces*, with *anterior* and *posterior margins* : the " margins " are really thick, rounded surfaces. This part has also an outer margin which articulates with the acromion.

The lower aspect of the clavicle can be recognised at once (Fig. 50), by the prominent ligamentous markings on it and by the presence of a groove in its middle part.

The bone presents a shaft and two extremities for examination. The *inner end* has an *articular surface* at its extremity for the sterno-clavicular fibro-cartilage, concavo-convex, and prolonged on to the under aspect of the bone for articulation with the first costal cartilage. Round this articular surface are markings for the sterno-clavicular ligaments : the Sterno-hyoid origin extends from these ligaments behind on to the back of the bone for a little distance.

* *Clavis* = a key : the Roman key was S-shaped.

66   Anatomy of Skeleton

The shaft (Fig. 50) has on its *lower* aspect, just external to and behind the costal facet, a rough *rhomboid impression* for the rhomboid ligament : outside this, the *groove* for the Subclavius muscle, bounded by ridges that give attachment to the costo-coracoid membrane : close to the back margin of its outer third a prominent *conoid tubercle* for the conoid ligament, and running outwards and forwards from this a *ridge* for the trapezoid ligament.

The upper surface of the shaft is comparatively smooth. On the inner third is the origin of Sterno-mastoid. An area for Pectoralis major occupies the front and lower aspect of the inner half of the bone. The Trapezius and Deltoid are attached to the back and front margins respectively of the outer third, and in the area of the last-named muscle a small *deltoid tubercle* may be apparent.

The *outer end* presents a narrow articular surface for the acromion, bevelled so that it looks downwards and outwards, with the result that the bone tends to ride above the acromion : it is normally on a higher level, and the direction of the articular

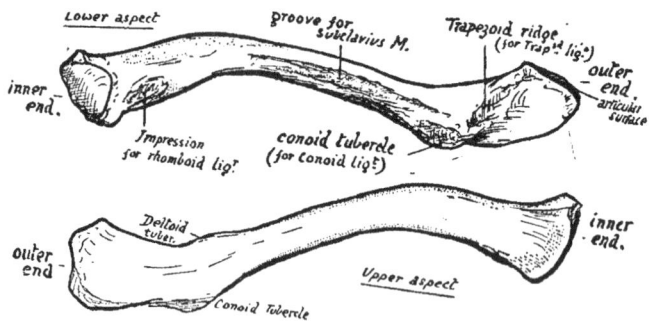

FIG. 50.—Left clavicle from below and from above. To tell right from left, have the subclavian groove below, the flat end externally, and the rhomboid and conoid impressions towards the back.

surface accounts for the difficulty in retaining the clavicle in place after dislocation at this joint. There are feeble ligamentous markings round the articular surface.

The inner end of the bone is separated from the sternum by a disc of interarticular fibro-cartilage, which is thinnest in the centre and may be perforated here.

The disc has strong attachment, as shown in the diagram (Fig. 52), by its back and upper part to the clavicle and at its lower part to the sternum and first costal cartilage, so that it forms a band of connection between this part and the clavicle ; but it must not be forgotten that the meniscus is also fastened round its periphery to the capsule of the composite joint, thus completely dividing the joint cavity into two save when it is perforated. The strong attachment to the clavicle above is associated with this capsular connection, because the capsule here is thickened and strengthened by the interclavicular ligament : the marking for this ligament can be recognised on the bone above and behind the articular surface.

The mechanical result of such fixation of the meniscus is that the clavicle is held down to the sternum and is kept from riding over its top edge, while at the same time the bone can be elevated and depressed, rotating round an antero-posterior axis passing through the attachment of the fibro-cartilage. The elevation of the clavicle on this axis must be accompanied by a sliding out of its lower surface on the costal cartilage, whence the necessity for the lower articular surface. The elevation is

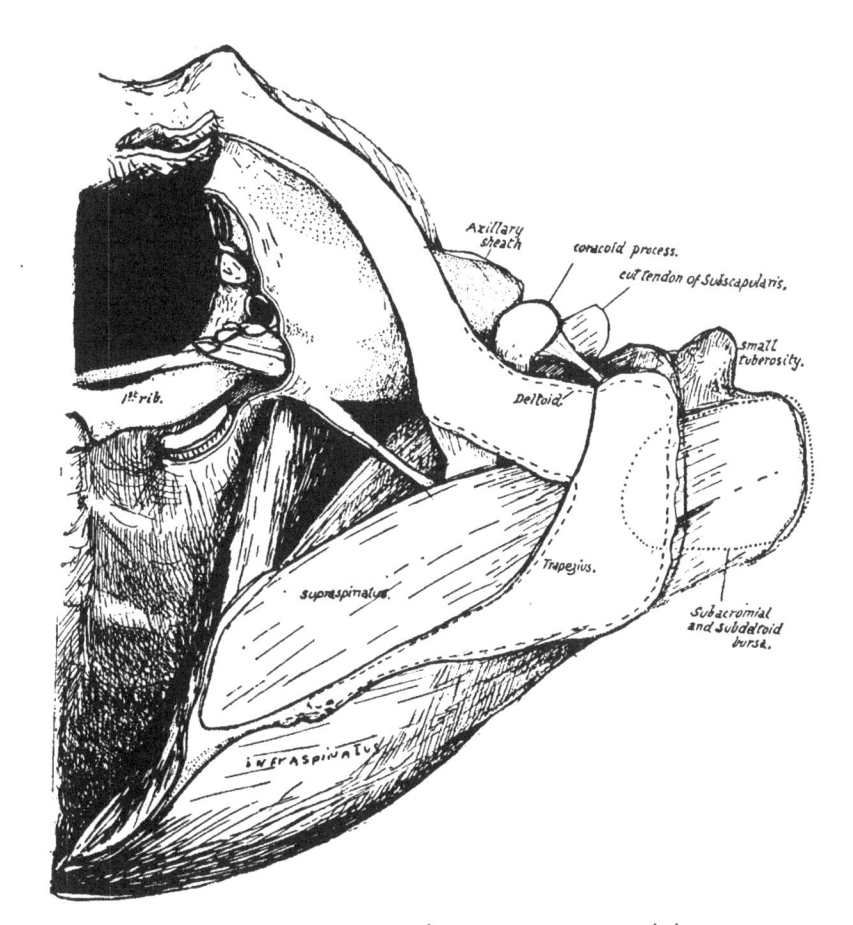

FIG. 51.—A view of the skeleton of the region of the shoulder seen from above, in which the muscles, vessels, etc., have been put in position to show their relations to the bones. The humerus is somewhat inverted, and the Supraspinatus has thus come well under cover of the outer end of the clavicle: observe that the subacromial bursa might also reach this bone if it were a little larger. Insertion of Supraspinatus is beyond the acromion and therefore under Deltoid, and the bursa covering it becomes the subdeltoid bursa. Subscapularis and Serratus magnus are seen, with the axillary space between them: evidently this space is continuous with the posterior triangle of the neck behind the axillary sheath. Capsule of shoulder-joint is not shown: it would, of course, be deep to the tendons. Attachments of Trapezius and Deltoid indicated, from which the deep relations of these muscles near the bones can be appreciated.

checked by the rhomboid ligament, and the amount of play allowed by the ligament depends on its nearness to the axis of rotation combined with the direction of its fibres: they run upwards and somewhat *inwards*, and thus a certain amount of movement outwards on the elevating clavicle is allowed before they become tense.

The security of the joint depends on these two connections, so that the remaining capsular fibres are comparatively weak and loose, particularly below and in front: the joint is supported in front by the sternal head of Sterno-mastoid passing over it, and the capsule is strengthened behind by the origin of Sterno-hyoid from it, extending out from it on to the back of the clavicle. A certain amount of antero-posterior movement, and even of very slight rotation, is possible in addition to elevation (see under " Scapula ").

The markings of the acromio-clavicular ligaments are slight because these ligaments are feeble, though the upper ones are strengthened by fibres derived from the Trapezius. The security of the joint really depends on the coraco-clavicular ligaments,

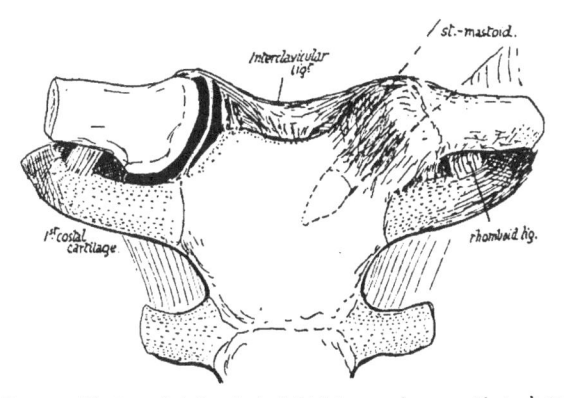

FIG. 52.—Diagram of sterno-clavicular joint, opened on one side to show the meniscus dividing it into two cavities.

and especially on the trapezoid fibres, whose direction enables them to withstand the upward and outward movement of the clavicle on the scapula that is necessary to allow dislocation of the bone. The conoid ligament is not placed obliquely, and it is necessary, when endeavouring to articulate the two bones and place them in their proper relative position, to ensure that the conoid tubercle on the clavicle is immediately over that on the base of the coracoid process of the scapula (see Fig. 51).*

There is a smooth area on the bone (Fig. 53) behind the trapezoid ridge, under which the Supraspinatus plays, particularly if the arm is inverted—this is shown in Fig. 51—and under certain circumstances the subacromial bursa may extend under this surface of the bone.

The insertion of Subclavius extends (Fig. 53) well in front of the conoid ligament. The muscle is enclosed by the costo-coracoid membrane, indicated in the figure by thick black lines, and this attachment of the membrane to the bone can usually be traced by the finger, extending from the rhomboid to the coraco-clavicular impressions: the membrane is attached at its margins to the corresponding ligaments.

* For action of the ligaments and movements in this joint, see under " Scapula." p. 78.

The extent of the muscle insertion is constant, and does not necessarily correspond with the length of the groove: the floor of the groove may present a ridge for an intramuscular tendon.

The Pectoralis major and Sterno-mastoid take origin from the clavicle by mixed tendinous and muscular fibres, so that their area can be found and followed with the eye and finger on any full-grown bone. The two areas are visible on the front (Fig. 54) aspect of the bone, where the lower pectoral surface is seen to extend further out than the upper one for the Sterno-mastoid: the latter surface is on the upper and front and the pectoral area on the front and lower aspect of the inner part of the shaft. Turn the bone up and follow the pectoral surface on its lower aspect, and it will be seen (Fig. 53) that it extends back to come into relation with the rhomboid ligament and costo-coracoid membrane.

The attachments of Trapezius behind and Deltoid in front are plainly visible on the outer third of the bone. A little interval exists between the Pectoral and Deltoid origins, and here a communication between the cephalic vein and the external jugular may cross the bone and may groove it: this is the persistence of a large venous channel that is present in this situation in the embryo.

The vaginal layer of deep cervical fascia, which encloses the Sterno-mastoid and Trapezius, is attached to the bone along a line joining the areas for these two muscles: there is no definite marking for it on the bone. In front of this line the clavicle is covered only by Platysma and superficial tissues, and crossed under this by the descending cervical nerves: some of these nerve filaments may groove the bone or even pierce it, and their relation to the bone accounts for the excessive pain and tenderness that sometimes accompany and follow slight lesions of the bone.

The inner curve in the clavicle arches over and protects the axillary sheath. The relation of the sheath to the bone is shown in Fig. 51: it lies behind and below it and not quite in contact with it. The omohyoid fascia comes down into the groove between the back of the bone and the sheath and obtains attachment to this aspect of the bone, low down on its posterior surface: the line of its attachment lies between the conoid and rhomboid impressions, and the double-layered fascia passes externally on to the conoid ligament and so reaches the base of the coracoid process of the scapula and the suprascapular ligament. Internally, the two layers separate at the rhomboid impression, the deeper one passing down the rhomboid ligament to reach the cartilage of the first rib *below* the origin of Sterno-thyroid, while the superficial layer runs inwards along the back of the clavicle *above* the origin of Sterno-hyoid to reach the top and back of the manubrium. In this way the central infrahyoid muscles are included between the layers as well as the (lateral) omohyoid, so that all the muscles of this group are in one sheet or morphological plane.

The suprascapular artery runs out behind the bone, held to it by the omohyoid fascia, and it gives off here the small nutrient artery which pierces the bone behind and below, running in an outward direction: the foramen may not always be present or it may be double. It is worth remarking that there is always an artery running behind the bone and derived from the thyroid axis, but it does not always reach the scapula: in this case the effective suprascapular artery arises from the third part of the subclavian artery, runs out more or less in company with the nerve, and passes with it under the suprascapular ligament.

The morphological value of the clavicle is uncertain. It has been customary to look on it as a bone that has been secondarily added to the pectoral arch, of which it is not really a constituent, in the sense that it does not represent one of the ventral segments of that arch already shortly described.

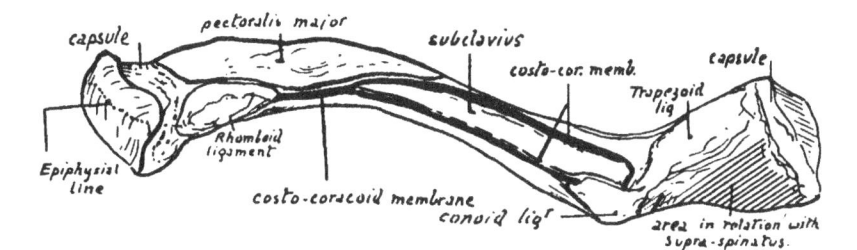

FIG. 53.—Lower surface of left clavicle. The various areas and ridges can be found at once on the bone Observe that Pectoralis major is in close relation with the rhomboid ligament and costo-coracoid membrane covering Subclavius. The outer part of Subclavius, however, is separated from Deltoid by an interval filled with soft tissue, to allow of the play of the latter muscle in movements of the arm. The costo-coracoid membrane reaches the chest-wall and Pectoralis minor by passing down the rhomboid ligament, while its outer part reaches the muscle by passing down the conoid ligament to the coracoid process. To understand the relation between Supraspinatus and the area behind the trapezoid ridge, see Fig. 51.

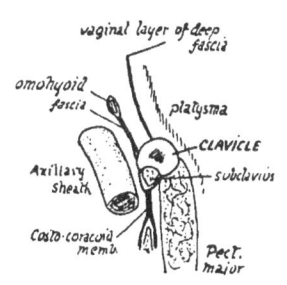

FIG. 53A.—A scheme of a section through the clavicle to show the different planes which come into relation with the bone. The descending cervical nerves cross it deep to the Platysma, the suprascapular vessels run behind it in the attachment of the omo-hyoid fascia, and the Sterno-mastoid and Trapezius are in the plane of the vaginal fascia.

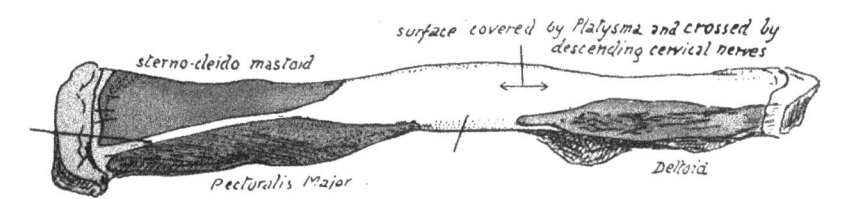

FIG. 54.—Left clavicle seen from the front. A is the blunt edge which lies between the areas for Sterno-mastoid and Pectoralis major; the presence of two large muscular surfaces gives the inner third of the bone a somewhat prismatic shape on section, moulded on the original cylindrical shape of the shaft. V indicates the interval between Pect. major and Deltoid areas, on which there may be a vein joining the cephalic and eXternal jugular; this is, in the embryo, one of the main venous channels which drain the limb-bud.

Limbs : Upper Limb

On the other hand, it has been held that the bone is really the much modified anterior ventral bar, corresponding with the precoracoid in the simple type of arch, and in support of this it can be shown that there is a cartilaginous structure in the anlage of the bone immediately preceding its ossification. Possibly the bone does not correspond with the " clavicles " that are found in some lower animals. Ornithorhynchus is the only mammal that possesses coracoid bars that reach the sternum, and if its " clavicle " is morphologically a separate bone it exists with a coracoid and a precoracoid—that is, if the epicoracoid of this animal represents the precoracoid of the simple type. It is possible that the human clavicle is partly a covering dermal bone, and partly a primary ventral bar which has been taken into the covering bone and is represented by the included cartilaginous core in the ossifying structure.

The thick compact bone of the shaft wall thins away toward the ends. The association with longitudinal pressure, due to the pull of muscles going from the trunk to the arm, is indicated by the well-marked longitudinal lamellæ in the cancellous tissue.

The clavicle is well developed in mammals in which the fore limbs are used for purposes beyond the mere to-and-fro movements of progression : thus functions that call for lateral motions in the limb require a clavicle to support the scapula.

This is seen in birds as well as mammals, for the flying birds have strong clavicles firmly articulated with the sternum, whereas the cursorial birds have only rudimentary bones.

Among mammals it is well developed in those that use their front limbs for prehension, as among the monkeys or the rodents, while animals like the horse and carnivora have none or only rudimentary suggestions of the bone embedded in the muscles of the shoulder.

Development.—At the end of the first month the bone is represented by a mesenchymal condensation lying obliquely below the future acromial region of the scapula, which is placed in the neck. In the fifth to sixth week a couple of centres of ossification appear in the rod of condensation, which is now in a peculiar condition of " pre-cartilage." The two centres seem to indicate the development of the shaft of the bone in two parts, and are preceded by the pre-cartilaginous change occurring also in two separate areas. In the seventh week the two areas and the two centres are continuous, and from now on the spreading ossification is preceded by cartilaginous changes in the mesenchyme.

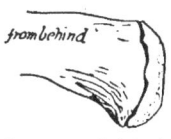

FIG. 55.—Epiphysis on end of left clavicle.

As a rule only one epiphysis develops, and this only forms a thin disc of articular surface (Fig. 55) at the inner end. The small size of this epiphyseal plate can be understood when it is stated that its centre occurs very late, about the twentieth year, when the bone is nearly fully grown, and fuses very quickly. On the other hand, the centres for the shaft are the earliest centres to appear in the body.

SCAPULA,

A " flat " bone, placed on the postero-external aspect of the upper part of the thorax, connected with the sternum by the clavicle, and carrying the humerus.

Triangular in shape, therefore with three borders and three angles, and a dorsal and ventral surface (Fig. 56). *Ventral surface* concave, forming the *subscapular fossa*, covered by and giving origin to Subscapularis : a raised rim along its inner border for insertion of Serratus magnus. *Dorsal surface* presents the prominent *spine*, projecting as a free process, the *acromion*, in a forward and upward direction externally :

70   Anatomy of Skeleton

the spine divides the dorsal surface into an upper *supraspinous fossa*, and a lower *infraspinous fossa*, giving origin to muscles of same name : the infraspinous fossa has a narrow area running along its outer side for the two Teres muscles. The spine gives insertion to Trapezius and origin to Deltoid along its *upper* and *lower borders*, and these borders are respectively continued into the inner and outer borders of the acromion : where the lower border of spine joins outer border of acromion is the *acromial angle*. An articular facet for the clavicle occupies the front portion of the inner border of the acromial process : in front of this the acromion presents a free extremity.

The angles are : (a) *Lower*, covered dorsally by Teres major and ventrally by Serratus magnus : often spoken of as " the angle." (b) Upper and *internal*. (c) Upper

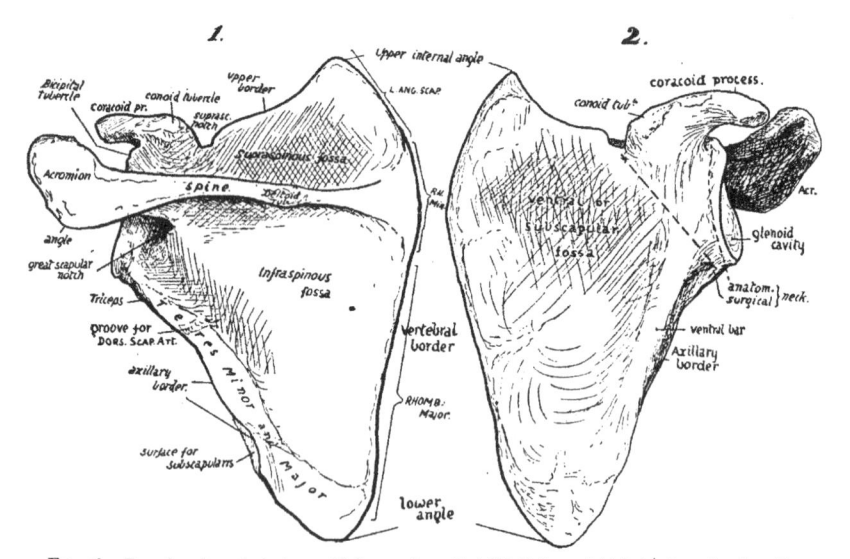

FIG. 56.—Dorsal and ventral views of left scapula. To tell left from right hold the spine dorsally, the acromion eXternally, and the pointed " angle " downwards.

and *externa :* this carries the *glenoid mass*, spoken of as the *head*, which has a *glenoid cavity* for articulation with head of humerus : the head is joined to the rest of the bone by the *neck*, which lies outside the free border of the spine : the rest of the main bone is then spoken of as the *body*.

Borders.—*Inner* or *vertebral border* thin, gives attachment to Levator anguli scapulæ opposite supraspinous fossa, to Rhomboideus minor opposite expanded *base* of spine, and to Rhomboideus major for the rest of its extent.

Outer or *axillary border :* sharp, may run somewhat on to dorsal surface between the Teres muscles : below neck has a rough area for origin of long head of Triceps.

Upper border : sharp internally, then presents a marked *suprascapular* notch for the *suprascapular* nerve, which is bridged by ligament occasionally ossified : immediately outside this supports a curved projection, the *coracoid process*, which arises from the border between the notch and the glenoid rim. Coracoid process projects forwards and outwards and turns slightly downwards, therefore has a concave *lower* surface,

Limbs: Upper Limb

inner and *outer* thick borders, marked by Pectoralis minor internally, and by coraco-acromial and coraco-humeral ligaments externally, and a rough *upper* surface, which presents a prominent *conoid tubercle* for conoid ligament behind, and, outside and in front of this, roughnesses for trapezoid ligament : further forward part of Pectoralis minor insertion, and, at the tip, origin of Coraco-brachialis and short head of Biceps. At the base of the process, just above the glenoid cavity, is the *bicipital* or *supra-glenoid tubercle* for the origin of long head of Biceps. The neck is sometimes termed the *anatomical* neck, lying outside the coracoid, and the term *surgical* neck is applied to the area crossed by a line drawn from below the glenoid to the notch internal to the coracoid. The lines of both these " necks " pass between the glenoid rim and the free border of the spine : the notch between the rim and spine is sometimes termed the great *scapular notch*.

The acromion lies above and rather behind the glenoid fossa, and does not come so far forward as the coracoid process : about an inch and a half separates the tips of the two processes. The upper surface of the acromion and the back surface of the spine are continuous and subcutaneous.

Ventral **Surface.**

Observe that the narrow rim on this surface close to the vertebral border, for the Serratus magnus, widens at the upper and lower angles—especially at the lower angle—and thus the origin of Subscapularis cannot reach these or the intervening border. These widenings correspond with the arrangement of the digitations of the muscle : the upper two are inserted into the upper angle, the next two into the border, and the remaining ones converge on the lower angle. The arrangement indicates that the main action of the muscle is on the lower angle, and such action is illustrated in the movement of raising the hand over the head. The arm is abducted by the fibres of the Deltoid acting from the outer margins of the acromion, and the effect of the contraction of this muscle is to approximate its points of attachment either by raising the arm or by rotating the scapula inwards on its upper attachments (Fig. 60). Naturally the latter movement would occur rather than the lifting of the heavy limb, if it were not for the Serratus magnus, which, pulling forward on the angle of the scapula, would tend to rotate that bone outwards and thus to neutralise the action of the Deltoid on the bone Thus the Serratus can be said to fix the scapula while the Deltoid raises the arm to a right angle : after this, since the humerus cannot be further abducted from the scapula owing to the conformation of the tissues, the Deltoid fixes the humerus on the scapula while the Serratus magnus, by its active contractions, rotates the scapula outwards and thus raises the arm higher : the Trapezius is associated with Serratus magnus in these actions, whereas the Rhomboids have the opposite effect when they contract.

Such action of the Serratus magnus on the scapula is best served by concentrating its force on the lower angle, as far as possible from the axis of rotation, and this calls for a modification in the structure of the scapula. It is evident that, with a weight tending to turn the glenoid cavity down on the one hand and a strong muscle pulling the angle forward and thus turning the glenoid up on the other hand, the bone requires strengthening in the line between the angle and the glenoid, for here is where it would " buckle " at once if it lacked resisting power. Consequently we find a very thick bar of bone in this line of greatest strain, more apparent on the ventral side owing to its concavity, but also visible dorsally where it is covered by the Teres muscles.

This thick ventral bar, then, is a " primary " ridge, concerned in the mechanical building of the bone, and has nothing to do with the axillary border, which is a secondary ridge carried out on its outer aspect. The covering of Subscapularis is really responsible for the axillary border as a ridge, and so we find that the origin of this muscle extends out to the axillary border, and follows this even when, as frequently occurs, it turns on to the dorsal surface for a little distance between the Teres major and minor (Fig. 57). The line of origin of Subscapularis extends along the border as far up as the origin of the Triceps. Its inner limit is made by the area for Serratus magnus, as already mentioned, and its margin reaches the top border of the bone in front of the upper and inner angle, and runs along this border as far as the suprascapular notch. The outer limit of its origin between this point and the axillary border is somewhat variable : the muscle marking given in the figure may be taken as an average type, and the concave irregular line passes inside the foramina found in the ventral fossa, which are for small vessels belonging to the suprascapular system. Observe now that the area of origin is marked by a variable number of secondary ridges, all converging in the direction of the fibres of the muscle : these are evidently for intramuscular tendinous slips, which increase the area of origin of the muscle so that it forms an exceedingly thick mass. If the bones are articulated and a line drawn to represent the lower border of the muscle, from the lowest point of its origin to the lowest point of insertion, it is evident that its inferior border comes down a considerable distance below the axillary margin of the scapula, as shown in the figures.

The outer third of the ventral surface is covered by the Subscapularis, which plays over it without arising from it. As it gets near the neck the subscapular bursa separates it from the bone, and the coracoid process arches forward over the top of its tendon : the subscapular bursa is frequently prolonged upwards over the tendon as a " subcoracoid bursa " between it and the concave lower aspect of the coracoid process under which it plays. The subscapular bursa is also between the tendon and the inner part of the capsule, and opens into the joint between the upper and middle glenohumeral ligaments, by a large aperture on and in which the upper part of the tendon lies as it passes out towards the humerus.

Dorsal Surface (see Fig. 57).

Notice that the floors of the supra- and infra- spinous fossæ are in different planes, depending on the relations of their contained muscles to the rounded end of the humerus. It is this difference in planes that really makes the subscapular hollow on the bone, for this hollow corresponds with the base of the spine, where the two planes meet at an angle (130 degrees) with each other as well as with the spine. It can be seen that the bone below the level of the spine is slightly concave, that is, it is slightly convex on the ventral side, and in the early embryonic condition the scapula is decidedly convex on its ventral aspect, and does not possess a supraspinous fossa : the Supraspinatus is at first a very small rudiment attached only to the upper and outer border of the bone, and as it extends it forms a fossa, and this in its turn makes an angle with the lower plate which is the subscapular hollow (see Fig. 58).

The two muscles that occupy the fossæ extend their origin to the vertebral border : here there is a definite rim which marks the attachment of their covering aponeuroses. Supraspinatus arises along the upper margin and in the fossa as far as the level of the notch, where the suprascapular vessels and nerve pass between it and the bone, turning slightly forward to reach the great scapular notch. Both muscles get an origin from the spine of variable length, but usually extending outwards as a narrow area considerably further than their origins from the floors of the fossæ. The presence of intramuscular tendons in the Infraspinatus is shown by the ridges on the bony surface. The aponeurosis covering Infraspinatus is strong, giving origin to muscular fibres and making

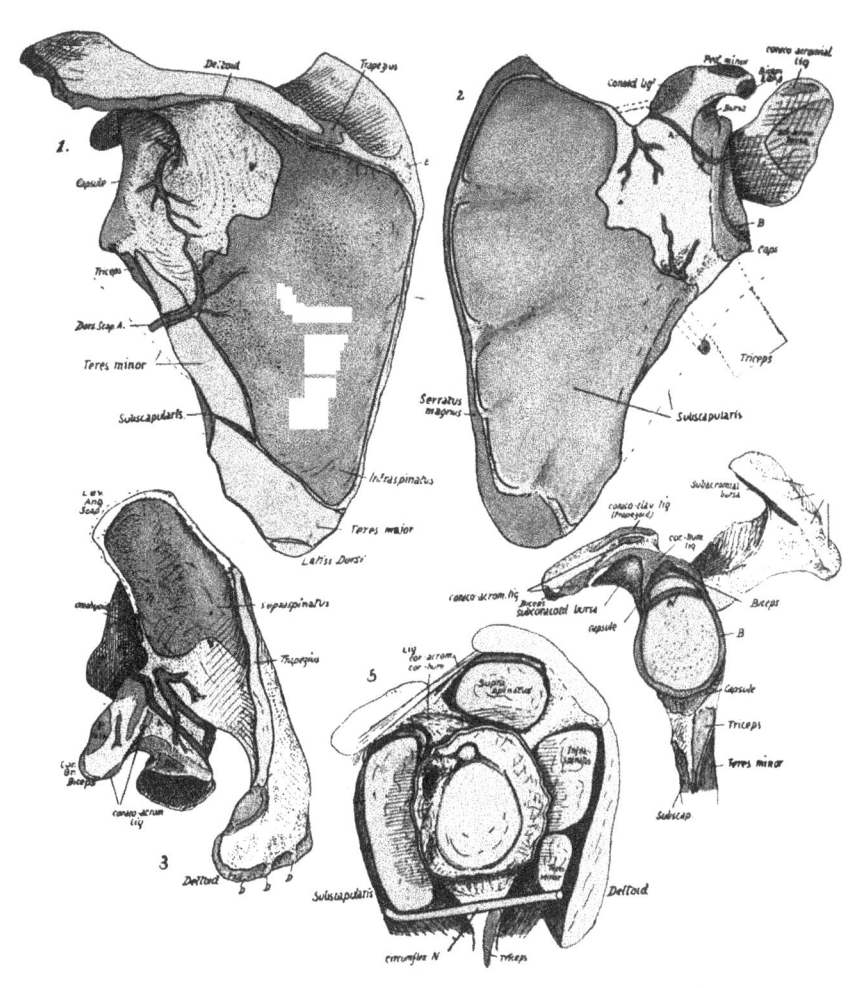

FIG. 57.—Left scapula: 1, from behind; 2, from the front; 3, from above; 4, glenoid region; 5, scheme to show relations of capsule of shoulder joint. A (dark green), line of junction of coracoid and scapular parts. B (light green), glenoid ligament round cavity. C, fibrous band under Subscapularis, across branch of subscapular artery into subscapular fossa. D,D, ridges marking tendons of origin of Deltoid (see Fig. 59). E, occasional bursa under Trapezius.

Limbs: Upper Limb

by its outer attachment the line separating the infraspinous and teres surfaces: therefore the origin of Infraspinatus reaches this line, and extends up along it for a considerable distance. The two Teres muscles lie against the aponeurosis on this line, and the Teres major in particular is very prone to extend its area of origin on to the aponeurosis, thus lying partly superficial to Infraspinatus. Origin of Teres minor extends to the neck of the bone, its area being interrupted by the passage of the dorsalis scapulæ vessels: these lie in a definite groove on the bone, and are covered by a fibrous arch from which muscle fibres arise. At the inner end of the groove there is a break in the ridge for the infraspinous aponeurosis, showing where the vessels pierce the aponeurosis to pass deep to the Infraspinatus, and this point can be used as the upper limit of origin of the outer border of Infraspinatus.

By articulating the bones it can be seen that the lower border of Teres minor comes down below the level of the axillary border of the bone, as in the case of Subscapularis, and almost to the same extent. In other words, the two muscles are only partly separated by the bone, and below this, from within outwards, by dorsalis scapulæ artery, long head of Triceps, and finally by the upper end of the humerus.

Teres major is a part of the Subscapular mass which has extended round the border of the bone to reach its dorsal surface, but, being part of that mass, it must lie in the same general plane—that is, it lies along the lower border of Subscapularis and passes with it in front of Triceps, instead of behind the plane of this muscle which would be the level of muscles belonging properly to the dorsal musculature of the scapula. It is not separated from its main mass as a result of its dorsal origin or the passage of nerves and vessels between them, but owing to its difference in function: this is the great cause of separation between muscles belonging to the same mass fundamentally.

The spino-glenoid ligament, which bridges over the great scapular notch, is simply a fascial band made by the fusion of the fascial sheaths of Supraspinatus and Infraspinatus: these coverings are at first separated by the spine, but they come together at the free edge of this process, and are attached to this edge and to the capsule, etc., further out. The fused sheaths thus form a "ligament" with these connections, and this must of course be over the vessels and nerve here, because they are deep to the plane of the muscles and fasciæ concerned.

Spine and Acromion.

This has a thick rounded bar at its free outer border, which is the only part of the process to be preformed in cartilage (Fig. 58). The remaining thinner portion is apparently ossified directly in membrane, which represents probably the distal attachment of part of the Trapezius.

The upper and lower borders are continued behind into the aponeurotic rims for the spinatus muscles on the vertebral edge: this indicates that the covering aponeuroses are attached to the spine. Examine the areas for Trapezius and Deltoid: in both cases they spread on to the superficial surface of the bone for some distance, but in the case of the Deltoid the area narrows to the line of the border in its inner half, and so corresponds with the fact that the most posterior fibres of the Deltoid are aponeurotic, lying on and fused with the infraspinous aponeurosis, and extending on this nearly to the vertebral border of the bone. The Trapezius area extends in along the spine till it lies above the deltoid tubercle, where it narrows and turns downwards and outwards, forming a hook-shaped roughened ridge. This arrangement of the insertion—which is really responsible for the existence of the "deltoid tubercle"—is due to the convergence of the fibres of Trapezius (Fig. 59), which would otherwise be twisted and overlap one another as they reached their insertion. The inner part of the spine is covered by Trapezius, which plays over it, and a small bursa may be found here very occasionally. The angle of the acromion overhangs the Infraspinatus,

74   Anatomy of Skeleton

which plays against its deep surface. In some animals, as in the rabbit, the angle is drawn out into a long " metacromial " process.

Observe that the facet for the clavicle occupies the greater part of the inner border of acromion, so that Trapezius only reaches a very little part of this border before passing on to the clavicle, whereas the Deltoid arises along the whole of the outer border and then along the front before it reaches the clavicle. Also observe that both muscles have comparatively broad markings for their attachments, which are evidently by mixed muscle and tendon, and moreover the Deltoid area along the outer border has four ridges on it which indicate the places of attachment of the intramuscular tendons in this muscle (Figs. 57 and 59).

The facet is obliquely cut, facing upwards and inwards, so that the clavicle is

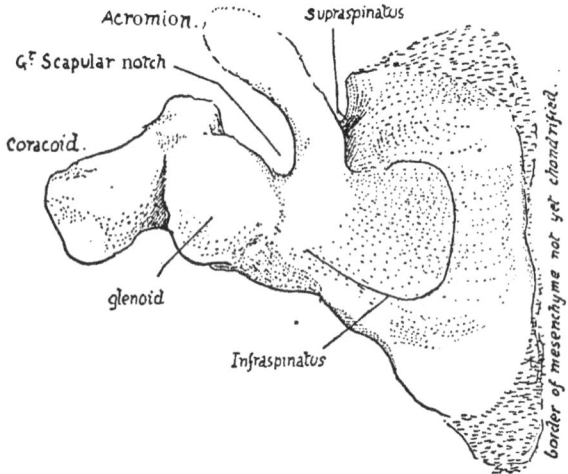

FIG. 58.—Dorsal surface of scapula in the sixth week.* From a model. There is no suprascapular notch because there is no supraspinous fossa. The cartilaginous acromio-spinous bar is distinct; the interrupted line at its end indicates that it is lost in a condensation mass which will chondrify later to form the acromion. The glenoid cavity is partly on the coracoid. Note the small size of the areas for Infraspinatus and Supraspinatus.

placed on a slightly higher level and tends to ride over the acromion : this explains the difficulty in retaining the bone in position when dislocation has occurred between them. The rough area outside the facet is for ligaments which are strengthened by fibres from Trapezius and Deltoid, thus being made stronger than the ligaments on the inferior aspect.

The remaining upper surface of the process is subcutaneous, and has ramifying on it the small arteries of the " acromial rete." Its lower surface is smooth and is lined in part by the subacromial bursa which separates it from the tendon of Supraspinatus (Figs. 51 and 57, 5) : near the tip, in front, this surface has a slightly marked roughness for the attachment of the coraco-acromial ligament. The tendon and bursa separate it from the capsule of the joint, over which it is slightly arched from before backwards.

The acromion is an epiphysis formed by centres occurring in the condensation in

Limbs : Upper Limb

the middle of which the embryonic " spine " ends (Fig. 58). Sometimes the epiphysis does not join the spine by bone, and in such a case the cleaned bone shows a separated acromion : the line of separation runs as a rule from in front of the angle to the back of the clavicular facet, but may be much further forward, suggesting that the epiphysis has two centres at least for its formation, so that one or both of the pieces so developed may fail to join.

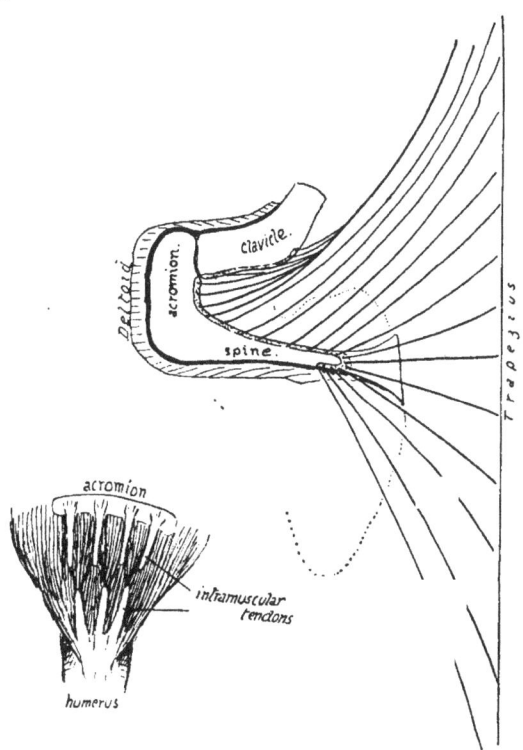

FIG. 59.—The larger drawing is a scheme to show how the disposition of the fibres of Trapezius accounts for the shape of its marking on the spine. The figure also shows the comparative attachments of this muscle and Deltoid on the scapula. The small figure is a scheme of the tendinous intersections or slips in the Deltoid at its origin and insertion, with the disposition of the fibres of the muscle ; this only affects the acromial part, on which most strain is thrown. The slips of origin make secondary markings on the acromion.

Glenoid Cavity (Fig. 57, 4).

The rim affords attachment all round to the glenoid ligament (B), which is partly continuous above with the origin of the long head of Biceps from the bicipital tubercle this is really on the base of the coracoid. The rim presents a notch in its upper and front part : this is better marked in the early state of the bone, indicating the junction of the " coracoid " and " scapular " parts of the articular surface, and the part above the notch has a separate centre of ossification. The oval cavity is very shallow, but in the recent state is deepened and enlarged by the glenoid ligament and by the fact that the cartilage lining it is thinnest centrally. The capsule is fastened to the outside of the rim and to the outside of the glenoid ligament, so that synovial membrane passes off this last on to the capsule with only a slight furrow intervening.

At the top, however, this arrangement is somewhat modified : the glenoid ligament not only joins the Biceps tendon, but is continued on the rim below it, and the capsule is attached in front of and behind the Biceps, passing up to become continuous with the coraco-humeral ligament on the coracoid process immediately above it.

Coracoid Process (Fig. 57).

This has already been described as arching forwards and a little outwards from the upper border of the bone over the tendon of Subscapularis : its tip is slightly depressed, showing a tendency to turn down in front of this muscle. We find, therefore, that its under surface is smooth and concave for the play of the tendon, from which it is separated by a subcoracoid bursa. Its other surfaces are roughened by the attachment of ligaments and tendons. Its upper surface ends behind in the rounded conoid tubercle from which the trapezoid ridge runs forwards along the outer part of the surface : in front of these is the slightly marked roughness for Pectoralis minor, on the inner border and upper surface. The acromial branch of the acromio-thoracic axis crosses the process between the ligaments and the insertion of the tendon, under cover of the Deltoid. The costo-coracoid membrane, on which this artery lies at first, is attached to the inner part of the upper surface between the ligaments and the insertion of Pectoralis minor. On the inner aspect are seen part of the attachment of this muscle and of the conoid ligament.

Looked at from the outer side a well-marked ridge is visible running forward toward the end of the process and lying immediately outside and below the trapezoid ridge. This outer edge shows two more or less definite markings, one near its posterior end (Fig. 57, 4) and a larger one near the front extremity, for the two main bands of the coraco-acromial ligament : between these the ridge affords attachment to the very weak central portion of the ligament, which is frequently absent.

Just below the level of the edge and the coraco-acromial ligament is the area of attachment of the coraco-humeral ligament, the apex of which extends far enough forward to lie under cover of the anterior coraco-acromial band : the area widens behind and externally to enclose the supraglenoid tubercle (where the Biceps has origin for its long tendon), and becomes continuous on each side of this with the capsular marking.

Whatever may be the comparative value of the coracoid process, there is no doubt that it is one of the primary parts of the bone, and is concerned in the formation of a part of the glenoid fossa, as already shown in Fig. 58. The portion that aids in forming the cavity is ossified from a distinct centre, which, from its occurrence before the true epiphyses, probably is also of primary value. This and the main centre for the process join, and are afterwards furnished with small epiphyses on the prominent tubercles. The coracoid element (ventral) is separated from the scapular or dorsal element before fusion along the dark green line A in Fig. 57, which passes through the notch on the rim of the glenoid as well as on the front wall of the suprascapular notch.

The suprascapular notch is bounded in front by the base of the coracoid process, to which the bands of the suprascapular ligament are therefore attached : these bands may be two in number, an upper stronger part bridging the notch and attached to its margins and dorsal sides of these margins, and a lower weaker and more ventrally situated band. Under the lower band runs a vein passing up from the subscapular fossa to join the suprascapular vessels, while the nerve and one of the venæ comites of the suprascapular artery pass between the two bands.

Limbs: Upper Limb

The omohyoid fascia is fastened to the conoid ligament and base of the coracoid process, passes from this on to the upper margin of the suprascapular ligament, and so gains the upper border of the bone behind the notch : here it encloses the Omohyoid muscle which arises from this part of the bone. Further back on the upper border, near the inner angle, fibres of Levator anguli scapulæ are inserted, and their attachment is continued round the angle on to the vertebral border as far as the " base " of the spine. Here the lesser Rhomboid finds insertion, and below the level of the spine the Rhomboideus major insertion extends down on the border nearly to the lower angle.

The axillary border is a secondary border, not present in the young bone, in which this margin of the body is formed directly by the ventral bar : the adult border is slowly extended over this and partly round it as the Subscapularis grows, and thus accurately marks the extent of that muscle.

The bone is laid obliquely on the upper, outer, and back aspect of the thorax, so that its ventral surface really looks inwards as much as forwards, and in addition slightly downwards. Put in this position it forms with its front muscles the posterior (postero-external) wall of the axilla, being held out through its clavicular support and slung up by its muscular attachments to the skull and neck above it. When the clavicle is broken the bone and arm are no longer held out, but swing inwards on the muscular attachments, being displaced downwards and inwards through a small arc of a circle that has its centre at the muscular origins : this leads to depression of the outer fragment of the broken clavicle below the inner, which is thus prominent under the skin. True raising of the inner fragment occurs to a certain extent from unopposed action of Sterno-mastoid, but it is limited by the costo-clavicular ligament, and a good deal of the upward displacement is only apparent, the result of the lowering of the outer piece. At the same time the inward and downward movement of the scapula can only take place on the dorsal aspect of the trunk, so that the outer fragment of the clavicle is also brought to a somewhat posterior plane, thus increasing the apparent prominence of the inner portion.

In the normal state the upper end of the vertebral border is nearer the middle line than the lower, being only about 2 to 3 inches from it. The lower angle varies, in this relation, with the position of the arm and the attitude of the shoulders. The scapula is separated from the ribs and intercostal muscles by Subscapularis and Serratus magnus, and in the ordinary position of rest the angle lies over the seventh rib or space. The angle is not kept against the chest by the Serratus magnus, but by the upper fibres of Latissimus dorsi (Fig. 39), which hold it like a strap against the chest-wall : under this it can move round the wall for a distance of about an inch and a half as a result of the action of Serratus magnus or Trapezius.

The upper end of the bone is usually over the second rib. The base of the spine can be located through the skin and is opposite the third dorsal spine : from this the spine can be traced upwards and outwards to the acromion. The spine and acromion are the only subcutaneous parts of the bone, but the inner and outer borders can be felt through the muscles, and the coracoid process can be recognised through the Deltoid about an inch below the junction of the middle and outer thirds of the clavicle.*

The scapula rotates on the clavicular attachment of the coraco-clavicular ligaments, particularly of the conoid ligament, but the first part of its movement is round an axis passing through the front part of the acromial joint. The partial upturning

* The glenoid is practically on the same level as the base of the spine when the arm is hanging by the side.

78   Anatomy of Skeleton

of the glenoid cavity that is the main object of such rotation is further increased by sterno-clavicular movement : it will perhaps give a clearer conception of these actions if the motion of raising the arm above the head is analysed. The arm is raised to a right angle by the Deltoid, acting from a scapula fixed by Serratus magnus and Trapezius (see Fig. 60). An arm held in full abduction by the Deltoid could be raised to a higher level by active contraction of the same muscles that would fix the scapula to enable the Deltoid to act—that is, the Serratus magnus and Trapezius, which rotate the scapula outwards, are also of service in holding it against the Deltoid and weight of the arm, which tend to rotate it inwards. The rotation appears to take place at first round an axis passing through the front part of the acromio-clavicular joint, but the consequent depression of the coracoid quickly tightens the conoid ligament and the axis is transferred to the upper attachment of this. Now the acromial surface slides up on the clavicular, but owing to the oblique plane of the surfaces this entails some outward displacement of the scapula, therefore tightens the inner part of the conoid band, and hinders further movement. By this time the alteration in direction of the glenoid cavity has only progressed a certain way, and further change is obtained by carrying the shoulder back through antero-posterior movements at the sterno-clavicular joint : this, when the angle is fixed as far forward as possible by Serratus magnus, has the effect of further rotating the scapula. At the same time the clavicle is elevated, and probably a very slight upward rotation occurs round its longitudinal axis, thus associating it with the rotating scapula and at the same time allowing elevation to the full extent, as the rhomboid ligament is attached behind this longitudinal axis.

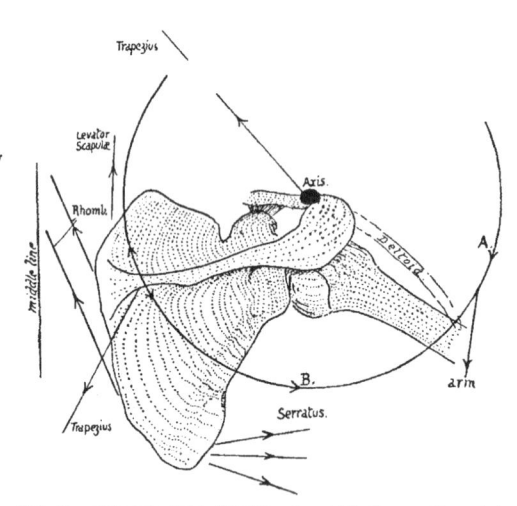

FIG. 60.—Scheme of the directions in which the muscles act in rotating the scapula. The centre round which the bone rotates is put approximately at the acromio-clavicular joint, and it is clear that any muscles pulling on the circle in the direction A. must rotate the bone in, while those pulling in the direction B. have the opposite effect. The upper fibres of Trapezius are mainly suspensory, the lower are external rotators, while the intermediate ones help in the retraction of the scapula which occurs in the later stages of elevation of the limb.

In life the various actions are combined, and go on more or less at the same time. On starting to raise the arm, the lower fibres of the Trapezius begin to rotate the scapula, by this action also, of course, holding it against the Deltoid. When the limb has come out a little distance from the side, the Serratus magnus joins in the movement, and so the raising of the humerus by the Deltoid goes on with simultaneous rotation of the scapula, though this last movement is not proportionately so rapid. In the same way the "terminal" movements of elevation and rotation of the clavicle, with drawing back of the scapula, begin to make their appearance fairly early in the whole action.

The movements are reversed as the arm comes down, and internal rotation of the scapula is checked by the trapezoid fibres. The rotators of the scapula may be again enumerated :—

    Turning the angle out are   Serratus magnus ;
                                        Trapezius.
    Turning the angle in are   Rhomboids ;
                                        Weight of limb (Deltoid)
                                        (? Levator anguli scapulæ).

When the limb is fixed on an external object, so that muscles may act from it on the scapula, these rotators are, of course, not used in this way.

In its build the bone is very thin in most of its extent ; in fact it may be perforated in old subjects. But though thin it is rendered very rigid by the stout bar of bone already mentioned as existing near the axillary border for supporting the strain of the Serratus magnus, by the thickening along the vertebral border, and by the strong spine placed across it. The strongest part of the spine is the thick outer edge, and this supports the acromion and resists the downward pull of the Deltoid from this process. In structure the bone only presents cancellous tissue in its thicker parts : in the young bone there is more of this tissue between two enclosing plates of compact bone, but the middle layer disappears in the adult except along the vertebral and axillary borders and in the spine.

The scapula is sometimes described as consisting of three bony plates diverging from a central axis : the spine is called the *mesoscapula*, the bone above the spine the *prescapula*, and the infraspinous plate the *postscapula*. These form angles with each other along the axis, which corresponds with the attached border of the spine : the angle between the meso- and pre- scapula is about 100 degrees, and those between pre- and post- scapula and meso- and post- scapula about 130 degrees.

A scapula is present in all mammals, whether they possess a clavicle or not : the human scapula is remarkable for the length of its postscapular part, which gives the bone its characteristic elongation. The proportionate length of this part increases after birth. The proportion of breadth to height is taken to form a " scapular index," and is lowest in the adult, highest in mid-fœtal life—higher in negroes than in Europeans.

Development.

The scapula is recognisable as a mesenchymal condensation early in the second month, and is preformed in large part in cartilage, and a centre appears in the perichondrium on its ventral aspect about its middle, before the end of the second month. From this the greater part of the bone is formed, so that at birth only the acromion, coracoid, vertebral border and lower angle, and glenoid cavity are cartilaginous. This is one of the primary centres for the bone, forming the mass of its dorsal or " scapular " element : the other primary division of the bone is the ventral or " coracoid " element, represented, in part at any rate, by the coracoid process and upper part of the glenoid cavity in the human skeleton. This second primary centre is delayed till after birth, and then appears within the first year in the thick part of the process, while another centre appears later, in the eighth to tenth year, for the base and the upper part of glenoid : these are consolidated with the main bone shortly after puberty.

About this time or a little later secondary epiphysial centres make their appearance —two for the acromion, one for the angle, one for the vertebral border, and one for the rest of the glenoid fossa : additional small centres may appear on the trapezoid ridge of the coracoid and at its tip.

All these coalesce with the bone between twenty and twenty-five.

Enough has been said already to show that the *body* can be recognised as the dorsal portion or scapular part of the girdle, while the *coracoid* belongs to the ventral division. But which of the two ventral bars is represented by it is not at all clear. If it is the caudal element, then the precoracoid has either disappeared, or has been partly taken up in the clavicle, and partly remains outside this as fibro-cartilage, etc. If it is really precoracoid, the true coracoid has disappeared, or may be represented by the supraglenoid centre : this last view accords with ossification periods, but leaves the cartilage in and around the clavicle unaccounted for. Those that have a fancy for homologising the two girdles can compare the coracoid process with either the pubis or ischium, according to the view they take of the value of the process.

Following the process of its development, it seems that the spine ought to be considered merely a muscular process, except perhaps as concerns its thick outer margin, which is preformed in cartilage : it is possible that this may have to be considered in the future as representing a ventral bar much modified. The ventral surface of the scapula has been homologised with the gluteal surface of the Ilium, on the assumption that rotation of the girdles has altered their disposition ; there seems to be no ground for this assumption.

HUMERUS,

A long bone forming the skeleton of the upper arm, articulating with the scapula above and with the bones of the forearm below : it has an elongated *shaft* with expanded *upper* and *lower ends* (Fig. 61). The *upper end* has an articular *head* directed upwards, inwards and very slightly backwards, joined to the rest of the bone by a slightly constricted *anatomical neck*. The remainder of the upper end presents in front a *small tuberosity* for insertion of Subscapularis, separated by a deep *bicipital groove* (for long tendon of Biceps) from the larger mass of the *great tuberosity* : this has upper, outer, front, and back aspects, being continuous with the bone on its inner aspect, and has three quadrilateral *facets on its upper and back aspects* for the tendons of Supraspinatus, Infraspinatus, and Teres minor.

The upper end joins the shaft by the *surgical neck*, a common site of fracture spoken of by surgeons as " fracture of the neck " of the bone. The *shaft :* front surface has, at the upper part, the continuation of the bicipital groove, bounded by *inner* and *outer lips*, continuous with small and front part of great tuberosity respectively : inner lip has a rough marking for Teres major, and outer lip has insertion of Pectoralis major while the floor of the groove gives attachment to Latissimus dorsi at a somewhat higher level. The outer lip runs down into the front part of a rough *deltoid impression* which is on the outer side of shaft. Nearly opposite this on the inner side, but as a rule a little lower, is the slight impression for *Coraco-brachialis*.

Upper half of shaft is more or less cylindrical, but the lower half widens to carry the lower end, so that the *posterior* surface is flattened in the lower part but rounded above : about half-way down, near the outer side behind, a slight twist in the bone gives the appearance of a badly-defined groove, the *musculo-spiral groove*, which passes spirally downwards towards a point a little distance below the deltoid impression : it has the musculo-spiral nerve and superior profunda artery in relation with it. The outer head of Triceps arises above and outside the groove, and the large inner head below and internal to it. The *front surface* of the lower half is not flattened, but presents a rounded, broad *anterior border* or ridge dividing it into *antero-internal* and *antero-external surfaces :* the ridge is continuous with the external lip of the groove above. This front surface of the lower part of the shaft gives origin to the Brachialis anticus which covers it, and is bounded at the sides by *internal* and *external supracondylar ridges* which lead down to the *internal* and *external condyles*.

The external supracondylar ridge has the outer intermuscular septum attached to it, and in front of this the Supinator longus (upper two-thirds) and Ext. carpi radialis longior (lower third) : the inner ridge is for the inner septum.

The *lower end* has the *condyles* on each side, a groove behind the inner condyle

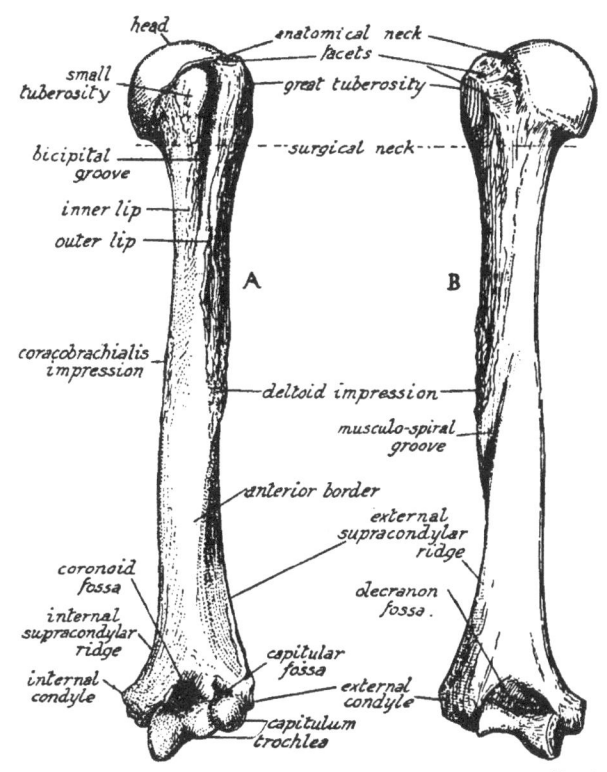

Fig. 61.—Left humerus. A. anterior aspect ; B. posterior aspect. To distinguish left from right, hold the bone with the rounded head above and with its articular surface looking inwards, while the bicipital groove is in front.

lodging the ulnar nerve : immediately outside this a hinge-like articular surface, the *trochlea*, for the ulna, and outside this the *capitulum*, a convex articular surface on the front and lower aspect to articulate with the head of the radius. Above the trochlea a deep *olecranon fossa* behind, and a shallower *coronoid fossa* in front, to receive the corresponding parts of the ulna : above the capitulum in front a very shallow *capitular fossa*.

The humerus is held to the glenoid cavity of the scapula during movement more by the surrounding muscles than by the tension of any ligaments. The ligamentous capsule as a whole is thin, attached to the glenoid ligament and glenoid rim on the scapula and to the anatomical neck of the humerus, extending to the surgical neck below: it is very loose below and in front and behind, but fairly tense above when the arm is by the side—for no further adduction is possible.

It has, however, one strong accessory band in the shape of the coraco-humeral ligament, which passes on one side beyond the glenoid margin on to the coracoid process, and on the other extends beyond the anatomical neck to the tops of both tuberosities, on either side of the commencement of the bicipital groove, so that its front part stands out as a fibrous edge between the Subscapular and Supraspinatus tendons, while it becomes continuous behind, under the latter tendon, with the plane of the capsule (see Figs. 57 and 62). The long head of Biceps passes over the head of the bone, in the articular cavity and surrounded by synovial membrane, to reach the groove, and is regarded as an additional factor helping to keep the head in the cavity: as it emerges from under the coraco-humeral ligament this is strengthened over the top of the groove by transverse fibres. The coraco-humeral band would tend to check extreme outward rotation. Three accessory *gleno-humeral* ligaments, of no apparent mechanical importance, are present *inside* the joint, making prominent folds of the synovial membrane on its front wall. They are directed down and out from the upper and front glenoid rim, and the upper band is attached at the *fovea humeri*, an impression on the anatomical neck just above the small tuberosity. The opening for the subscapular bursa is between the upper and middle bands, and here, therefore, the subscapular tendon comes into relation with the articular surface. The head of the humerus is almost a third of a sphere, set at an angle of 140 degrees with the axis of the shaft, and covered by hyaline cartilage that is thicker in the centre, thus contrasting with the opposing cartilage that lines the glenoid cavity. The area of its articular surface is more than twice that of the glenoid surface, so that a large part must always lie in contact, not with the glenoid cavity, but with the surrounding capsule: when the arm is by the side this capsular contact is mainly above and behind, but is below when the arm is raised.

The want of security evident in such a shallow articulation is a consequence of the necessity for having free movement in all directions, and the two bones cannot be firmly held together by ligaments, for this would restrict the range of motion. So the security of the joint must depend on the surrounding muscles, and the closer these are attached to the articular surfaces, and especially to the more movable one, the more efficiently can they perform the function of bands which are always tense in any position or action of the joint. This is probably the reason why we find strong tendons grouped closely round three surfaces of the capsule and inserted into the humerus immediately outside the anatomical neck to which the capsule is attached (see Fig. 62).

They are the Subscapularis, Supraspinatus, Infraspinatus, and Teres minor: the two spinatus muscles are inserted into their corresponding facets, but the other two extend beyond their evident markings on the bone, as usually seen.

The Teres minor (Fig. 63) is inserted by *tendon* on its facet, but in addition is attached by *muscle* fibre to the shaft for $\frac{1}{2}$—$\frac{3}{4}$ inch below this, thus not causing a secondary marking here. The two modes of insertion probably depend on the fact that the portion on the tuberosity, being raised, is under pressure from the Deltoid, whereas that below the prominence escapes such pressure: occasionally the lower portion is also tendinous, and so there is occasionally a marking for it on the bone.

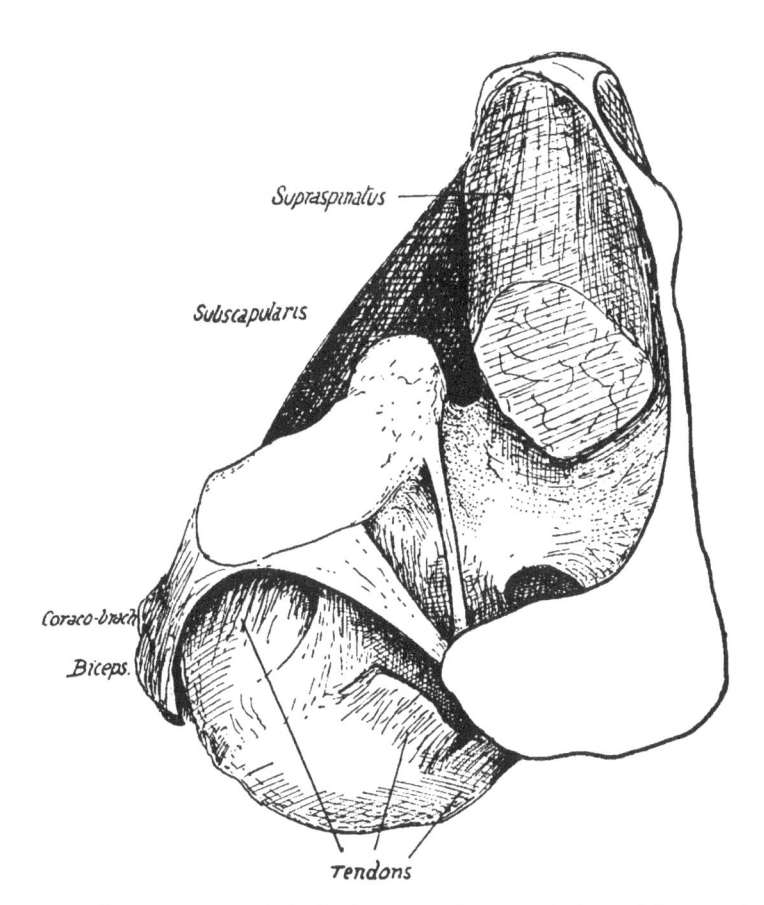

Fig. 62.—View from above of the left shoulder-joint, exposed by removal of part of the Supraspinatus. The cut tendon of this muscle is seen on the humerus, and just in front of it the coraco-humeral ligament makes a prominent band between it and the tendon of Subscapularis; the edge of this band overhangs the latter tendon, but the ligament is directly continuous behind with the capsule under cover of Supraspinatus. The Supraspinatus plays under acromion and coraco-acromial bands (especially when the humerus is inverted), with a bursa interposed. For attachments of ligaments, see Fig. 57, 4. Observe that the short head of Biceps is in part continuous with the anterior coraco-acromial band, and it and the Coraco-brachialis are separated from the capsule by Subscapularis.

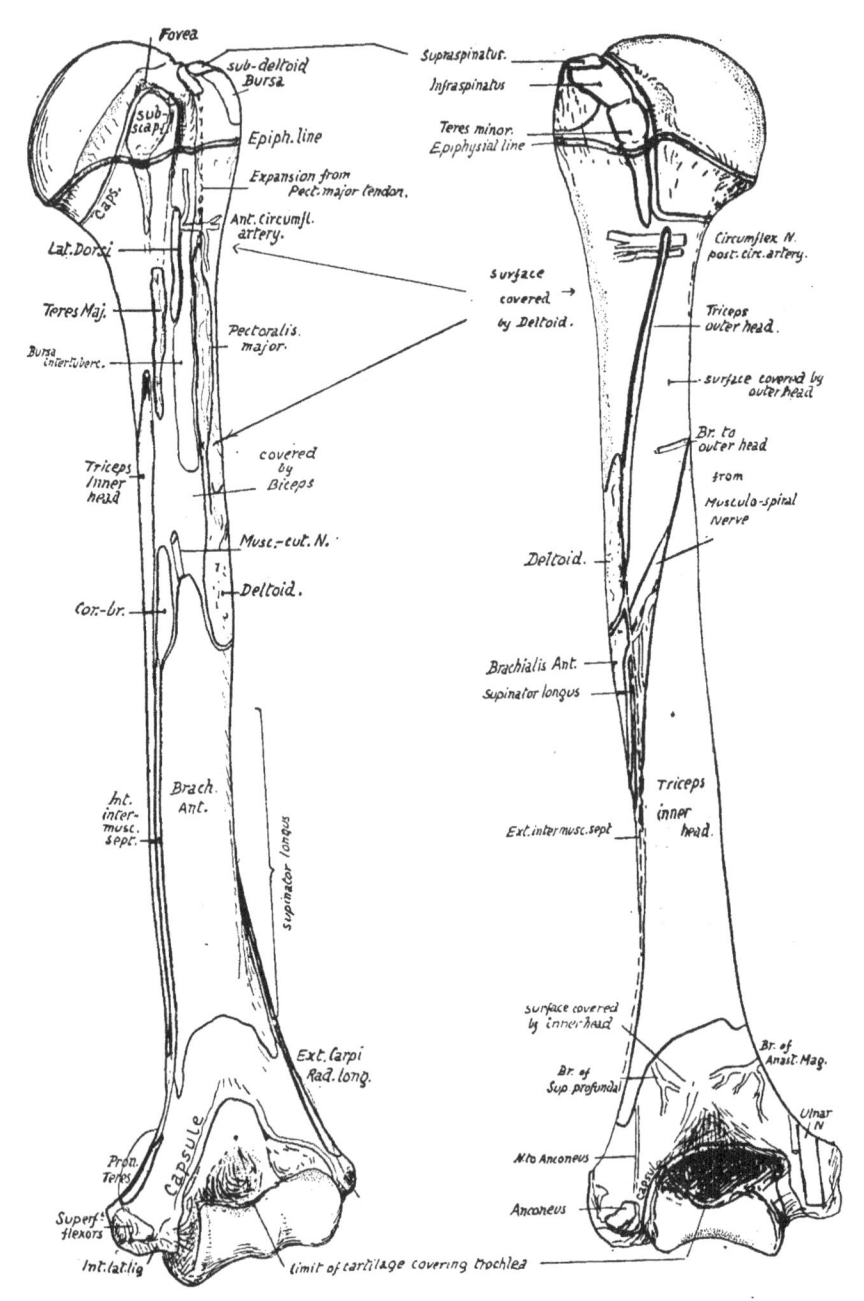

Fig 63.—Anterior and posterior views of left humerus, showing muscular attachments and relations of the bone. Compare with Figs. 65, 67, and 68.

Limbs : Upper Limb

The insertion of the Subscapularis also comes down well below the small tuberosity on to the shaft, and may here be either muscular or tendinous, with absence or presence of a corresponding marking. Bearing these facts in mind, it is easy to show the line of capsular attachment on the bone (Figs. 63 and 68). Starting behind, it passes down close to the inner side of the three facets, so leaving the neighbourhood of the anatomical neck as it runs along the Teres minor. It still passes down along the insertion of Teres minor, thus reaching the level of the surgical neck, but near the lower end of this insertion the line turns suddenly inwards and forwards along the surgical neck and inclines upwards to come again on to the anatomical neck internal to the Subscapularis on the small tuberosity : it sometimes reaches the lower part of this insertion. So the line passes up to the Supraspinatus facet, but it should be noticed that the inclusion of the coraco-humeral ligament in the capsule has led to the line of humeral attachment of capsule being extended outward on both tuberosities at each side of the upper end of the bicipital groove : thus the coraco-humeral ligament is visible between the Supraspinous and Subscapular tendons (Fig. 62) and covers the issuing tendon of the Biceps, being strengthened here by the transverse ligament.

It is not quite apparent why the capsule and muscles should leave their association with the anatomical neck in the lower part, or why there should be no strengthening tendon below the joint as on the other aspects, but probably all these modifications are connected with the marked downward displacement of the head that occurs when the arm is abducted : at any rate, the lower part of the joint thus becomes its weak point mechanically, and when the arm is abducted the head of the humerus can be fairly easily dislocated downwards, the only resistance being offered by the Supraspinatus tendon. The long head of the Triceps, however, provides some support when the arm is abducted.

Summing up these observations on the region of the head of the bone, we can obtain a conception of the immediate relations of the capsule (Figs. 57 and 62). It has Subscapularis tendon in front (separating it from Coraco-brachialis, etc.), Supraspinatus above (separating it from acromion and Deltoid and corresponding bursæ), Infraspinatus and Teres minor behind (separating it from Deltoid). Below it the circumflex nerve and vessels run back close to the neck of the bone, in contact with the lower margins of Subscapularis and Teres minor : the long head of Triceps is below and internal, but not in immediate relation, except when the arm is abducted.

Articulate the bones and observe that the greater tuberosity projects (Fig. 51) far out beyond the level of the acromion. This implies that the tendon of the Supraspinatus passes to its insertion under cover of the Deltoid, and the " subacromial bursa " which intervenes between the tendon and the acromion becomes " subdeltoid " as it covers it further out, to extend beyond it on to the outer side of the upper and more prominent part of the tuberosity.

It is a matter of some practical interest to be able to place the upper epiphysial line. This is the " growing end " of the bone, and the upper epiphysis is a compound of three centres, one for each tuberosity and one for the head : these appear within the first few years of life and coalesce at about six years. The epiphysis formed in this way can be marked out as in the figure (Fig. 63), the line passing just below the facets for Teres minor and Subscapularis and just touching the lower lip of the articular surface. It is apparent that the lower third or more of the capsule is attached to the diaphysis, the remainder being on the epiphysis.

The compound nature of the epiphysis has led to the introduction of the term *morphological neck* to indicate the true junction of the head and shaft only : this

84   Anatomy of Skeleton

(Fig. 64) corresponds with the lower half or less of the attached surface of the head centre, the upper part being that which is fused with the tuberosities.

The structures attached to the epiphysis are :—

(a) Upper two-thirds of capsule (with coraco-humeral, transverse, and upper gleno-humeral ligaments).

(b) Supraspinatus, Infraspinatus, and upper parts of insertion of Teres minor and Subscapularis : the extreme upper part of upward expansion from Pectoralis major.

The bicipital groove starts as a deeply-cut notch between the upper parts of the tuberosities, and is bridged across here by the transverse ligament. As the tuberosities fade away below, the groove becomes shallower, but is then deepened again somewhat by the increasing prominence of the outer lip that marks the area of Pectoralis major insertion, and of the inner lip, to a lessor degree, where the Teres major is attached.

The Pectoral insertion is well marked : it can be followed down into the front part of the deltoid impression, showing the connection between the two tendons. The finger can usually detect a fine secondary line running up from the pectoral marking to the front part of the great tuberosity : this indicates the line of attachment of an upward expansion from the Pectoralis major to the capsule, and this expansion lies in front of the long tendon of the Biceps in its groove and of the short head further in, and between these it covers part of the insertion of the Subscapularis (Fig. 65).

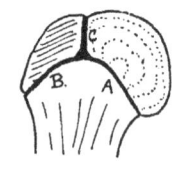

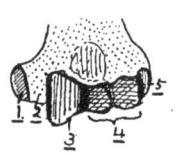

FIG. 64.—The uppermost figures are to illustrate schematically the meaning of the " morphological neck." In the first figure the head is joined to the shaft along the epiphysial line A., and a muscular tuberosity along the line B.; the neck is therefore along A. This type is seen in the femur. In the second figure, however, the tuberosity has also come into relation with the head along the line C.; it is evident that C cannot be regarded as part of the " neck," because it does not mark the junction of head with shaft, and the same remark applies to B. Hence the morphological neck is represented by A. only. This type is seen in the humerus. The lower figure shows the five different ossifications which are concerned in forming the lower end of the humerus. The dotted area is shaft, which reaches the end at 2 ; 1, 3, 4, and 5 are centres of inner condyle, trochlea, capitulum, and outer condyle respectively. Notice that the capitular centre is also responsible for the outer part of the trochlea.

It is seen in the figure also that the tendon and its expansion are covered by the front part of the Deltoid. This muscle is only in the same plane as the great pectoral at its origin, becoming superficial to it as it passes down.

The long tendon of the Biceps is surrounded by a synovial covering, the *bursa intertubercularis*, derived from the joint ; this again is enclosed in a fibrous sheath thrown round it from the deep aspect of the Pectoral tendon and its upper expansion, just before they are attached to the bone, and this sheath is more or less fixed to the floor of the grove above and below the insertion of Latissimus dorsi. Thus the bursa does not come into direct relation with the bone in the groove, but is separated from it by the fibrous sheath as well as by the Latissimus dorsi and anterior circumflex artery. The roughnesses generally produced by these attachments in the floor of the groove are therefore more extensive than that due to the insertion of the tendon of Latissimus dorsi only, though this expands a little as it reaches the bone.

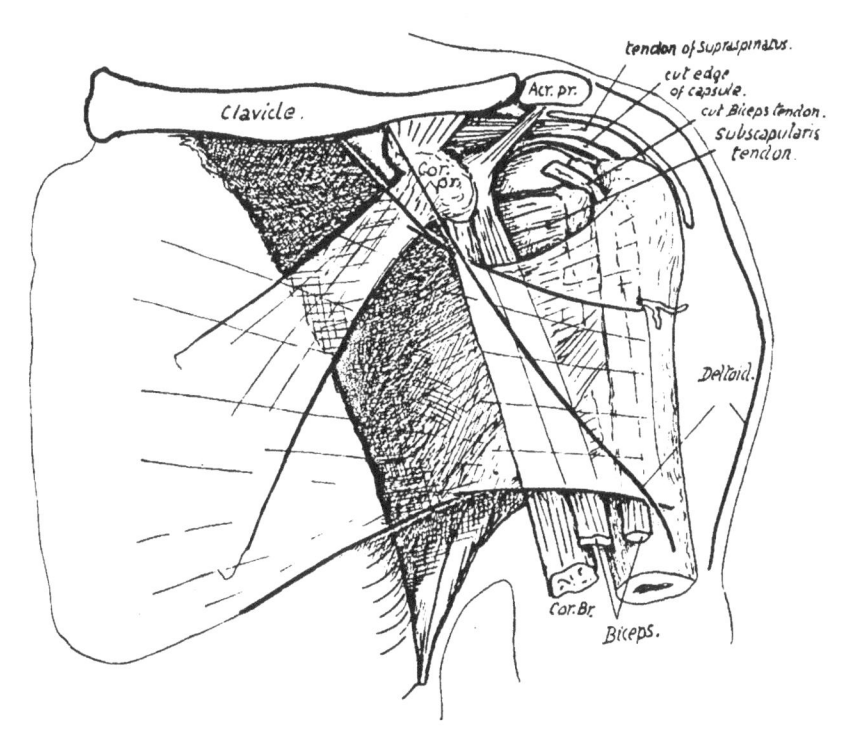

FIG. 65.—Diagrammatic drawing of the relations of structures in the anterior humeral region. Pectoralis major and Deltoid are indicated as transparent, their margins showing as thick black lines, and the latter muscle is seen to lie on the former, covering in its insertion. The Pectoralis major gives an eXpansion upwards under cover of Deltoid, which reaches the capsule and is attached to the outer lip of the groove above the insertion of the main tendon. The two heads of Biceps and Coraco-brachialis are in the neXt (deeper) plane, therefore in contact with the tendon and its eXpansion, and the tendon of Subscapularis is immediately deep to them, so that, when the humerus is everted and the long head of Biceps carried outwards, the Subscapular tendon is "eXposed" between the heads of the Biceps and is covered by the eXpansion—as seen in the figure. The structures which form the posterior wall of the aXilla go to the bone in the neighbourhood of the inner lip, and those of the front wall reach the outer lip of the bicipital groove, and of the structures placed between them, the long tendon lies against the bone between these lips, and the Coraco-brachialis forms the outer wall of the aXilla.

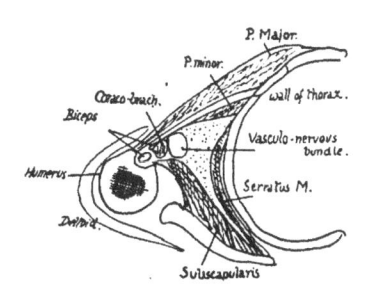

FIG. 66.—A scheme of a section through the aXilla at about a level half-way down the preceding figure. Both figures are for comparison with Figs. 51 and 63.

Limbs: Upper Limb

The tendon of Teres major has a clearly-cut marking on the inner lip, which terminates at its lower end.

The two lips of the groove mark the continuation on to the humerus of the planes of the front and back walls of the axilla. The outer wall must therefore lie between the structures going to

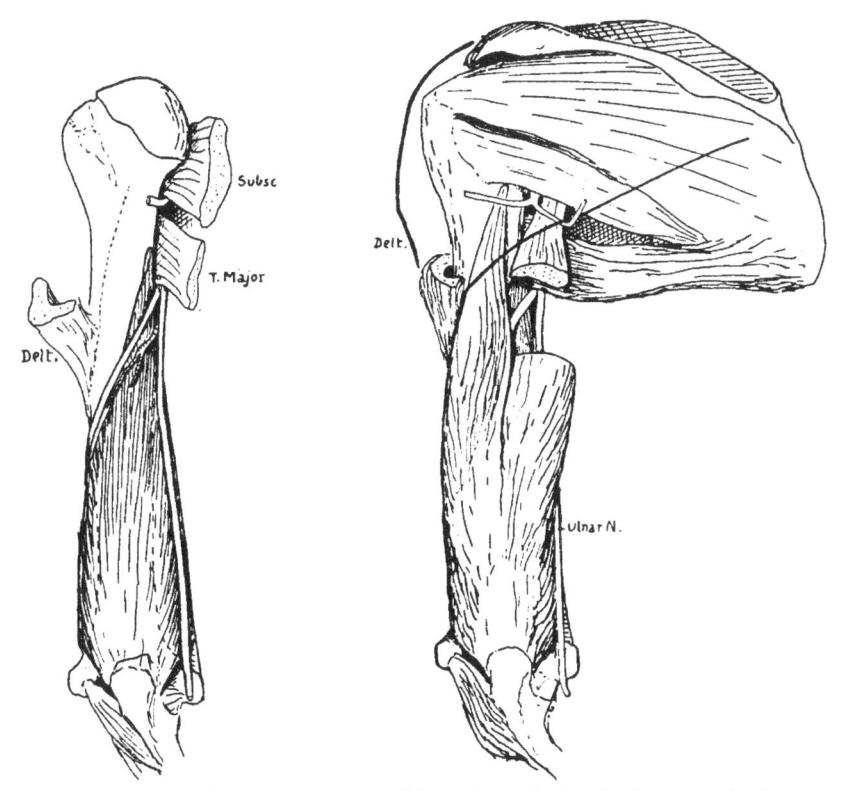

Fig. 67.—The two planes of the Triceps. In the first figure the inner head s seen as the deep plane of the muscle; observe the musculo-spiral nerve crossing the inner head to reach the bone externally, a consequence of the height to which the origin of the head extends. In the next figure the other heads are seen lying on the deep plane, and a piece of the long head is cut away to show the musculo-spiral nerve lying at first between the long and inner heads before passing between the inner and outer. The ulnar nerve is applied to the inner head from the time it enters the arm, only separated from it by the musculo-spiral and superior profunda. The Deltoid is indicated by thick lines; the circumflex nerve appears between long and outer heads deep to the muscle, so that each of its divisions must cross one of these heads under cover of the Deltoid.

these lips: the innermost of these intermediate tissues is the Coraco-brachialis, which thus comes to form the outer wall of the axilla (Fig. 66).

The anterior circumflex artery runs outwards from the axillary sheath in the neighbourhood of the upper part of the Latissimus dorsi tendon, thus reaching the floor of the bicipital groove on this tendon. It runs at first deep to the median and musculo-cutaneous nerves, then passes behind Coraco-brachialis and short head of Biceps, and in the groove is covered by the long head in its fibrous and serous coverings : it gives its bicipital or articular branch upwards in the groove, and some small foramina in the upper part of the groove receive minute twigs from this.

The artery itself proceeds transversely, crossing the outer lip above the pectoral tendon, and therefore piercing the upper expansion of the tendon : it now finds itself on the surface of the bone covered by the Deltoid.

Deltoid.—The insertion is by strong tendon on the outer and front aspect of the shaft, but the posterior fibres tend to turn under the more external ones, and to reach the back and outer side of the bone at a slightly higher level : the roughness they cause on the bone may extend some distance up the back of the shaft. The front part of the muscle altogether covers the tendon of insertion of Pectoralis major (Fig. 65).

Triceps.—Outer head: attached below to back of Deltoid tendon, and extending up to get just under cover of the lower border of Teres minor, a little distance internal to its insertion. If a thick line is drawn from this point to the back of the deltoid impression it will mark the attachment, for the origin is by compressed tendon.

Inner head : extends upwards internally, passing under cover of Teres major for some ½—¾ inch (Fig. 67). Sometimes these fibres are aponeurotic, making a mark on the bone just behind the Teres ridge, but they are frequently muscular and cause no roughness : the position of the muscle, however, is constant here. The area of origin is as shown in Fig. 63, and extends down farther on the outer than on the inner side. When the head is in position it is evident that the musculo-spiral nerve and superior profunda artery must lie on it and cross it to reach the groove, which is really outside this head of the muscle : this is the only part where these structures run in contact with the bone (see Fig. 67).

There are only two planes in the Triceps. The inner head, lying on and arising from the bone, is deep to the other two heads, which lie side by side behind it. Thus the musculo-spiral nerve, sinking back below the Teres major, disappears from view on the inner side of the arm by passing between the *long* and *inner* heads : it crosses the latter head, covered by the former, and then passes under cover of the *outer* head and reaches the bone outside the inner head. The circumflex nerve, passing back above the Teres major, goes between *outer* and *long* heads, and its anterior division crosses the former under cover of Deltoid while the posterior division crosses the long head, under the Deltoid to reach its lower border and turn round it. The ulnar nerve runs in a straight line from the Teres major along the inner side of the inner head just behind the internal intermuscular septum, and the septum reaches the prominent internal condyle, so the nerve passes behind the condyle, coming into contact with the bone here and passing on to the internal lateral ligament below.

External Intermuscular Septum is attached to the outer supracondylar ridge. If the ridge is traced up, it is found to lead toward the back of the deltoid impression, but is lost before reaching it owing to the depression of the spiral groove. The septum may be looked on as derived from the back of the Deltoid tendon, or, perhaps more truly, as a continuation down of the line of origin of the outer head of Triceps from the point where it is attached to the back of the tendon. Brachialis anticus arises from the bone in the angle between the line of the septum and the Deltoid tendon, but is separated from the septum below this by the origin of Supinator longus (upper two thirds) and Extensor carpi radialis longior (lower one-third) from the front of the external ridge in front of the septum. The last-named muscle has a prominent marking on the bone.

Brachialis anticus shows individual variation to some degree in the distance to which its origin extends behind the Deltoid : also in its upward extension between

46

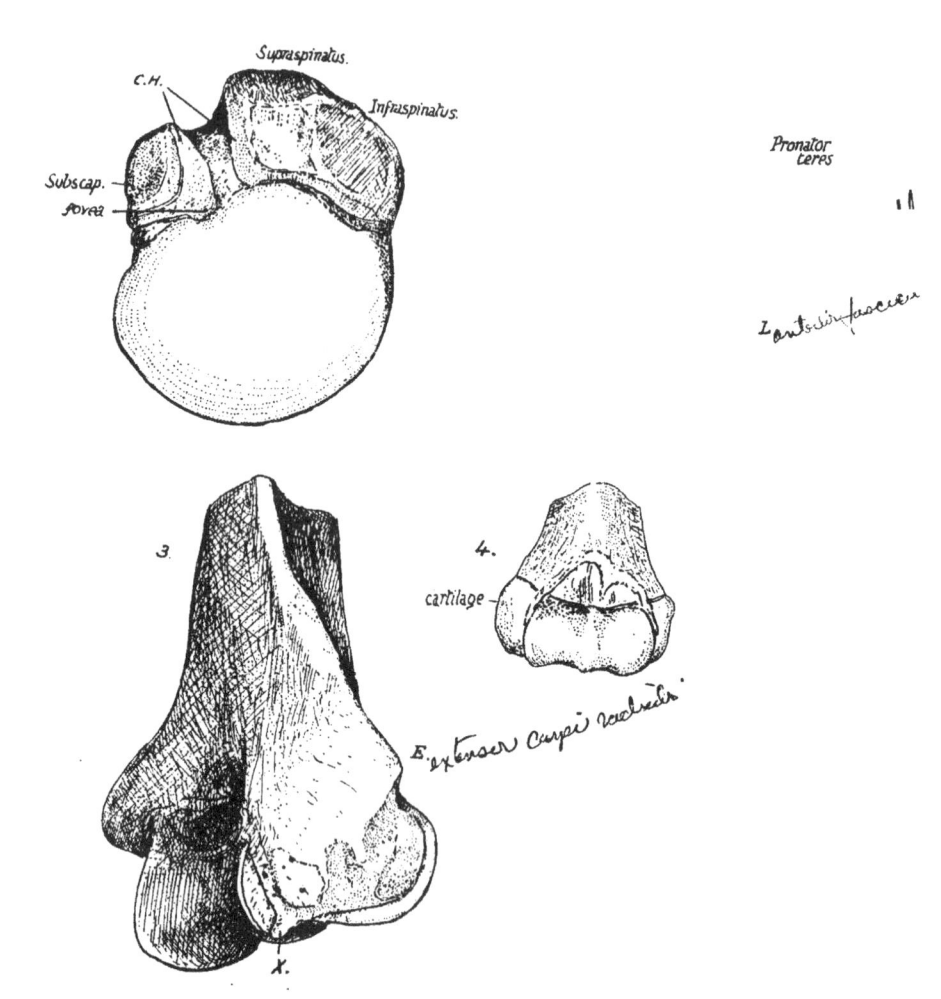

FIG. 68.—1. Upper end of right humerus from above. Shows the attachment of the coraco-humeral ligament, C.H., along the upper end of the bicipital groove or notch, and how it is continuous with the general capsule at the anatomical neck; at this continuity in front the thin fibres of the superior gleno-humeral ligament, deep to the proper capsular fibres, reach the bone at the fovea humeri. 2. Front and inner view of lower end of right humerus. F. is the area of origin of the superficial flexors of the forearm. L. is the area of attachment of the anterior fasciculus of the internal lateral ligament; this is continuous with the fibres of the anterior capsule at the edge of the trochlear eminence. The ligamentous structures do not encroach on the trochlea, which is therefore altogether covered by synovial membrane or cartilage; the extent of the cartilaginous area is shown in Fig. 63. It can easily be appreciated that the anterior band of the lateral ligament is attached at L. in what is practically the axis of the movement of the ulna on the humerus, and hence this strong band can be tight in ordinary humero-ulnar movements and serve to maintain the bones in position. 3. The external condyle of the right humerus. The point of bone E. gives origin to the Extensor carpi radialis brevior and superficial Extensors of the forearm; these also arise from the ligaments which are attached to the rough area immediately below E. The ligaments are shown *in situ* in Fig. 72. The whole rough area is for the ligamentous tissue. The origin of Anconeus is generally larger than is shown in the figure. X. marks the attachment of the cruciate ligament. 4. Lower end of left humerus at birth, from the front. The whole of the lower end is cartilaginous. The capsule is attached partly on bone and partly on cartilage, so that separation of the cartilaginous end would involve opening of the joint. Observe the definite indication of the two parts of the composite joint cavity, radial and ulnar.

Limbs : Upper Limb

Deltoid and Coraco-brachialis, and if this is not great the musculo-cutaneous nerve may lie in contact with the bone for a short distance here before passing on to the muscle.

At its lower part, moreover, the area of origin of the muscle seems to vary in extent, but it always ends some little distance above the capsule. There is a large origin from the internal septum, but there can be none from the outer septum, except at its upper end.

The *capsule* of the elbow-joint makes a distinct marking on the bone, well above the coronoid and capitular fossæ in front, but across the floor of the olecranon fossa behind. The upper part of the olecranon fossa contains loose fatty tissue which is continuous, through deficiencies in the capsule leading to the subsynovial plane, with intra-articular fatty pads.

Condyles.

Inner condyle (or epicondyle) is the more prominent, in an inward and backward direction,* though the outer supracondylar ridge is better marked and sharper than the inner. Two surfaces are visible (Fig. 68) on the inner prominence when looked at from the front : one is antero-internal, and is for the superficial flexor muscles of the forearm, while the other extends along the whole length of the lower aspect, and marks the attachment of the internal lateral ligament. The most anterior fibres of this ligament are those that arise from the extreme mesial end of this lower facet and are therefore the most superficial, and it is on to this that the Flexor Sublimis extends from its humeral origin, which is from the deepest part of the upper facet, to reach its ulnar origin on the coronoid process (see Fig. 77).

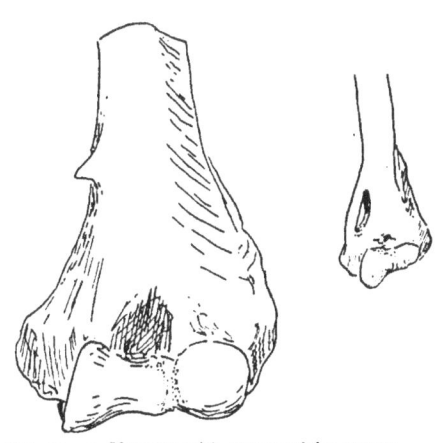

Fig. 69.—1, Humerus with supracondylar process. Humerus of cat, showing the supracondylar foramen for nerve and artery.

Immediately above the upper part and close to the supracondylar ridge is the origin of Pronator teres. This is not a rounded origin, but has an area in the shape of an elongated and narrow triangle with base downwards, which can be easily seen on the bone, marked out by the diverging lines produced by the thin aponeurotic covering. The internal brachial ligament of Struthers is attached to the tip of the process behind the lower end of the pronator origin.

This is a thin fibrous cord attached above to the coracoid process, and possibly represents the lower portion of the Coraco-brachialis sheet : it may be connected with the internal septum or with the supracondylar process if this is present. The *supracondylar* process (Fig. 69) is found now and then, and is a hook-like point of bone continuous with a fibrous band, from which the Pronator teres may arise and under which the median nerve and brachial artery may run : it is looked on as an atavistic structure, representing the front wall of the bony foramen found normally in carnivora and some other animals and transmitting the nerve and artery.

* It may be pointed out here, as a fact of some surgical importance, that the inner condyle points in the same inward and backward direction as the head of the bone.

The concavity on the back of the condyle contains the ulnar nerve and inferior profunda artery. Observe that, owing to the disposition of the front fibres along the lower border of the condyle, the internal lateral ligament has a sort of concavity or groove on its superficial posterior aspect, opening out to the forearm, and the groove on the back of the condyle leads the nerve into this, so that it enters the forearm on the internal ligament, behind the anterior fibres. This portion of the bone on which the nerve lies is formed by an extension of ossification from the shaft, so cutting off the epiphysis of the epicondyle from the three centres that form the "epiphysis of the lower end" (Fig. 64).

The external condyle (Fig. 68) has a rough surface looking outwards for the external lateral ligament of the elbow: the superficial extensors of the forearm are fused with the upper part of this, and so take origin here from the upper portion of the rough surface and from the point of bone above. The surface traced back becomes rapidly but irregularly smaller in area, being only linear along the margin of the capitulum where the fibres are those of the posterior capsule : from this the line of capsular attachment turns up sharply at a right angle along the margin of the trochlea. Included in the angle is the origin of Anconeus, while the point of the angle projects slightly (Fig. 68) between capitulum and trochlea, marking the attachment of the cruciate ligament, a band running in the capsule to the back margin of the lesser sigmoid cavity of the ulna and marking the remnant of the original division between the radial and humero-ulnar joints.

The trochlea and capitulum are placed in a plane somewhat in front of that of the shaft : the outer supracondylar ridge is turned forward with the condylar mass to correspond with this disposition, but the inner ridge, though it shows a slight forward tendency in its upper part, is turned rather backwards below in conformity with the slight backward direction of the condyle on to which it is continued.

The trochlea is only covered with articular cartilage over part of its extent, as is easily seen on the bone (Fig. 63) : its cartilage-covered surface is broadest behind, in accordance with the comparative breadth of the olecranon that articulates with it here. When, therefore, the forearm is flexed, the broad olecranon overlaps by its outer border the inner part of the capitulum, and the ulna tends to move radialwards, moreover, as a result of the disposition of the surface of the trochlea.

The inner part of the trochlea is deeper than the outer, so that the axis of the whole surface made up of these two parts is set obliquely across the vertical axis of the shaft. The result of this is seen when the forearm is extended to the full and it is apparent that the line of the forearm does not coincide with that of the arm : the angle is modified by various factors in the forearm, but that made between the axes of trochlea and humerus is about 105 degrees in male and 108 degrees in female bones.

The inner vertical surface of the trochlea is covered by synovial membrane and has no ligamentous fibres attached to it ; in fact the whole structure might be described as intra-articular in position.

The capitulum is altogether covered by hyaline cartilage, and is spheroidal to permit of a gliding movement of the head of the radius during flexion and extension, and a rotating movement of the head on it during pronation and supination. Its centre is the first to appear at the lower end of the bone.

The upper end of the bone lies below and, for its greater part, external to and in front of the end of the acromion, and can be appreciated by the fingers through the Deltoid. In a thin arm it is possible to feel the small tuberosity through the anterior fibres of this muscle : when the arm is hanging by the side, the bicipital groove lies vertically below the tip of the acromion, but, by eversion of the arm, this groove and

Limbs: Upper Limb

its tendon is carried out (away from the short head, which is fixed to the coracoid process) and the small tuberosity becomes apparent, covered of course by the Subscapularis tendon. The articular head can be felt through this tendon by passing the fingers up into the axilla, and its direction and continuity with the rest of the bone demonstrated by rotation : when there is wasting of the muscles, the relation of the head to the line of the axillary border of the scapula can also be appreciated. Examine from the front, by rotating the arm, the position of the head with reference to the coracoid process, which is just internal to it. The shaft of the bone can be palpated better from the sides than from the front or back, owing to the thickness of muscles on these aspects. The shaft is easily felt just below the insertion of the Deltoid, and the external supra-condylar ridge is directly palpable lower down : the inner ridge is not easily examined. The two condyles are evident. Examine their relations to one another, and notice that a line drawn between them has the top of the olecranon on or just above it when the arm is straight, but below it when the elbow is bent : it is placed much nearer the inner condyle. It is possible to examine the back of the bone just above the outer condyle, but not so easily on the inner side.

Ossification.
The bone is preformed in cartilage. A centre for the shaft appears in the sixth to seventh week, and at birth the ends are cartilaginous, but in about a fifth of the total cases a small centre is present in the articular head at birth.

The upper is the growing end, and three centres, for the head, greater, and lesser tuberosities, appear in that order between the first and fourth or fifth year, quickly fusing with each other to make the upper epiphysis, so that this is present as a single mass in the fifth or sixth year. It does not unite with the shaft till between the twentieth and twenty-fifth years.

The lower end has four special centres, three of which are fused to form the lower epiphysis, while the fourth, for the internal condyle, is separated from these by an extension downwards of the shaft : this extension forms the bone between the condyle and trochlea (the part in relation with the ulnar nerve behind), so that the lower margin of the bone may be said to be made from five centres, of which four are epiphysial. The areas formed by these centres are shown in Fig. 64.

The centre for the capitulum comes first, in the second to third year, followed by that for the inner condyle in the fifth to eighth year : the trochlear centre appears at about eleven, and the outer condylar nucleus about a year later. Thus these centres for the lower end are all present a few years before puberty. After puberty the extension of the ossification of the shaft takes place, and the three centres of the lower epiphysis fuse with each other and with the shaft about the age of seventeen, followed by junction of the inner condyle with the shaft within a year.

(See Fig. 68 for the appearance of the cartilaginous lower end and its relation to the capsule.)

ULNA AND RADIUS.

These two bones, connected by an interosseous membrane, constitute the skeleton of the forearm. They articulate with the humerus above and carry the hand below. The ulna is the inner bone, and forms the main articulation with the humerus, while the radius, placed on its outer side, can rotate its broadened lower end round the ulna in supination and pronation, and articulates directly with the carpus at the wrist : the ulna is only connected with the carpus by ligaments.

Thus the hand is carried by the radius and moves with it.

90   Anatomy of Skeleton

ULNA (Fig. 70).—This has a shaft, an upper end, and a lower end or head. The upper end is its thickest part, and presents a deep notch, the *great sigmoid cavity*, that receives the trochlea of the humerus: it is walled behind by the hook-like *olecranon* process, and in front and below by the smaller *coronoid* process. On the outer side of the coronoid process is the *lesser sigmoid cavity*, which receives the head of the radius: a triangular area below this has a sharp posterior margin for origin of Supinator brevis. The coronoid process is prolonged downwards in front, having *inner* and *outer margins* enclosing a rough tuberosity for insertion of Brachialis anticus: the margins give origin to muscles, the inner has a prominent tubercle at its upper end for internal lateral ligament and Flexor sublimis, and the outer reaches the front margin of the lesser sigmoid cavity above, where the orbicular ligament is fastened.

The shaft has a prominent sharp *outer* or *interosseous* border, a well-marked sinuous *posterior* or *subcutaneous* border, and a badly-marked rounded *inner* border. These separate three *surfaces*: the *internal* and *anterior* surfaces are continuous and afford origin to Flexor profundus digitorum, but the *postero-external* surface is sharply defined by the interosseous and subcutaneous borders, and shows markings for some of the deep extensor muscles. The lower fourth of the front surface is marked off from the rest by an oblique rough line, the *pronator ridge*, and gives origin to Pronator quadratus.

The lower end or *head* is slightly enlarged, has a smooth articular surface on its rounded periphery, externally and in front, for the radius, a prominent *styloid process* for internal lateral ligament of wrist, behind, below, and internally, and a smooth articular surface external to this, on the lower surface of the head, that is continuous with the outer surface and articulates with the triangular fibro-cartilage: a groove on the back, between the head and the styloid process, is for the tendon of Extensor carpi ulnaris: a notch on the radial side of the base of the process is for the attachment of the apex of the fibro-cartilage.

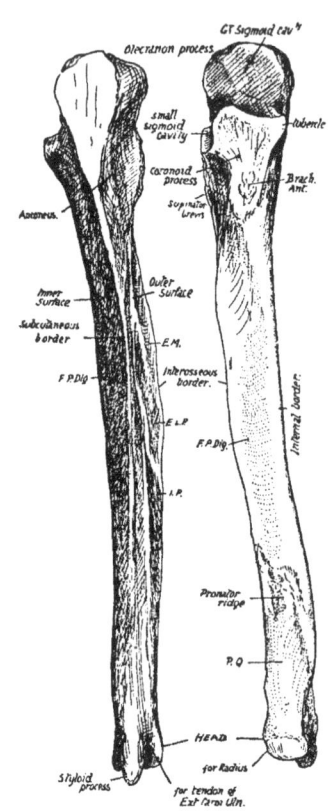

FIG. 70.—Right ulna. *F.P. dig.* origin of Flex. prof. dig.; *E.M., E.L.P., I.P.* origins of Ext. ossis met., Ext. Long. poll., and Ext. indicis; *P.Q.* origin of Pronator quadratus. To tell right from left, hold the bone so that the olecranon is above and behind, and the small sigmoid cavity (for the radius) on the outer side.

The shaft tapers from above, and is cylindrical in its lower part: it is concave forwards, especially in the upper half or so.

RADIUS.—This bone increases in size from above down, is somewhat concave in front, and decidedly concave internally, so that a broad area is provided in the middle of the forearm, by it and the ulna and the interosseous membrane, for the origin of muscles. It presents an *upper* end or *head*, a *shaft*, and a *lower end*. The head is like a thick disc, with a concave articular surface above for the capitulum of the humerus, and a peripheral articular surface for the lesser sigmoid cavity of the ulna and the orbicular ligament which surrounds it and holds it in this cavity. The head is connected by the narrow *neck* with the shaft, and this bears, just below the neck and on the inner side, the *tubercle* for the insertion of the Biceps: here the shaft is somewhat cylindrical, giving insertion externally, in front, and behind to Supinator brevis, and this area is limited in front and behind by *anterior* and *posterior oblique* lines: Flexor sublimis arises from the anterior line. On the outer and back surface, where these lines approach each other, and about half-way down, is a rough Pronator teres impression. Below the oblique line the shaft has a hollowed *anterior* surface for origin of Flexor longus pollicis, bounded internally by the *interosseous border*, externally by the *outer border* continuous with the anterior oblique line, and below by a *pronator ridge* and a *quadrate surface*, comparable with that on the ulna, for Pronator quadratus. The *posterior surface* is rounded from side to side, and shows markings for Extensor ossis metacarpi pollicis and Ext. brevis pollicis near the interosseous border. Lower down this posterior surface widens into the lower end, and here presents grooves for tendons. The lower end is enlarged by the addition of a mass on its outer side, which projects down in the *styloid process*, for the external lateral ligament. As a result the lower end has an *outer surface* on the styloid process for thumb extensor tendons, a *postero-external* behind it, carrying the radial extensor tendons, a *posterior* internal to this, marked by an oblique groove for Ext. longus pollicis, and a broad and shallow one for the common and index extensors; an *internal*, carrying a concave

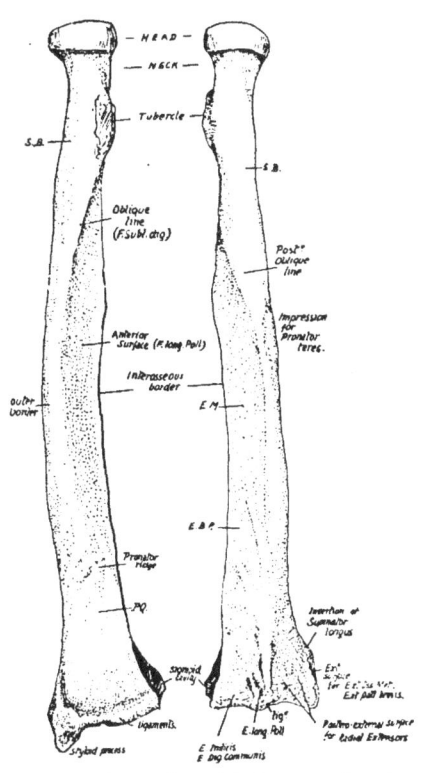

Fig. 71.—Right radius. S.B. insertion of Supinator brevis; E.M., E.B.P. origins of Extensor ossis and brevis p s; P.Q. Pronator quadratus insertion. Toolled right from left, hold the bone with the head up, the tubercle internally, and the concave aspect of the lower third forward.

sigmoid cavity (which plays round the head of the ulna) on its lower part and a non-articular triangular area above this, where some fibres of Pronator quadratus are inserted: also an *anterior* rough for ligamentous attachments, and an *inferior*, which articulates with the carpus. The articular surface on the inner side is not continuous with that below, being separated from it in the recent state by the attachment of the base of the triangular fibro-cartilage to the intervening border.

The radius is held to the ulna by certain bands of connection, which, while keeping it in constant contact, yet do not in the least impair its power of rotation round that bone: these can be considered under three headings:—

(1) The interosseous membrane and oblique ligament.
(2) The lower radio-ulnar joint, especially the triangular fibro-cartilage.
(3) The upper joint: the orbicular ligament.

(1) *Interosseous Membrane.*—Aponeurotic, becoming thin and fascial at its lower end. Fibres chiefly directed downwards and inwards, from radius to ulna: some bands, however, on posterior aspect running in the opposite direction. The interosseous borders on the bones fade away below, showing a tendency to reach the posterior aspect: here the thin membrane becomes continuous with the annular and other ligamentous structures on the back of the wrist.

It has a free edge above, corresponding with the upper border of Extensor ossis metacarpi pollicis, and the posterior interosseous artery passes over this. The anterior interosseous artery pierces it near the lower fourth.

The main function of this sheet is to afford additional surface for the origin of deep flexor and extensor muscles, but, owing to the direction of the majority of its fibres, it can also, when taut, transmit to the ulna some of the pressure that is more directly exerted on the radius through the hand and at the same time will tend to keep the bones together in proportion as the pressure increases. That this is not the chief function of the membrane, however, is evident when we find that it is not taut during pronation, which is the position most commonly assumed when pressure is transmitted from the hand.

The *oblique ligament*: not a part of the interosseous membrane: a band directed down and out from the outer margin of the coronoid process to the radius just below its tubercle, and therefore slightly in front of the plane of interosseous membrane as well as altogether above it. On the ulna it is, when broad, continuous at its upper end with the orbicular ligament, but it is as a rule a feeble band; its mechanical action, if of any value at all, would be to check any downward sliding of radius on ulna.

(2) *Fibro-Cartilage of Lower Radio-Ulnar Joint.*—This is responsible for keeping the bones apposed at their lower ends, for the feeble ligaments of this joint are of necessity very lax to permit of the extensive rotation of the radius round the head of the ulna. The structure is triangular in shape, the apex fastened to the ulna at the outer side of the base of the styloid process, in the axis of rotation, and the base attached to the inner margin of the lower end of radius: it follows, therefore, that the radius can rotate round its axis, made fast to it by the fibro-cartilage, which is thus like a broad spoke, so to speak, in the wheel in the circumference of which the radius moves. It is usually only definitely cartilaginous in its outer part.

Synovial membrane is loose like the capsule, and is sometimes termed the *Membrana sacciformis*. The capsule is attached to the edges of the fibro-cartilage, so that, unless this is perforated, the radio-ulnar cavity is quite separate from the radio-carpal joint.

(3) *The Orbicular Ligament* holds the head of the radius in the lesser sigmoid cavity of the ulna: it is a part of the general capsule of the elbow-joint, from which it cannot be separated (Fig. 72).

Attached to the front and back margins of the lesser cavity, the orbicular fibres surround the radial head, to which they are not fastened, so that it is free to rotate within the circle formed by them and the cavity. It is lined by synovial membrane which is reflected from its lower border on to the neck of the radius. The fibres not only surround the head, but also the upper part of the neck, and here they are somewhat constricted so that the upper end of the bone is more firmly grasped by them.

Limbs: Upper Limb

The *elbow-joint* contains, enclosed in a common capsule and with a common synovial cavity, the humero-ulnar, radio-ulnar, and humero-radial articulations. There are indications in the joint that the radial articulations were at one time distinct from the humero-ulnar, but for practical purposes the three may be considered as forming one compound joint, and the capsule is attached to the humerus above, while it reaches the ulna below and is also attached to the orbicular band round the head of the radius, to which bone it is not fastened.

The inner and outer parts of the capsule, consisting of radiating fibres from the two condyles, are termed the lateral ligaments: the inner passes to the inner margins of the olecranon and coronoid processes, and the outer goes mainly to decussate and blend with the orbicular fibres, and partly to the ulna behind this (Fig. 72). Between the lateral ligaments are the anterior and posterior capsular fibres: posterior part is thin, consisting of a single and incomplete layer of vertical fibres, while the anterior portion of the capsular contains three layers; deepest, thin and incomplete, transversely disposed: intermediate thick, vertical, and complete: superficial, one or two bands of oblique fibres running downwards and outwards to be lost on the front of the orbicular ligament.

We have already seen (p. 87) how the front capsule is attached

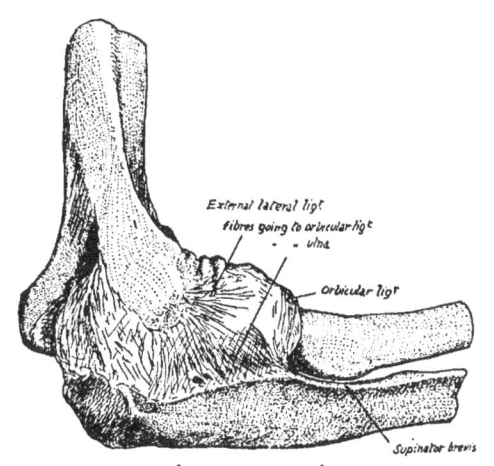

FIG. 72.—Capsule of elbow-joint seen from the outer side.

above the fossæ on the humerus, while the posterior fibres gain the floor of the olecranon fossa: below, the anterior capsular fibres make a marking on the coronoid process a little distance below its upper border (Fig. 75), and the posterior fibres join the olecranon a little way from its articular margin.

Evidently the anterior and posterior fibres of the capsule can have little to do with maintaining the articulation of the bones of the forearm with the humerus, and a cursory examination of the articular surfaces will suffice to show that the greater sigmoid cavity, deep as it is, does not grasp the trochlea to a sufficient extent to hold the bones together without external help, so we are compelled to seek for the provision of such assistance in the lateral ligaments.

Fig. 73 shows the internal lateral ligament as a coarsely fasciculated group of fibres radiating from the internal condyle; the front fibres pass to the prominent tubercle on the inner side of the coronoid process and the neighbouring bone, while the posterior fibres go to the side of the olecranon. The intermediate fibres are not strong, and the notch between the coronoid and olecranon is bridged by transverse fibres (oblique ligament of Cooper). The anterior band is very strong, and is attached above to the lower and front part (Fig. 68) of the internal condyle as far as its tip—that is, practically in the axis of rotation of the ulna on the humerus; therefore this band is in a state of nearly constant tension in all positions of the humero-ulnar joint.

94   Anatomy of Skeleton

The more posterior fibres, on the other hand, are relaxed on extension, and do not seem to be of much value as strengthening the joint : their position and direction make a ligamentous groove or concavity with the more prominent front band, into which the ulnar nerve passes from the back of the condyle (see Fig. 76).

The external lateral ligament radiates from the outer condyle, for the most part to decussate with and join the orbicular fibres, but some of its anterior and posterior fibres reach the coronoid and olecranon respectively : the posterior group is of especial importance, for this is attached to the external condyle (Figs. 68 and 72) in the line of the axis of movement, so that it is also in a state of tension. It is a well-marked set of fibres, passing from the condyle to the strong ridge on the ulna running from the olecranon along the back of the lesser sigmoid cavity, and for a little distance below this on the prominent rim of the triangular area below the cavity : it is partly fused with the orbicular ligament, and accounts for the greater length of the posterior border of the lesser sigmoid notch (Fig. 72).

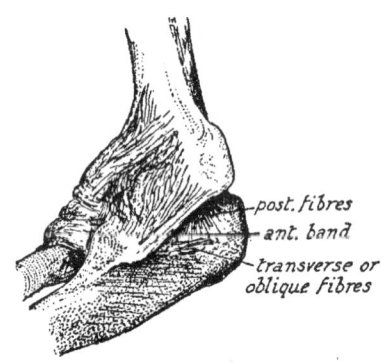

FIG. 73.—Inner aspect of elbow-joint. Note the strong anterior band from front of inner condyle to inner tubercle on coronoid process. This stands away from the posterior fibres, thus making a concavity in which the ulnar nerve lies.

The upper end of the radius is on the whole in front of the plane of the ulna : its head is against the coronoid process, but the ulna is bowed backwards below this prominence, and this, in conjunction with the presence of the olecranon and heavy shaft below the olecranon, gives the radius its rather anterior position : the articular surfaces for the two bones on the humerus are nearly on the same level, but the radius articulates with that bone by the upper surface of its head, whereas the ulna carries its articular surface on its front aspect. When the forearm is pronated, the radius turns over the front of the ulna altogether, with the exception of its head.

This movement takes place round an axis that can be drawn, for practical purposes, from the middle of the head of the radius to that of the ulna, just outside the base of its styloid process, as shown in Fig. 74.

ULNA.

Observe the groove that crosses the floor of the **great sigmoid notch,** separating the olecranon and coronoid : the articular cartilage is also interrupted here by a line of fibrous tissue. At each end of the groove is a notch in the margin, bridged on the inner side by the transverse fibres of the lateral ligament, but covered in externally by the outer ligament, which may, however, show a slight deficiency here (Fig. 72) for vessels and fatty tissue.

The smooth central ridge that runs down the sigmoid notch corresponds with the hollow of the trochlea, and, in further correspondence, we find that the outer moiety of the notch thus divided is greater on the olecranon and smaller on the coronoid : compare with a humerus.

Olecranon.— Upper surface (Fig. 75) from its margin backwards presents synovial, ligamentous, bursal, and tendinous areas. Its posterior surface has a broad, triangular *subcutaneous* area, in relation with which are a couple of small subcutaneous bursæ. Its inner side has rough markings for the internal lateral ligament and, behind and below these, the attachment of the fibrous arch that carries the origin of Flexor carpi ulnaris from the inner condyle to reach the inner margin of the subcutaneous area, and thus to pass to the subcutaneous border. Externally is a definite ridge for the external lateral ligament, and behind this a slightly hollowed surface for the upper part of the Anconeus insertion.

Coronoid.—Cartilage covering upper surface is continuous with that in lesser sigmoid notch. Front surface (Fig. 75) is an elongated triangular area, with a rough tuberosity for tendon of Brachialis anticus, above this an occasional bursal area, then marking for anterior capsular fibres, then a marginal synovial area. The inner and outer margins afford attachments to structures as shown in Fig. 75. The ulnar artery crosses the lower part of the inner margin, running off the Brachialis anticus in a downward, inward, and backward direction on to the Flexor profundus; it therefore passes here deep to the deep head of Pronator teres, which separates it from the median nerve (Figs. 76 and 77).

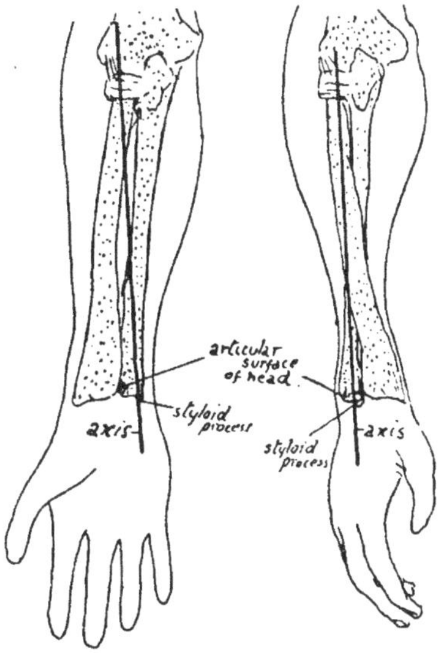

Below the lesser sigmoid cavity is a triangular hollow area: articulate the two bones and rotate the radius, and it becomes clear that the tubercle of this bone with its attached tendon requires the room afforded by this hollow to enable it to rotate. There is no space here for the attachment of muscles, so the ulnar origin of Supinator brevis is by thin aponeurosis from the back margin only of this area: a few muscle fibres may arise, however, just in front of this.

The head has really two articular surfaces which are continuous, a lower one for

FIG. 74.—The axis round which the radius turns is seen to pass through the head of that bone above, and through the head of the ulna below; these two points are the centres of the circles of which the circumferences of the heads are segments. Thus the radius rotates *in* the small sigmoid cavity of ulna above but moves *round* that bone below. Observe that in pronation the part of the head of the ulna which projects behind is covered by the radius in supination.

the triangular fibro-cartilage, and a circumferential one for the radius: owing to the extent of movement of the radius, this last surface is on two aspects of the head. When the hand is pronated, the postero-lateral articular surface on the ulna causes the prominence visible on the inner part of the back of the wrist (Fig. 74).

The capsular fibres are lax and make little marking on the bone. The fibro-cartilage is fastened by fibrous bands to the depression on the outer side of the base of the styloid process: external to this the head of the bone is articular. The styloid process has the internal lateral ligament of the wrist attached to it: as this is almost in the line of the axis of rotation, the ligament does not interfere with pronation and supination.

In mid-fœtal life the styloid process articulates with the cuneiform bone in the carpus, but this connection has become ligamentous at birth.

96   Anatomy of Skeleton

The remaining relations of the bone are shown in the figures, and a few short remarks may be made on these (Figs. 76, 77, and 79).

F. Profundus Dig.—Arises by muscle fibre, therefore a smooth area on bone; arises higher on inner side, extending as far as the hollow smooth surface goes, up to side of olecranon bordering on internal lateral ligament : thus the ulnar nerve can pass directly from the ligament on to the muscle. A part of the inner surface of the shaft is free from muscular origin, between the attachments of the little finger and ring finger portions.

Pronator ridge is made partly by fibres of the interosseous membrane extending

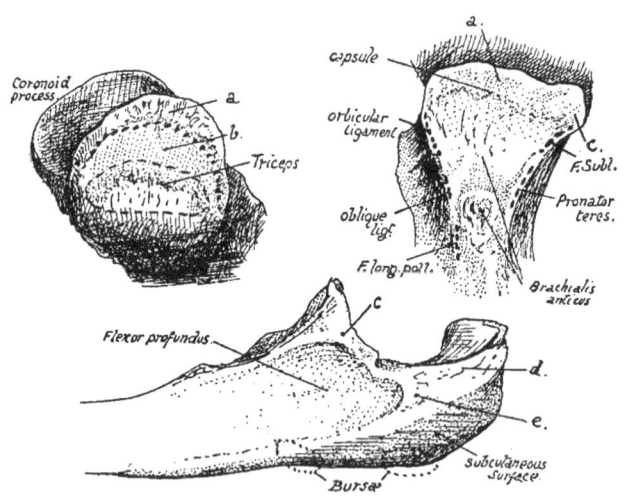

FIG. 75.—1, Upper aspect of olecranon : *a*, surface covered by synovial reflection ; *b*, area for bursa deep to Triceps : the dotted line in front of *b* marks the capsular attachment. The bursal area is frequently much smaller or absent, with consequent approximation of Triceps and capsular markings. 2, Front surface of coronoid, showing the positions of attachments on its margins and surface. Internal lateral lig. is attached to the tubercle *c*, and F. sublimis reaches the bone along this band (see Fig. 77). Synovial membrane covers the surface *a*. 3, Inner side of upper end of ulna. The origin of F. profundus reaches the base of olecranon. *c*, tubercle on coronoid for ligament and F. sublimis ; *d*, posterior fibres of int. lat. lig. ; *e*, F. carpi ulnaris (see Fig. 76).

on to the front of the bone, and partly by aponeurotic fibres on the surface of the muscle.

Pronator Quadratus.—Observe the limited area of origin, and that the inner half of the muscle is covered by the broad tendon of Flexor profundus, while the outer portion is exposed in part by the narrower tendon of Flexor longus pollicis : this explains why the radial artery lies in one part of its course on the muscle, whereas the ulnar artery does not do so.

The *extensor surface* is subdivided by a " vertical ridge " into an area between the ridge and the interosseous border for deep muscles and a smooth strip between the ridge and the subcutaneous border which is for the play of Extensor carpi ulnaris, not affording origin to it. The vertical ridge, traced up, leads to the back of the Supinator brevis origin and outer border of olecranon.

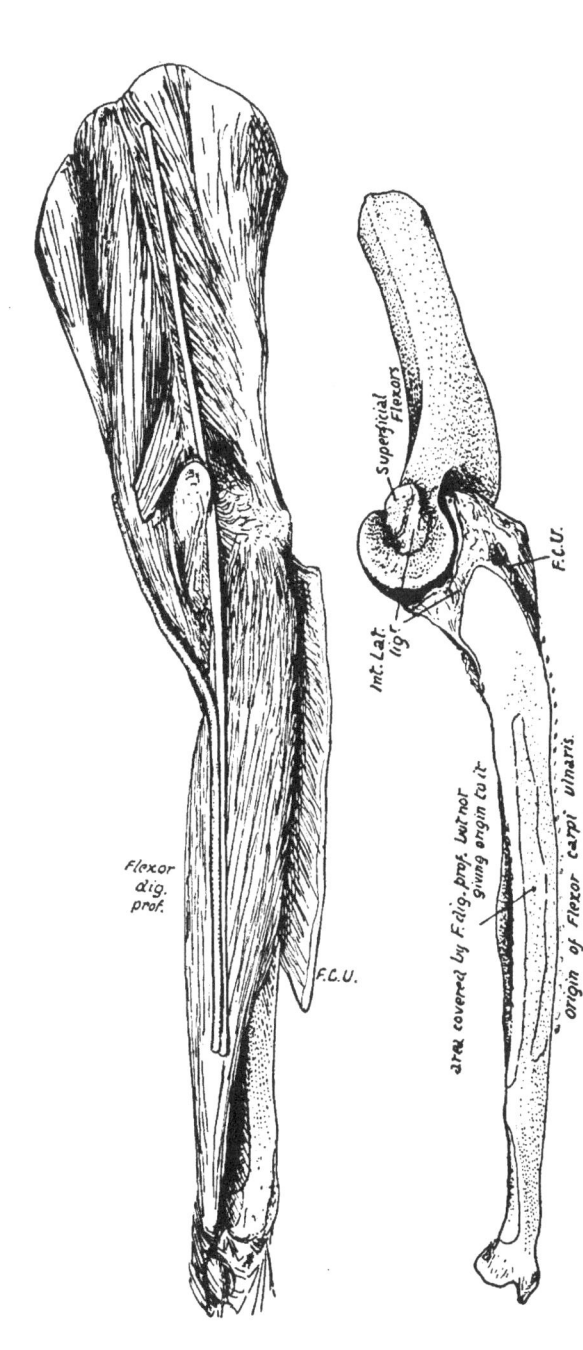

FIG. 76.—Shows the attachments on the inner side of the ulna. F.C.U. is the line of origin of Flexor carpi ulnaris rom the o ecrano which it reaches from the internal condyle by means of a fibrous arch thrown over the ulnar nerve; from the olecranon its origin extends down the subcutaneous border as shown by the dotted line. Immediately deep to this muscle, the Flexor profundis arises from the inner side of the shaft and base of olecranon. The various structures are shown in position in the upper figure, the Flexor carpi ulnaris being turned back and separated from the condyle to expose the nerve lying on F. profundis. The nerve is lying in the groove between the anterior and posterior fibres of the lateral ligament and runs straight from there to the front of the forearm and hand; this is possible because of the position of the ulna, which carries the articular surface for the humerus on its front aspect and is also curved backwards in its shaft, so that it lies behind the plane of the internal condyle and the nerve passing behind the condyle. The ulnar artery, as a result of the direction of the Brachialis anticus passing to its insertion on the bone runs off tie muscle with a backward direction and thus comes nto relation with the nerve n the upper third o the forearm. Compare with Fig. 77.

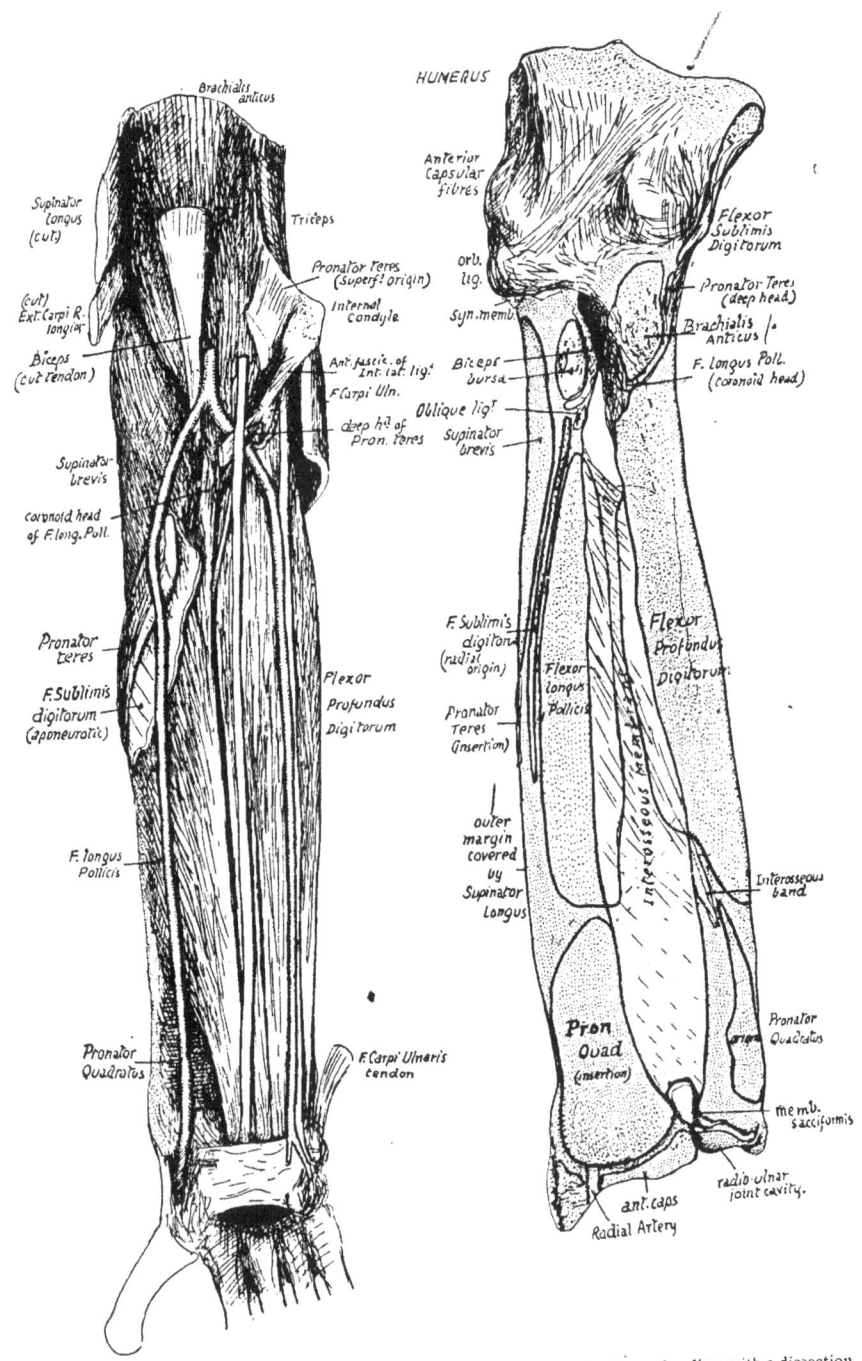

FIG. —The attachment of structures on the anterior surfaces of the right ulna and radius, with a dissection

The ridge is made by a thin aponeurotic fascia which covers the deep muscles and is continued on to the surface of Supinator brevis above : the origins of the deep muscles are mapped out by their aponeurotic fibres.

Extensor Carpi Ulnaris.—Origin is prolonged from the external condyle and capsule to the subcutaneous border by an aponeurotic sheet which is attached to the line (Fig. 79) below the insertion of Anconeus, and is often described as an "intermuscular septum" on this aspect of the Anconeus. The origin is from the *middle third* of the subcutaneous border by aponeurosis, and thus the muscle comes to lie on and play over the smooth area on the extensor surface : the continuation of this surface downwards leads to the groove for the tendon by the styloid process, where there is a synovial sheath.

Subcutaneous Border.—This only extends to the lower fourth of the shaft : below this it cannot be found. The explanation is that it is made by the attachment of the deep fascia, rendered aponeurotic internally by the fusion with the aponeurosis of origin of F. carpi ulnaris, and externally with that of Ext. carpi ulnaris : between these two aponeuroses the ridge is subcutaneous. At the lower end, however, the deep fascia plays over the bone without firm attachment, as this would interfere with the radio-ulnar movements to some extent.

Fig. 76 shows how the position and shape of the ulna are such as to allow the ulnar nerve to come at once into a plane in front of the bone without turning forward : articulate the arm and forearm and extend the latter, and it can be seen at once that the internal condyle is in front of the plane of the ulna, so that the nerve can run straight down to its position on the anterior annular ligament. At the same time the ulnar artery, in coming inwards and downwards off the Brachialis anticus, must have a backward inclination, and this brings it into relation with the nerve after a very short course.

The female bone is smaller and lighter, with less marked muscular ridges : the head of the bone is more globular and knob-like in the female than in the male.

The whole length of the bone can be felt by the finger from the olecranon, along the subcutaneous border, to the styloid process. Observe the relation of the olecranon to the condyles of the humerus (*q.v.*) when the elbow is straight and bent, and notice that its distance from the inner condyle increases as flexion progresses. The styloid process is found on the back of the wrist when the forearm is supinated, but, on pronation, the rounded mass seen and felt here is the articular head itself, uncovered by the forward and inward movement of the radius, and the styloid process appears now to be on the inner and posterior aspect of the wrist : the apparent change is really due, of course, to the swinging round of the hand with the radius.

RADIUS.

The concave upper surface of the *head*, covered by cartilage, is in contact with the nearly spherical capitulum of the humerus, and has a double movement on this, gliding during flexion and extension and rotating during pronation and supination. The circumference of the head is also covered by cartilage, and is altogether intra-articular and in synovial contact with the ulna internally and orbicular ligament for the rest of its extent : the synovial membrane comes down for a little distance below the level of the orbicular ligament, to be reflected on to the neck of the bone.

The *tubercle* has a rough posterior part into which the tendon of the Biceps is inserted, and a front part somewhat facetted, against which the tendon would play in pronation and flexion, and which is therefore guarded by a bursa that is partly enclosed behind by the tendon. The oblique ligament runs downwards and outwards just below and internal to this structure to be attached to the radius. Articulate the two bones, and it will be seen that the tubercle of the radius lies in front of the ulnar origin of Supinator brevis : this can be seen in Fig. 78. During pronation the tubercle turns back and with the attached tendon glides on the front of these aponeurotic fibres, so that another bursa is placed on this aspect of the tendon and extends forward to come also between it and the oblique ligament : this bursa is frequently replaced by smooth "semi-bursal" areolar tissue.

The oblique ligament is possibly a modified part of the ulnar origin of Flexor longus pollicis :- the interosseous border of the radius commences just below and rather behind the insertion of the ligament, but as the fibres of the membrane arising here run in the opposite direction an interval

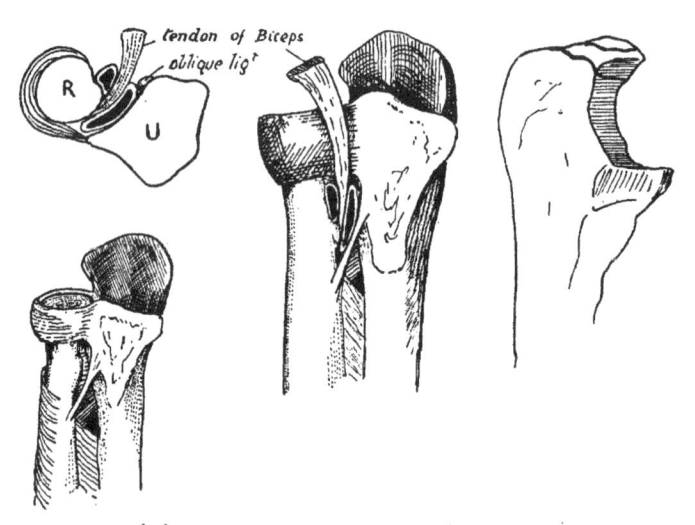

FIG. 78.—The right-hand drawing is from an ulna to show the small portion of the olecranon which is formed from the two epiphysial centres, and the others illustrate the position and relations of the bursae on the tendon of Biceps. One of these is between tendon and radius, the other between tendon and oblique ligament internally and Supinator brevis behind, occupying the greater part of the triangular area below the lesser sigmoid cavity. This second bursa is often wanting.

is left between them covered in behind by the lower fibres of Supinator brevis, and the posterior interosseous artery runs through this interval to pass under the lower margin of the muscle.

The relations and attachments on the shaft of the bone are shown in Figs. 77 and 79. Examination of the shaft soon makes it evident that it has been moulded in a general way, like so many bones, by the influence of the different muscle-groups that can affect it : thus the extensors and flexors, which act in planes more or less parallel with the bone, have hollow surfaces that afford origin to them, whereas the direct pressure of the Supinator brevis has a stronger convex surface of cylindrical shaft to withstand it. In this way the oblique line on the front surface may be looked on as a primary ridge, but there is also a secondary marking to be found on it, for the aponeurotic origin of Flexor sublimis : the origin is aponeurotic owing to the pressure of the overlying Pronator teres, and it may be very thin, and the secondary marking may not be apparent at first, though the finger as a rule is able to detect it. The origin of the

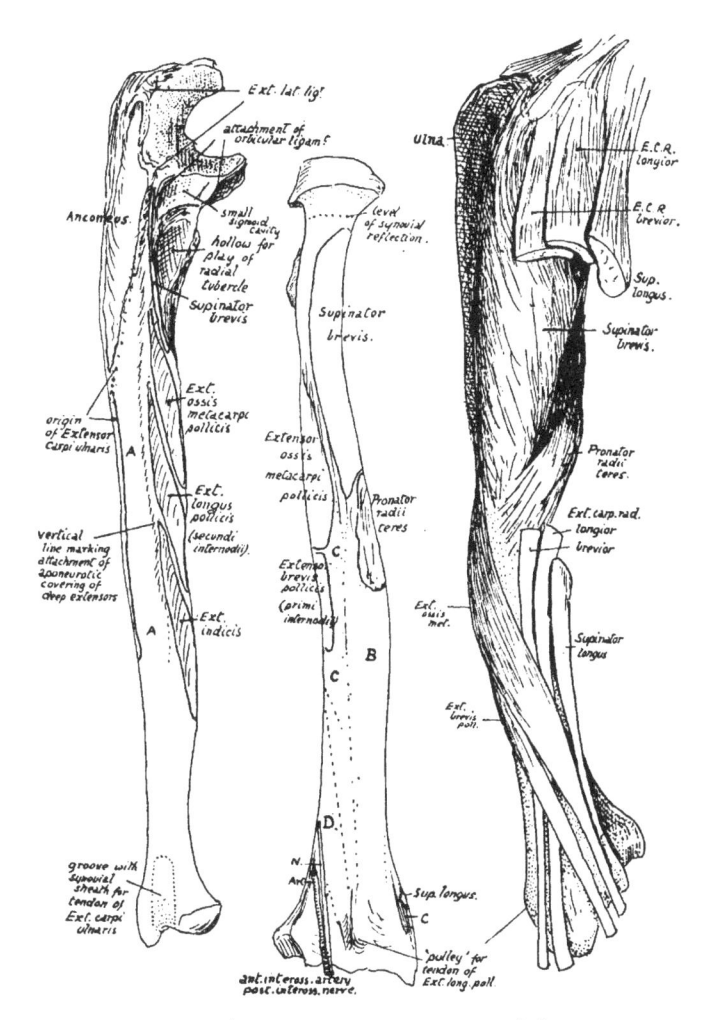

FIG. 79.—1. Extensor surface of right ulna, divided by a vertical line which is made by the aponeurotic covering of the deep muscles; these, therefore, lie between it and the interosseous edge. Thus the surface *A* is for superficial muscles; it is covered by Ext. carpi ulnaris, but does not afford origin to it, and is continued below into the groove for this tendon. 2. Right radius from behind. *B* is covered by Radial Extensor tendons; *C* by Ext. oss. met. and Ext. br. poll., which cross the radial tendons further out. *D* is for Ext. long. poll., which crosses the radial tendons in the hand (Fig. 93). 3. Outer side of radius with muscles and tendons in position. Compare with Fig. 80.

muscle may be prolonged down to the outer border to within 2 or 3 inches of the lower end of the bone, and may be muscular here.

The oblique line behind may be also considered a primary structure with a secondary line on it that is often not appreciab'e.

The Supinator brevis is inserted as shown in the figures, but makes no marking on the bone save near the lower part of its area, for this is as a ru'e the only part where tendinous fibres are to be found in it. It extends down to the top of the impression for Pronator teres, which is more on the back than the front of the outer side of the bone.

The flexor surface has a very faint quadrate ridge, due only to attachment of aponeurotic fasciæ of the muscle, and not, as in the ulna, to implantation of fibres of the interosseous membrane.

The Pronator quadratus has a large area of insertion, reaching out to the outer margin and down to the rough ligamentous ridge: internally it extends to the inner margin and round this on to the triangular area that lies above the sigmoid notch.

The extensor area gives origin to two of the deep muscles, and a faint indication of the position of the muscles running downwards and outwards may be found by the finger and sometimes by the eye. Below the surface for origin of Ext. brevis pollicis is an area covered by Ext. longus pollicis and Ext. indicis, but not giving origin to them: deep to these the anterior interosseous artery and posterior interosseous nerve lie on the bone.

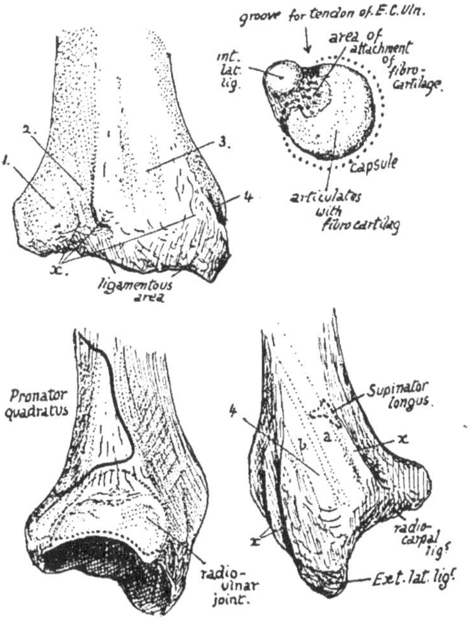

FIG. 80.—The upper right figure is of the head of the ulna seen from below. The others are of the lower end of the radius from behind, from the inner side, and from the outer side. 1, groove for Ext. comm. dig., Ext. indicis, ant. inteross. artery and post inteross. nerve; 2, for Ext. sec. internod. (longus poll.); 3, for radial extensors; 4, for Ext. oss. metacarpi (a) and Ext. primi internod. (brevis pollicis) (b). x marks attachment of posterior annular ligament. On the inner side the dotted line shows the extent of the synovial cavity; this goes up to the insertion of Pronator quadratus above but does not reach the margin below because the triangular cartilage is attached here. On the outer side notice the insertion of Supinator longus partly covered by tendon of Ext. ossis metacarpi pollicis.

The outer aspect of the radius is covered by Supinator brevis in its upper half, then by the tendon of Pronator teres, and below this by the radial extensor and Supinator longus tendons which have passed down over the Supinator brevis and Pronator tendon. These are shown in Fig. 79. At the lower end of the radius the mass that ends in the styloid process projects outwards, nearer the front than the back of the bone, so that the outer surface now becomes postero-external, while a new outer surface is made on the prominent mass. Thus the Radial Extensor tendons come to lie in the postero-external groove on the lower end, while the deep muscles which have crossed them lie on the new outer surface, further out and further forward. The tendon of Supinator longus, which is rather in front of the two radial extensors in the marginal group, runs into the base of the projecting mass, spreading out a little and making a definite secondary mark on the bone : it is partly crossed here by Ext. ossis met., but not by the tendon of Ext. brevis pollicis. Each of these two surfaces on the lower end contains in its groove a synovial sheath for its two tendons.

The posterior surface has an oblique narrow groove for the tendon of Ext. longus pollicis, and a broad shallow groove that has in it the anterior interosseous artery and posterior interosseous nerve, covered by Ext. indicis and Ext. communis tendons. The oblique groove has a prominent outer margin separating it from the postero-external surface and ending below in a bony "pulley," on which the tendon turns more outwards as it leaves the radius to pass to the thumb.

Fig. 81 is a diagram of the six surfaces on the lower end of the radius.

The posterior annular ligament is attached to the bony ridges that separate these grooves and holds the tendons in place. The triangular fibro-cartilage has its base attached to the inner and lower margin of the lower end of the radius and thus connects it with the ulna and separates the articular surface of the sigmoid notch from that of the lower or carpal aspect.

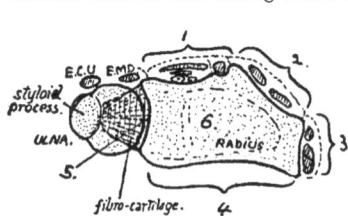

FIG. 81.—Scheme to show the six surfaces on the lower end of the radius. These are : 1, posterior; 2, postero-external; 3, external; 4, anterior; 5, internal, articular for ulna; 6, inferior, articular for carpus. Compare with Fig. 80. Extensor tendons lie on surfaces 1, 2, and 3, and also over the radio-ulnar joint (Ext. min. dig.) and ulna (Ext. carpi ulnaris).

The concave carpal articular surface has a ridge running across it from before backwards : this divides it into an outer area for the scaphoid and an inner for the semilunar bone. The outer area is triangular and extends on to the inner surface of the styloid process : the inner does not cover the whole surface of the semilunar, for this is partly in contact with the fibro-cartilage, but in adduction of the hand the semilunar moves under the radius.

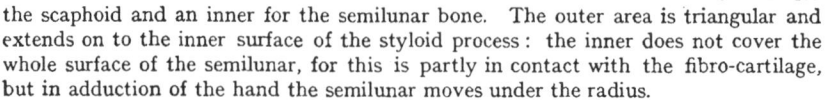

The attachments of ligaments, etc., on the lower end are shown in Fig. 80.

The female bone is said to exhibit a comparatively greater curve than the male bone : otherwise the sexual differences are merely general.

The head of the radius can be felt in the hollow below the external condyle, where, covered only by ligamentous and aponeurotic structures, it can be recognised by the finger on rotating the forearm : examine it here, and note its changing relation to the condyle as the elbow is bent. The muscular mass of Anconeus lies between it and the olecranon. The tubercle can be felt in thin arms just below the head when the forearm is well pronated. The upper part of the shaft of the bone is not easily followed in a well-developed limb, but the lower half can be palpated indirectly on its outer aspect, back and front. The lower end can be felt through and between the tendons behind, externally, and to some extent in front. Notice the level of the styloid process, lower than the posterior border and situated behind the tendons going to the base of the thumb.

Ossification of bones of forearm.

The bones are laid down in cartilage in which shaft centres appear during the seventh week, that for the radius a few days before the ulnar centre.

Limbs: Upper Limb

At birth the ends of the bones are cartilaginous, and also the tubercle of the radius · the coronoid process has been formed as part of the *shaft* of the ulna.

The additional epiphysial centres appear first in the radius, perhaps on account of it being the direct supporting bone for the hand.

The " growing ends " are distal, and in these centres appear, in the radius early in the second year, and in the ulna about the sixth or seventh year. They unite with the shaft between eighteen and twenty-one, the ulnar head joining a year or two before the lower end of the radius.

The upper ends begin to ossify, in the radius about the sixth year, and in the ulna during the tenth year, and unite between seventeen and twenty, the ulnar fusion again preceding that in the radius.

The centre for the radial tubercle appears about twenty and joins a few years later. There is frequently an additional centre for the tip of the olecranon : observe in the figure that the olecranon centres do not correspond with the whole of the descriptive process (Fig. 78).

Occasional centres have been described for the two styloid processes. The centre in the head of the radius has been said to be formed by the junction of several small ones, but this is probably not a correct interpretation of the observations.

The above account can be taken as representing the normal history of ossification in the male, but for the female there seems to be a marked modification in the direction of earlier appearance of epiphysial centres at the growing ends of the bones : thus in the radius the lower centre is found six months earlier than in the male, while that in the ulna is a year earlier.

HAND.

The skeleton of the hand consists of twenty-seven bones arranged in three divisions or groups. The proximal part, carried by the Radius, is the *Carpus*, consisting of eight small irregular bones arranged in two rows, *proximal* and *distal :* the distal row supports the *metacarpus*, a set of five long bones numbered from without inwards, articulating by their *bases* with the bones of the distal (or second) row of the carpus, and carrying on their distal rounded extremities, or *heads*, the proximal parts of the skeleton of the digits : this is composed of three *phalanges* in each finger numbered from the metacarpal as *first, second,* and *third* (or ungual) phalanx respectively, but the thumb only possesses two phalanges. Certain " sesamoid " bones are also normally found in the hand, but these do not belong to the bony skeleton, being modifications in tendons or ligaments.

CARPUS.—The constituent bones are closely applied to one another and firmly secured in their rows by strong ligaments, so that the carpus as a whole forms a strong bony and fibrous mass, concave on its palmar aspect and convex dorsally : the flexor tendons of the fingers occupy the concavity.

There are four bones in each row. These are shown, somewhat diagrammatically, in Fig. 82, and are named as follows :—First row : (1) *Scaphoid* (or *Navicular*), (2) *Semilunar*, (3) *Cuneiform*, (4) *Pisiform*.

Second row : (5) *Trapezium*, (6) *Trapezoid*, (7) *Os Magnum*, (8) *Unciform*.

Each of these bones, with the exception of the pisiform, articulates with its neighbours in its own row, and in addition with bones above and below it. The articulations can be followed in the figure, where it can be seen that every one of the elements articulates with its neighbour, on both sides save in the case of the marginal bones ; but the other contacts are rather more complicated than this.

Anatomy of Skeleton

(1) *Scaphoid* has one side contact, with the semilunar. But it articulates with three bones below, trapezium, trapezoid, and os magnum ; above, with the radius.

(2) *Semilunar:* two side facets, for scaphoid and cuneiform: below, with os magnum and a small area of unciform above, with radius and a small area of fibro-cartilage.

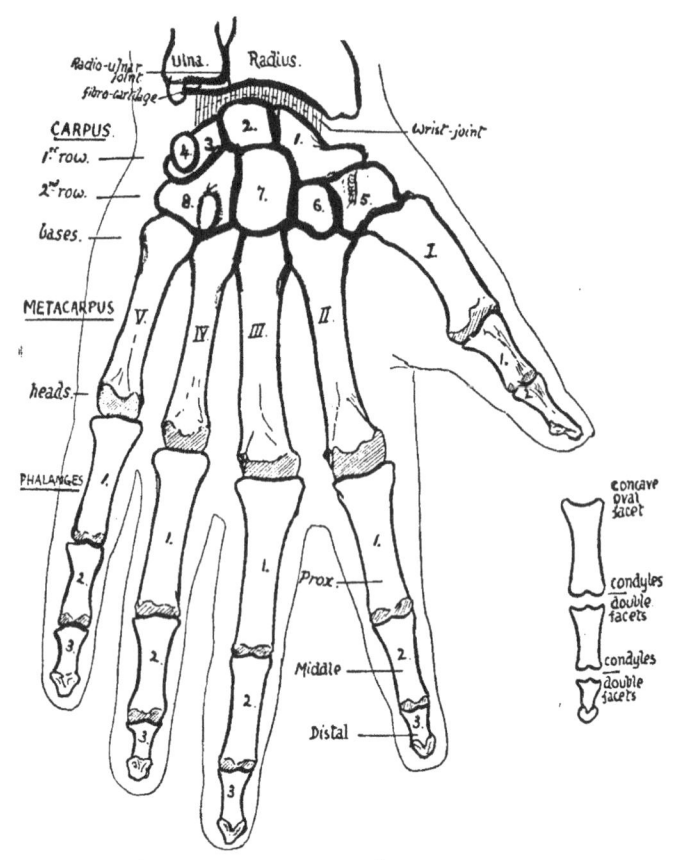

FIG. 82.—Semi-schematic outlines of bones of hand from the front. Description in text. Observe that the carpal and metacarpal bones are numbered from without inwards while the phalanges are numbered from above downwards. The smaller figure shows the distinctive features of the different phalanges.

(3) *Cuneiform:* with semilunar and pisiform, the latter really being in front of it, in conformity with the formation of the anterior concavity of the carpus: below, with unciform: above, with the triangular fibro-cartilage of the wrist.

(4) *Pisiform:* a very small bone, only articulating with the cuneiform.

The articulations of the distal row are—with their neighbours, with the proximal row above, and with the elements of the metacarpus below.

(5) *Trapezium:* at the side, with trapezoid : above, with scaphoid : below, supports first metacarpal and also, owing to small size of trapezoid, helps to support second metacarpal.

(6) *Trapezoid:* at the sides, with trapezium and os magnum : above, with scaphoid : below, with second metacarpal.

(7) *Os Magnum:* at the sides, with trapezoid and unciform : above, with scaphoid and semilunar : below, with middle metacarpal and also second and fourth.

(8) *Unciform:* at the side, with os magnum : above, with cuneiform and semilunar : below, it supports the fourth and fifth metacarpals.

This account shows that only one bone, the trapezoid, articulates with a single metacarpal.

The appearance of the *carpo-metacarpal articulation* is enough to show that the bones are practically fixed there and that no appreciable movement can take place except in the thumb,* where the form of the joint surfaces indicates that movement can take place from side to side or from before backwards : this joint has a separate synovial cavity, whereas the others have a common cavity.

The proximal row of carpal bones is immediately concerned in the wrist joint, and the curves of the articular surfaces on its upper aspect show that its range of movement at the wrist extends from before backward as well as from side to side. Observe that the joint surface on the scaphoid and semilunar is largely dorsal, more so than is necessary to correspond with the lower surface of the radius, which looks downward and slightly forward. It follows from this that the carpus can be extended on the radius to a greater extent than it can be flexed. The upper surface on the cuneiform is also somewhat dorsal in position, though to a less extent, thus agreeing with the plane of the fibro-cartilage.

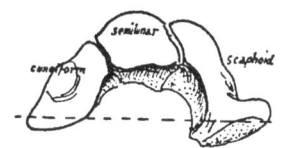

FIG. 83.—The bones which make the upper surface of the mid-carpal joint. The axis of movement is approximately indicated by the interrupted line.

When the hand is abducted or adducted the carpus moves round an antero-posterior axis passing through its centre—that is, when the hand is abducted or moved to the radial side, the first row of the carpus moves towards the ulna, the semilunar passing partly under the fibro-cartilage ; but when the opposite and more extensive movement takes place the bones are moved outwards, and the cuneiform may come partly under the radius.

Observe that the scaphoid and semilunar form a socket for the reception of the os magnum (Fig. 83) : the cuneiform is kept from this by the articulations between semilunar and unciform. In this joint flexion and extension are possible between the two rows of carpal bones, and in fact the so-called flexion and extension of the wrist really takes place in the more advanced stages at this *mid-carpal joint*. The back rim of the socket is deeper than the front so that, in contrast with the wrist, extension is the more limited of the two movements.

Lateral movement in the mid-carpal joint is quite impossible owing to the locking of the bones at the sides. The movement of the os magnum necessitates proper surfaces between the other bones of the second row and those of the first row, and we find accordingly that the curve of the head of the os magnum is carried on to the neighbouring part of the unciform, which fits into a concavity on the cuneiform, but, further in, this curve on the unciform becomes a concavity—that is, in the

---

* Among the others the fifth metacarpal has most freedom, and that is not much : the articular surfaces suggest this, and the movement is used in opposing the thumb.

104   Anatomy of Skeleton

line of the axis of movement of the second row on the first the shape of the articulating surfaces is appropriately changed, and at the same level on the outer side a similar form of curve is formed between the scaphoid and the trapezoid and trapezium.

All the bones of the carpus, with the exception of the pisiform, are affected by the anterior concavity of the whole structure, and this form is maintained by strong ligaments that are divisible into dorsal and palmar : the latter are the stronger set, as more strain is thrown on them in many of the functions of the hand, but it must not be forgotten that the anterior annular ligament, passing between the extremities of the curve, is the most effective factor in holding up the concavity of the carpus.

The bones of each row are joined together by interosseous ligaments, so that non-articular rough areas can be found on their sides, bordering on their articular surfaces, for these bands. The ligaments of the first row cut off the cavity of the radio-carpal joint from that of the mid-carpal joint, while this is again separated in part by the interosseous ligaments of the second row from that of the carpo-metacarpal system.

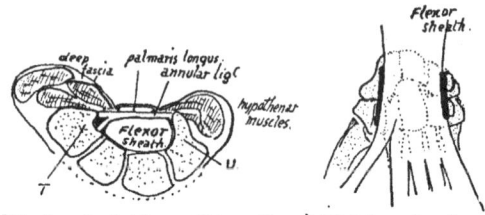

FIG. 84.—1, A scheme of a section showing how the flexor sheath occupies the carpal concavity, how the front wall of the sheath is thickened as the annular ligament, and how the carpal bones projecting on each side beyond the sheath are utilised and covered by thenar and hypothenar muscles. *T.* trapezium ; *U.* unciform. 2, Seen from the front, showing the flexor sheath lying on the carpus and metacarpus. The thick lines mark the side limits and attachments of the annular ligament : the marginal parts of the carpus project beyond these. The muscles arising from the projecting parts run down to the digits along the sides of the sheath, so that there are two side compartments and a central one for tendons and lumbricals. The superficial side muscles overlap the sheath at their origins, *i.e.*, they arise from the annular ligament.

In addition to these, the bones of the lower row are joined by anterior and posterior transverse intercarpal ligaments, which also extend on to the metacarpals. The transverse fibres that cover the first row are mainly connected with the radio-carpal fibres of the wrist-joint.

The transverse fibres are partly covered in front by the radiating fibres of the ligaments that help to connect the two rows with one another. These run from the os magnum to the neighbouring bones, the scaphoid, cuneiform, unciform, trapezoid, and the four inner metacarpals : some of these fibres also reach the radius, and a band goes to the base of the styloid process of the ulna, thus not interfering with rotation of the radius and hand round that bone. These radiating fibres are continuous in part with the fibres of the anterior capsule of the wrist. The two rows of bones are also connected at their margins by internal and external lateral ligaments which are continuous with the fibres of the lateral ligaments of the wrist, and by ligamentous or tendinous bands connecting the pisiform with the hook of the unciform and with the bases of the inner two or three metacarpals.

On the back of the carpus the first row is covered by radio-carpal bands passing downwards and inwards, and with these some fibres between scaphoid and cuneiform : some ligamentous slips from these pass to the bones of the lower row. In addition are vertical fibres between the ulna and cuneiform, and cuneiform and unciform, on the inner side, and from radius to scaphoid and scaphoid to trapezium on the outer side.

If we now analyse the carpal concavity, we see that the projection forwards on the radial side is due to the position and shape of the bones themselves, so that the trapezium and outer part of the scaphoid are turned forward, carrying the thumb in front of the rest of the metacarpus. On the inner side the cuneiform and unciform lie more in line, and the projection is brought about by the pisiform placed in front of the cuneiform and by a hook-like process—whence its name—projecting forward from the unciform. But the carpus between these margins is concave forward, so that we find

Limbs: Upper Limb

the various bones have, in general, larger dorsal than palmar surfaces, a feature well exemplified in the trapezoid.

The flexor tendons, lying in the concavity of the carpus, are covered by the anterior annular ligament, which is a purely carpal structure and is attached to the prominent margins of the concavity—that is, it is fastened externally to the trapezium and tuberosity of scaphoid, and internally to the pisiform, hook of unciform, and piso-uncinate ligament which connects these two. When the anterior annular ligament is in position the tunnel under it is floored by the whole front surfaces of os magnum, trapezoid, and semilunar, by a large part of scaphoid, and by very small margins of trapezium, unciform and cuneiform: *all these bony surfaces are of course covered by ligaments.* The remainder of the trapezium and scaphoid project beyond the line of outer attachment of the ligament, and the remainder of unciform and cuneiform project beyond its inner attachment, and the projecting areas thus provided, covered by ligamentous tissue, are utilised for the origin of muscles of the thenar and hypothenar eminences respectively. The figure (Fig. 84) shows this, and likewise indicates how these muscles also arise from the annular ligament. The ligament is really the thickened front wall of the "flexor sheath," and the general structure of the palm can be understood from the figure, for it shows three compartments, a central one that is the flexor sheath, and two side ones that contain the thenar and hypothenar muscles.

On its dorsal surface the carpus is covered largely by ligaments and crossed by the posterior carpal arch and other arteries under cover of the extensor tendons running to the digits and metacarpus. These tendons are covered as they lie on the radius by the posterior annular ligament—this is directed downwards and inwards, having no attachment to the ulna, but reaching the inner part of the carpus, to which it is attached, some of the fibres turning round this inner border to reach the pisiform and anterior annular ligament. This arrangement is in accordance with the fact that the carpus moves with the radius in supination and pronation, and any attachment to the ulna except in the axis of rotation would interfere with such movement.

THE METACARPUS.—Each of the five metacarpal bones has a *shaft*, a proximal *base*, and a distal *head*. With the exception of that of the thumb, each base presents side articulations for its neighbouring metacarpals in addition to the basal carpal facet or facets.

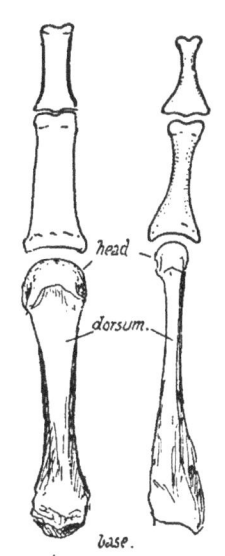

FIG. 85.—To illustrate the contrast between metacarpals and metatarsals, and between phalanges of hand and foot; slightly exaggerated. Notice the strong shaft of the metacarpal widening toward the head, which is heavy and square-cut, and compare this with the narrow, compressed shaft and head of the metatarsal, with its heavy base. The hand phalanx is broad, strong and flattened in the shaft, whereas that of the foot has a rounded feeble shaft with heavy ends.

Each shaft is thick and strong, and, except in the thumb, widens on its dorsal aspect as it is followed to the strong, square-cut head; this appearance of the shaft and head is one of the characters in which these bones differ from the corresponding metatarsals in the foot (Fig. 85). The five bones make a radiating series with intervening spaces that are filled by the Interossei muscles which arise from the shafts of the bones.

PHALANGES.—*All the first* phalanges have oval concave facets at their proximal ends for articulation with their corresponding metacarpals : the first phalanges are the only long bones in the body that possess such facets, so that any such bone must be a first phalanx of either hand or foot (Fig. 82).

The *second* phalanges all possess double facets on the proximal ends for articulation with the two "condyles" that are on the distal ends of the first phalanges, and the distal ends of the second phalanges are somewhat similarly furnished, so that the proximal ends of the *third* phalanges are also doubly facetted ; but the last phalanges have flat expansions at their ends that distinguish them at once—this expansion supports the fibro-fatty pulp of the finger tip. The two phalanges of the thumb

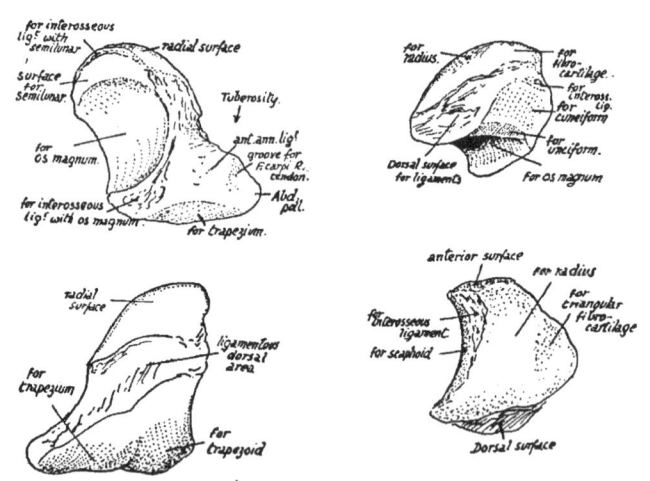

FIG. 86.—On the left, a left scaphoid : top figure, from the front ; lower figure, dorsal view. On the right, left semilunar : upper figure, from behind and internally ; lower figure, from above.

resemble the first and third phalanges in the other digits : the terminal phalanx probably is compounded of a true third and remnants of a second phalanx.

SEPARATE BONES OF CARPUS.

The separate bones of the carpus bear individually the general characters that have already been indicated as resulting from the building up of the whole mass, but it may be pointed out here that the effect of the carpal concavity in enlarging the dorsal surface of the bones is more than neutralised in the three main bones of the first row by the somewhat dorsal position of the upper articular surfaces that is correlated with the forward position of the hand with reference to the plane of the forearm.

Scaphoid (Fig. 86).—Upper radial facet convex : inner facet is crescentic for semilunar, and below this a large concavity for head of os magnum. A small rough area between radial and semilunar facets is for interosseous ligament connecting it with semilunar. Lower surface rests on trapezium and trapezoid, and is divided into

Limbs: Upper Limb

corresponding areas by a badly-marked ridge. Dorsal surface is encroached upon by the lower and upper articular surfaces, so is only represented by a narrow rough strip for dorsal ligaments of the wrist.

Anterior surface presents a concavity or notch, below which is the prominent *tuberosity*. The upper parts of the deeper fibres of the anterior annular ligament are attached to the tuberosity, which may, as in the figure, show a ridge for this, with a groove external to it for the tendon of F. carpi radialis : further in is a ligamentous area   d this may be enlarged in the angle between the trapezoid and os magnum surfaces by the attachment here of an interosseous band connecting the bone with the os magnum.

Abductor pollicis gets some slight origin from the tuberosity outside the groove for the radial flexor.

Semilunar.—Rather obliquely placed, in an upward and inward direction. A deep concavity below for head of os magnum, a flattened four-sided facet on inner side for cuneiform, and between these two, on the lower and inner angle, an elongated strip

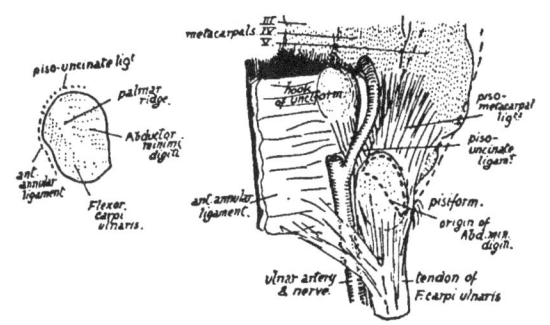

Fig. 87.—Palmar aspect of pisiform. On the right the bone is shown in position ; the Abductor minimi digiti is indicated by interrupted lines in its position lying on the ligamentous fibres connecting pisiform with bases of metacarpals. Observe that the oblique and raised anterior ridge is directed downwards and outwards ; this enables one to place the bone on its proper side.

of articular surface for unciform. Outer facet for scaphoid is elongated and crescentic, with a roughened area above it for the interosseous ligament between the bones : a similar band is on the ulnar side, connecting it with the cuneiform and attached above the facet for that bone, on the edge that separates it from the upper articular surface, but a definite marking is frequently wanting (see Fig. 86).

Upper articular surface extends on to dorsal surface in its outer part : this is for the radius ; but the inner part of the surface, more on the upper aspect of the bone, is for the triangular fibro-cartilage, and the two areas, though continuous, are sometimes separately recognisable on the bone. As a result of this the dorsal surface is diminished, particularly its outer part, and is a roughened and purely ligamentous one. Anterior surface, larger than the dorsal, has only weak ligamentous bands attached to it, and is not much roughened.

Cuneiform.—Three-sided pyramid with its base resting against the semilunar, its apex directed downwards and inwards, an inferior surface concavo-convex and resting on unciform, a postero-superior surface carrying a small facet for the fibro-cartilage, but otherwise rough for ligaments, and an antero-superior (anterior) surface ligamentous, but with a facet on its inner part for the pisiform.

Pisiform.—A small nodular bone with an articular facet on its dorsal surface, which rests on the front aspect of the cuneiform. Its long axis runs from above downwards and outwards, directed toward the hook of the unciform: the piso-uncinate ligament is attached to the distal extremity, sometimes making a marked tubercle. The tubercle is continuous with an oblique line visible on the radial aspect of the bone: the line gives attachment to the anterior annular ligament and some fibres of F. carpi ulnaris, the remainder of the tendon running on to the proximal and palmar surfaces of the bone. A slightly hollowed surface lies between the radial line and the dorsal articulation, increasing in depth as it is traced in a proximal direction, and making, at this end, a characteristic hollow on the radial side which enables the observer to recognise the upper end—and the outer aspect. The inner or ulnar surface is marked by faint striations from fibres of the posterior annular ligament which reach it. In front of these, between them and the tubercle for the piso-uncinate ligament, is an area for attachment of piso-metacarpal fibres. The Abductor minimi digiti arises from the front of the bone between these various areas, and extends on to the different ligamentous fibres occupying the surrounding surfaces. The deep hollow at its proximal (radial) side is occupied by a fibrous pad connecting the flexor sheath with the posterior annular ligament.

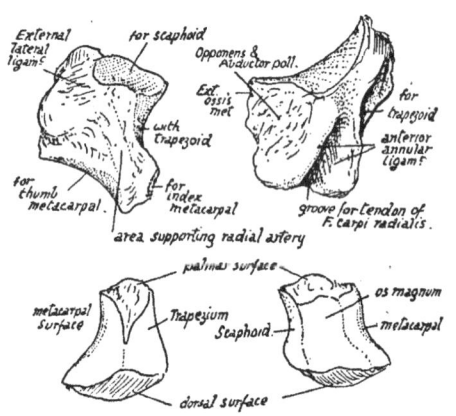

FIG. 88.—Upper figures, left trapezium: 1, postero-external aspect; 2, palmar aspect. To distinguish left from right, hold the bone so that the deep groove for Fl. carp. rad. is in front and near the proximal part of the bone, running towards the index finger, while the concavo-convex facet for first metacarpal looks toward the thumb. Lower figures, left trapezoid: 1, outer and 2, inner views. To distinguish the sides, place the smaller non-articular surface in front, with the pointed downward prolongation from it toward the outer side; of the two articular areas between which this prolongation lies the larger saddle-shaped one should look toward the index finger, and the smaller convex one outwards. (See text.)

Trapezium.—Upper surface has facet slightly concave from before backwards for scaphoid, and outside and behind this a ligamentous area for external lateral ligament. An internal surface concave from above down for articulation with trapezoid. The bone has its lower back and inner part prolonged down to articulate with the outer side of the base of the metacarpal of index finger, so that its lower aspect shows this facet looking downwards and inwards, and further out and proximally a concavo-convex surface for the metacarpal of the thumb—looking downwards and outwards.

The front surface is on the whole somewhat flattened, giving attachment to ligaments and to Opponens and Abductor pollicis, and insertion at its outer part to Extensor ossis metacarpi. Its inner margin forms a prominent *ridge* for attachment

Limbs: Upper Limb

of superficial fibres of the anterior annular ligament, and this overhangs a well-marked groove containing tendon of F. carpi radialis and bounded below by a smaller ridge for deeper fibres of the ligament : thus the tendon might be described as passing through the outer attachment of the annular ligament (see Fig. 84). The surface carrying the groove looks inwards and forwards towards the palm, and the groove is directed towards the base of the index metacarpal.

The postero-external surface is roughened by ligaments and supports the radial artery as it runs to the first intermetacarpal space.

Trapezoid.—Difference in size between anterior and posterior surfaces is well marked : increase in dorsal surface particularly in a downward and inward direction, while the rough anterior surface is drawn into a lower and outer point, fitting as a roughened angle between the facets for trapezium and index metacarpal (Fig. 88).

This bone only supports one metacarpal, and its lower surface is moulded into a concavo-convex form to fit the deeply-grooved base of this bone. Outer facet for trapezium is convex, and upper one for scaphoid is slightly concave, and these two are directly continuous, as is the upper one with that on the inner side for os magnum,

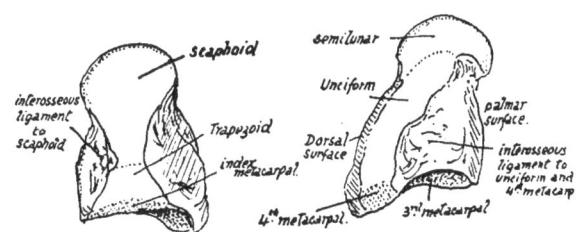

FIG. 89.—Left os magnum. From the outer and inner sides. To tell left from right, hold the round head proximally, the larger non-articular surface dorsally, and the abruptly flattened side inwards.

and this again with the lower facet. In fact all the facets form a continuous articular surface round the bone, only partly interrupted in front and externally by the pointed prolongation of the anterior surface.

Os Magnum.—A large dorsal surface coming obliquely to a point directed downwards and inwards and narrowing proximally to give the appearance of a neck supporting the upper articular head : front surface smaller, irregular, roughened for ligaments. Outer surface with two facets, partly continuous, for scaphoid and trapezoid, but with a hollow area between them on the palmar side for an interosseous ligament. Inner surface rather flattened, with facet for unciform extending along its posterior part, but roughened in front of this for interosseous ligament : this is the strongest of these ligaments in the hand, and has the largest surface of attachment.

The lower surface has a main facet for the middle metacarpal, and on each side of this a smaller one for the neighbouring metacarpals : that for the fourth metacarpal is small and confined to the back part of the surface, for the front portion is taken up by the large interosseous band, which not only ties the os magnum to the unciform, but also to the fourth metacarpal.

The obliquity of the distal border of the dorsal surface is a consequence of the presence of the styloid process on the middle metacarpal, and the dorsal surface may present a bevelled-off portion to receive this.

Unciform.—Recognised at once by the prominent hook on its palmar aspect, from which its name is derived. This process forms part of the inner wall of the flexor sheath, so is concave externally. The concavity is in relation with the synovial covering of the flexor tendons of the little finger : the tip gives attachment to the annular ligament : the convexity outside the tip affords origin to the Opponens distally and Flexor brevis minimi digiti proximally : the upper border has the piso-unciform ligament attached to it, and the lower border has fibres of Opponens arising from it. The inner and lower part of the process is grooved by the deep ulnar nerve turning round it : the artery takes a wider sweep, and so changes its relation to the nerve (see Fig. 87).

Above the hook the front surface is for ligamentous attachments, and further in is covered by the fibres that connect the pisiform and inner metacarpal bases.

Outer surface with facet behind for os magnum and roughened in front of this for interosseous ligament. Lower aspect has a continuous articular surface showing two facets separated by a low crest, for the fourth and fifth metacarpals. Upper or superointernal surface carries a concavo-convex facet for cuneiform, and, external to this, is prolonged proximally to an angle which articulates with semilunar. Dorsal surface large and roughened for ligaments. Inner surface, nearly reduced to a point, gives attachment to ligaments.

Observe that the unciform process marks by its lower edge the line of the carpometacarpal articulation.

Much of the position of the carpus can be appreciated during life, and many of its bones examined. On the front aspect, the tuberosity of the scaphoid is felt at the proximal end of the thenar eminence, deep and external to the radial flexor tendon where this disappears. The trapezium is covered by the thenar muscles, but can be indistinctly felt through these just distal to the scaphoid : metacarpal movement to and fro does not affect the mass thus felt. The line which curves round and limits the upper end of the thenar eminence passes over the scaphoid-trapezium joint, so that the scaphoid tuberosity is found between this line and the distal one of the two deep creases which cross in front of the wrist. The pisiform is easily felt a little below this last line, at its inner end and in the base of the hypothenar eminence. Between the pisiform and scaphoid, the proximal convexity of the curved upper margin of the carpus reaches the middle third of the upper of the two transverse deep lines, sc that this line can be taken as giving that of the wrist-joint : this part of the carpus cannot be felt, of course, being covered by flexor tendons, but it may be noted that the semilunar lies in the middle, between the two transverse creases. The hook of the unciform can be felt on pressing below and external to the pisiform, and a line drawn out from its lower edge marks the lower level of the carpus.

The presence of tendons interferes with recognition of the separate bones on the dorsum, though they may sometimes be made out in thin hands. The carpus, as a whole, lies between the bases of the metacarpus and the lower end of the radius, both of which lines can be found. The joint between os magnum and semilunar lies about halfway between the base of the third metacarpal and the end of the radius (see Fig. 93). Trapezium and part of scaphoid lie in the floor of the area between the long and short extensor tendons of the thumb.

SEPARATE BONES OF METACARPUS.

With the exception of the first, the metacarpals articulate with each other at their bases, so bear facets on their sides here, and also impressions for interosseous ligaments :

Limbs: Upper Limb

they are, moreover, attached to each other at their distal ends by the transverse metacarpal ligament, which lies on the palmar aspect of the heads of the inner four bones and is attached to them through the capsule of the metacarpo-phalangeal joint. They are arranged in a somewhat radiating fashion, and make a palmar concavity, and the intervals between them are filled up by the interosseous muscles: the concavity contains the flexor sheath and, deep to this in the outer part of the palm, the adductors of the thumb, while the deep palmar vessels and nerve cross it near the bases of the bones. Their dorsal surfaces are covered by the plane of extensor tendons, and the dorsal interosseous fascia extends from bone to bone between them.

The different metacarpals can be easily recognised, but it is well first to observe the general distinctions in appearance between the metacarpals and metatarsals, as shown in Fig. 85. The proximal ends of the metacarpals are distinctive.

First, or Thumb, Metacarpal (Fig. 90). — Sometimes confounded with phalanges: distinguished by its simple rounded head and its basal concavo-convex surface for articulation with trapezium. Shaft is broad and rather flattened. A small facetted tubercle on the outer side of the proximal end is for insertion of Extensor ossis metacarpi, and may be used to aid in placing the bone on its proper side; but the tubercle is often indistinct, and there may be a large ligamentous tubercle on the inner side. The saddle-shaped surface for trapezium is divided by a low ridge running antero-posteriorly, and the larger subdivision is external, as shown in Fig. 90.

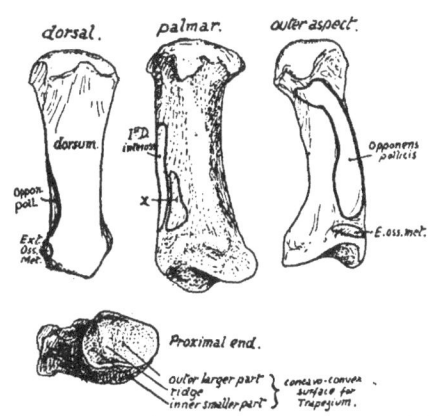

FIG. 90.—First right metacarpal. The side to which the bone belongs can be ascertained by looking at the proximal articular surface; the outer slope of the saddle-shaped surface is larger. x, origin of deep head of F. brevis pollicis.

Second, or Index, Metacarpal.—Recognised at once by the deep cleft or groove running antero-posteriorly across its carpal surface, as shown in Fig. 91, for articulation with the trapezoid: it is very apparent when looked at from the dorsal side. The bone touches the middle metacarpal, and therefore has a metacarpal facet on its inner side, and this is separated from the basal surface by a bevelled-off area for the os magnum. On the outer side there is no metacarpal facet, but a small rounded or four-sided surface looking forwards, outwards and upwards for the trapezium: immediately behind this is a roughened facetted dorsal area for the insertion of Extensor carpi radialis longior.

Third, or Middle-Finger, Metacarpal.—Recognised by its styloid process, which also is on the outer side of its basal end behind: from this the bone can be placed on its proper side. There is only one carpal facet, but metacarpal facets are on both sides. Observe that these are divided into anterior and posterior parts, partly on the outer side and altogether on the ulnar side, by rough areas for interosseous ligaments.

The styloid process may be short, separating the index metacarpal from the os magnum, or may be larger and longer, extending between these to reach the trapezoid. It may be a separate free ossicle, or this may have fused secondarily, not with the metacarpal, but with the os magnum or, more rarely, with the trapezoid.

Fourth, or Ring-Finger, Metacarpal.—This can be recognised because it manifestly has not the character of the bones already considered, yet has metacarpal facets on each side, and not on one side only, as is the case in the fifth metacarpal—in fact it can be recognised mainly by its negative characters. The ulnar metacarpal facet is single, the radial double. The carpal facet, for unciform, is continuous with the single inner facet, but only indirectly with the posterior of the two outer ones, through the medium of an obliquely-cut surface for the os magnum. It is cut off from the anterior one by an area for the strong interosseous ligament connecting os magnum, unciform, and the fourth metacarpal, and this ligament may be enlarged at the expense of the anterior facet.

Fifth, or Little-Finger, Metacarpal.—Distinguished by its possession of a metacarpal facet on one side only : the non-articular inner side presents a prominent tubercle near the dorsum for the insertion of the ulnar extensor. The carpal and metacarpal facets are continuous round the outer margin of the base.

The *shafts* of the bones can also be distinguished from each other. That of the first metacarpal has been considered already, and the markings on the others can be examined now.

The shafts are moulded by the interosseous muscles that are so closely applied to them. A glance at the dorsal aspect of the metacarpus shows that the greater breadth of the distal part of this surface in each bone is due to the encroachment of the dorsal interossei on the proximal portion, and it is interesting to observe that the extension of these muscles dorsally is most marked when there is a corresponding palmar interosseous on that side of the bone : thus the proximal part of the dorsal surface or ridge is pushed towards the outer side on the index metacarpal and to the inner side in the fourth and fifth bones, an obliquity that is very marked in some specimens.

Examine the sides of the shafts. The interossei have short tendinous fibres mixed with their fibres of origin, and therefore there is a very slight roughness visible over their areas ; but the finger and eye will have no difficulty in finding the fine cleanly-cut line that marks the separation between dorsal and palmar interosseous, and in well-marked bones the areas can be almost completely mapped out in this way. Fig. 92 shows these origins : the upper row is a view of the radial aspect of the bones ; in the lower row they are seen from the inner side.

It is evident that no two of these bones are alike in their markings for interossei, and it is possible to put the bones in their right order from their shafts alone, provided that the sides to which they belong are known. Observe also that the nutrient foramina are on the radial side in the three inner bones, and on the ulnar side in the outer two : the canal descends in the first but ascends in the other metacarpals, so being in accord with the rule that the canal is directed away from the " growing end " of the bone, for there is normally only one epiphysis on these bones, and it forms the base of the first metacarpal but the distal ends of the others.

The inner side of the fifth shaft receives the insertion of the Opponens of the little finger : this belongs morphologically to the group of palmar interossei, being the ulnar belly of the metacarpal flexor of the finger, which has extended its insertion proxi-

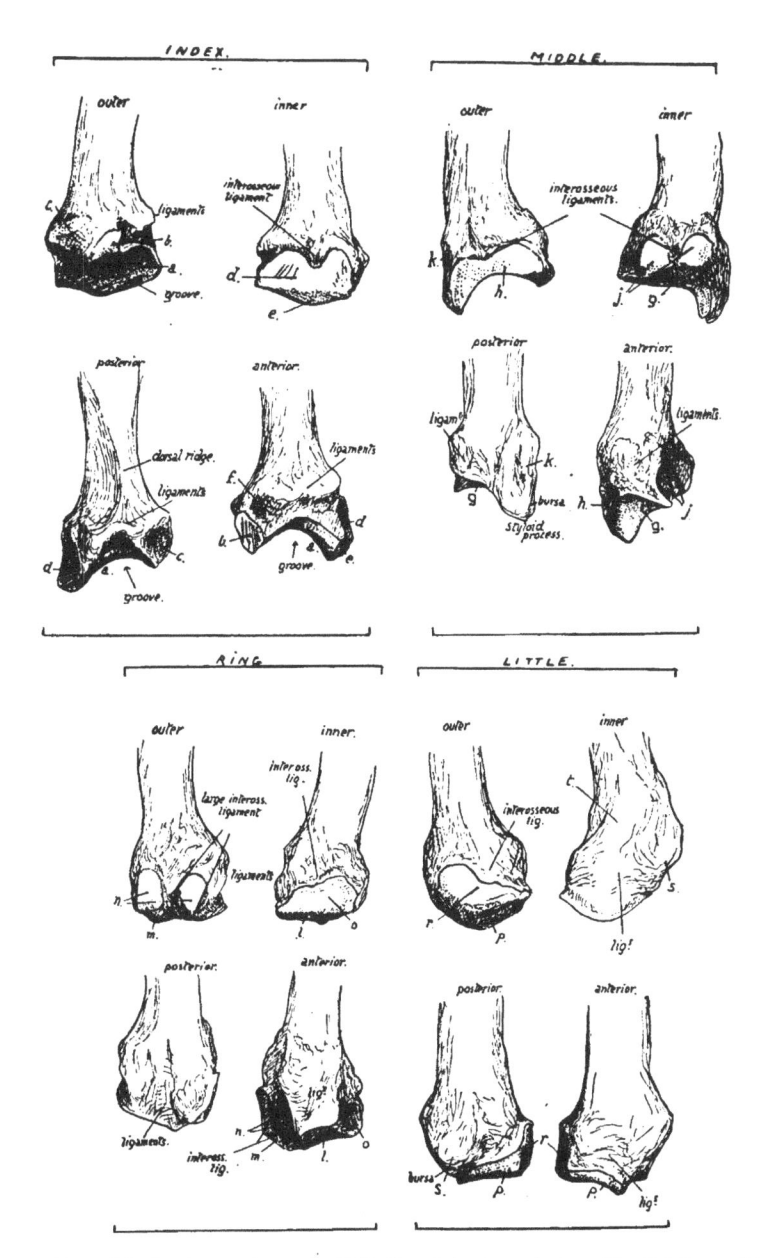

FIG. 91.—Bases of the four inner metacarpals of left side. Outer, inner, back, and front views are given of each bone. The cartilage-covered articular surfaces are coloured blue. *a*, surface for trapezoid; *b*, facet for trapezium; *c*, insertion of Extens. C. radialis; *d*, surface for middle metacarpal; *e*, surface for os. magnum; *f*, insertion of Fl. C. rad. and anterior ligaments; *g*, surface for os magnum; *h*, surface for index metacarpal; *j*, surface for fourth metacarpal; *k*, insertion of Ext. C. rad. brev.; *l*, surface for unciform; *m*, surface for os magnum; *n*, surface for middle metacarpal; *o*, surface for fifth metacarpal; *p*, surface for unciform; *r*, surface for fourth metacarpal; *s*, insertion of Ext. carpi ulnaris.

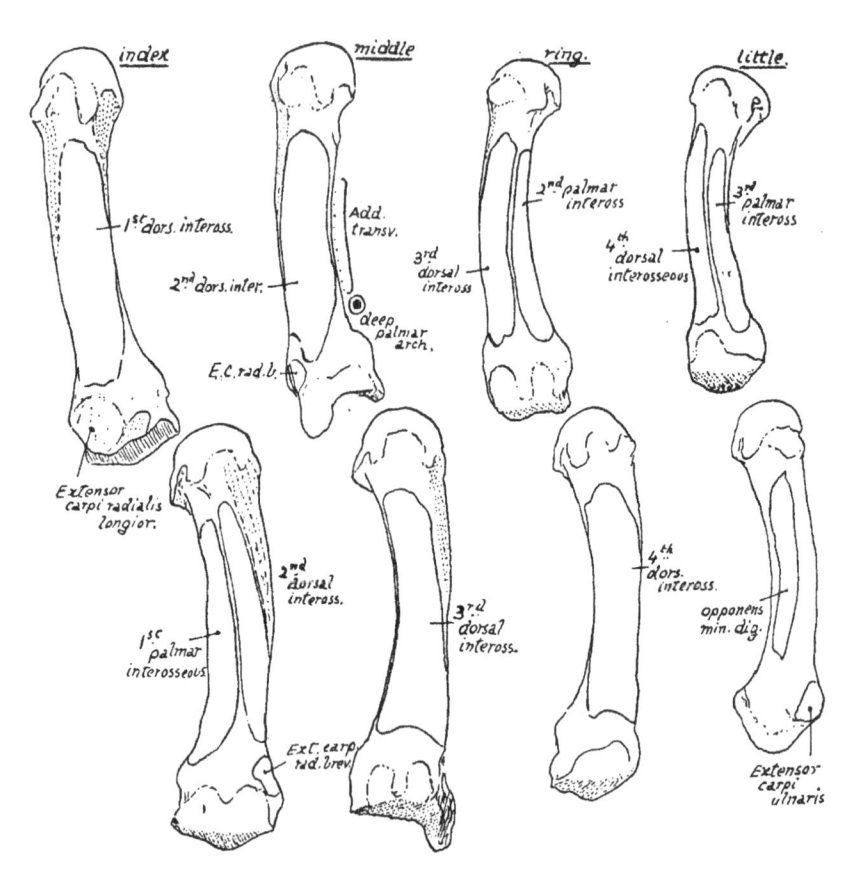

FIG. 92.—The four inner metacarpals, showing the attachments of the Interosseous muscles and of certain other muscles. The upper row represents the bones from the outer side, and the lower row shows their internal aspects.

mally from the phalanx to the metacarpal while its origin has grown round the flexor sheath on to the annular ligament.

A *palmar ridge* is seen running along each of the shafts, made by the meeting of the interosseous surfaces on the palmar aspect: it may be slightly displaced to the inner side in the fourth and fifth bones * as a result of their having a palmar interosseous on the radial side, but a displacement in the opposite direction is not apparent on the index metacarpal.

The ridge is widened in the middle metacarpal by the origin of Adductor transversus from it. On this bone, and on that of the Index, the ridges are not as a rule so clearly carried to the bases, but are interrupted by an ill-defined groove that marks the situation of the deep palmar arch : on their bases, proximal to this, the Adductor obliquus has part of its origin.

It is apparent from the markings for the interossei (Fig. 92) that these muscles form thick masses which practically cover in the bones on their palmar aspects and, with the Adductors in the outer part of the palm, separate them from the flexor sheath. Dorsally, however, the distal two-thirds or more of the shafts are not covered by the muscles, and here the shaft is broad and flat and in relation with the extensor tendons of the fingers.

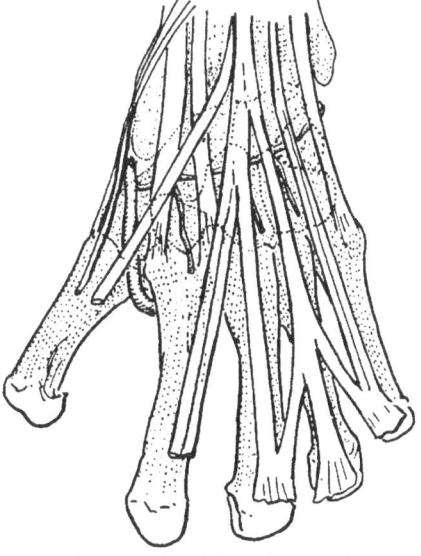

FIG. 93.—To show the relation of the tendons, etc., on the dorsum to the underlying carpal bones and to each other. The radial artery and posterior carpal arch are also shown.

A summary of the muscles that take origin from the different metacarpals, and the tendons inserted into them, may not be out of place here.

METACARPAL.	MUSCLES.	TENDONS.
First	Flexor brevis (deep head)	Ext. ossis metacarp.
	Opponens (insertion).	
	First dorsal interosseous.	
Second	First and second dorsal inteross.	Flexor carpi radialis.
	First palmar inteross.	Extensor c. rad. long.
	Adductor obliquus	Extensor c. rad. brev.
Third	Second and third dorsal inteross.	Flexor c. rad.
	Adductor transversus.	Ext. c. rad. brev.
	Adductor obliquus.	
Fourth	Third and fourth dorsal inteross.	
	Second palmar inteross.	
Fifth	Fourth dorsal inteross.	Extensor carpi ulnaris.
	Third palmar inteross.	
	Opponens (insertion).	

* Like the other characters described in these bones, these varieties should be studied in well-marked and good-sized male bones : they are not striking, or even apparent, in small and feeble bones.

F.A.

The radial flexor is partly fused with the ligaments in its attachment to the middle metacarpal. The main insertion of Ext. c. rad. brevior is on to the base of the styloid process, but a contiguous impression on the index metacarpal receives part of it : bursæ lie under the two radial extensor tendons at their insertions. The ulnar flexor is not directly inserted into the metacarpus, but the piso-metacarpal ligaments may be looked on as a continuation of its tendon to the inner bones.

The Ulnar extensor may have a small bursa between its tendon and the proximal part of the tubercle on the inner and dorsal aspect of the base of the fifth metacarpal.

The *heads* are somewhat square-cut, rounded off in an antero-posterior direction. Observe that the articular surfaces are slightly more vertically extensive in front than behind in addition to being broader, and that they show a transverse constriction owing to the existence of the depressions on the sides of the head : the constriction is best marked in the index metacarpal and least apparent in the fifth. Each surface is prolonged proximally in front and at the sides into two cornua, on which the sesamoid bones or cartilages play that are found in the palmar capsular structures (see later).

These sesamoids are always present and well developed in the thumb, and frequently, if not usually, in the index finger on the outer side and in the little finger on the inner side, and corresponding differences in the size of the cornua can be noticed in the metacarpals. The sesamoids are more numerous in the fœtus, especially on the radial side of the index and middle fingers, and on the ulnar side in the ring and little fingers.

The depressions on each side of the head have a facetted surface behind them, and to this also the lateral ligament of the joint is attached. The anterior part of the capsule is thickened and firmly attached to the base of the phalanx, but only by areolar tissue to the metacarpal, so that there is no very definite marking to be found for it on these bones.

The extensor expansion covers in the posterior aspect of the joint and takes the place of a capsule there. Thus the head of each bone may be said to be surrounded by ligamentous and fibrous structures : these are all connected together, but the mass can be analysed into its constituent parts as shown in Fig. 94.

1. The deepest structure is the proper capsule, which extends on the palmar aspect of the joint from one lateral ligament to the other : it has the sesamoid bones or cartilages when present embedded in it, and is firmly attached to the base of the phalanx, but lies loosely on the metacarpal, so that it does not limit extension of the joint.

Each lateral ligament has an extensive attachment to the metacarpal (Fig. 94) and the phalanx, so that its front fibres become tight in extension and its posterior part in flexion : the interosseous tendons are partly attached to the lateral ligaments.

2. The transverse metacarpal ligament lies on the palmar surfaces of the capsules, a strong fibrous band stretching across the four inner joints and only indirectly attached by their capsules to the bones : its function is evidently to limit divergence of the bones.

The interosseous tendons lie behind it, while the plane of the Lumbricales and digital vessels and nerves is in front of it.

3. The proximal end of each digital sheath (theca) lies on the transverse band, and thus thickens still more the structures in front of the joint.

The vessels and nerves pass down between the commencement of the sheaths.

4. The extensor expansion throws a fibrous covering downwards and forwards on each side of the joint, covering it in and passing over the lateral ligaments and base

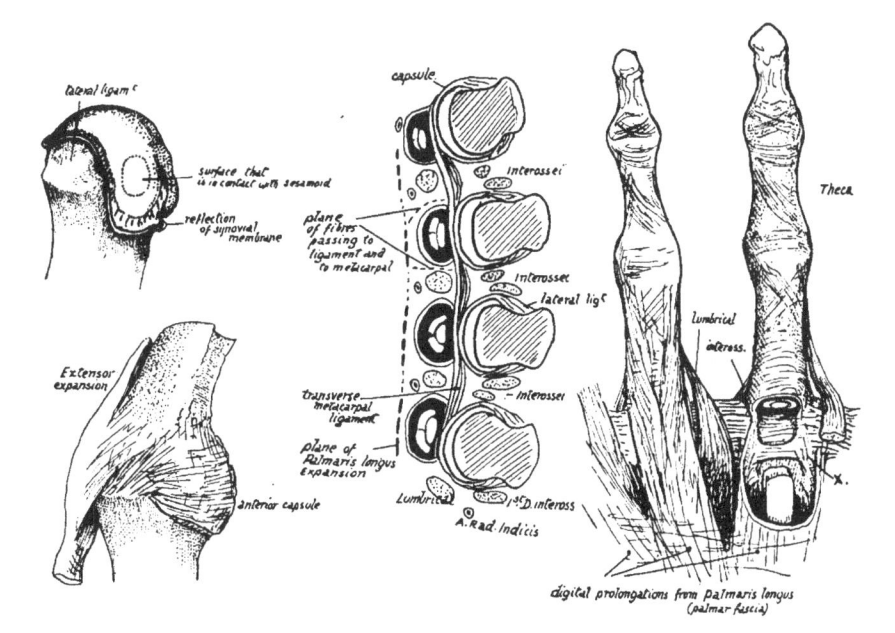

Fig. 94.—The central figure is a schematic analysis of the structure of the metacarpo-phalangeal joints as seen on transverse section. The plane of the transverse metacarpal ligament lies between the capsule and the tendons of the flexors of the digits and Lumbricals. The eXpansion of the tendon of Palmaris longus which forms the palmar fascia covers the flexor tendons, and in the lower part of the hand sends in fibres between the tendons to reach the metacarpal shafts and transverse metacarpal ligament. Thus the tendons are in a kind of incomplete compartment before they enter the proper theca of each finger. The Lumbricals pass down between the tendons, in front of the ligament, which separates them from the tendons of the Interossei. On the right the relation of these structures to the theca is shown in two fingers; observe that the fibres from the palmar fascia eXtend down as far as the base of the second phalanX, a detail which eXplains the flexion of the two proXimal phalanges found in pathological contraction of the fascia. On the left is the head of a metacarpal bone, with the synovial and ligamentous areas shown upon it. Below this the metacarpo-phalangeal joint is figured, the transverse ligament not being represented; notice the direction of the lateral and front fibres, and the eXpansion of the eXtensor tendon which covers the joint behind and postero-laterally.

of phalanx: it has some attachment to these and to the interosseous tendons, so that the direct action of the extensor tendon on the finger is mainly to extend the first phalanx through this attachment.

Observe that the articular surface in the metacarpo-phalangeal joint of the thumb is flatter and broader but less extensive from before backwards than those of the other bones: associated with this is the much lessened range of flexion and practical absence of power of lateral movement in this joint. Its ligaments are the same with the exception of the transverse metacarpal band: the sesamoids are large and always present and the comparative slightness of the grooves for them depends on the smaller extent of the articular surface.

Phalanges.

The distinguishing characters of these bones have already been noticed (p. 105).

Looked at from behind the phalanges are rounded and cylindrical in appearance, enlarged at their ends, but on their palmar surfaces they are flattened and bounded by sharp margins on each side. The rounded dorsal surfaces are covered by the expansions of the extensor tendons, while the flat, anterior surfaces form the bony bed of the flexor tendons, against which they are held by the fibrous sheath that is fastened to the margins of the bones.

The first phalanx is attached to the metacarpal as already described, and the interphalangeal joints are formed on the same plan, save that there is of course no representative of the transverse metacarpal band. Thus we find rounded facets on the sides of the distal extremities for the lateral ligaments and ligamentous markings on the proximal ends for these and for the capsules in front.

The presence of condyles with intervening depressions in the interphalangeal articulations indicates that there can be no lateral movements in these joints, and the lateral ligaments are attached at what is practically the centre of movement in flexion and extension, so that they are constantly tense.

The dorsal expansion covers the articulations behind and is attached to the bases of the last two phalanges, making a well-marked roughness there, which is therefore not ligamentous.

The theca which holds down the flexor tendons, is not a continuous fibrous layer, but is made up of bands (Fig. 94), so that the markings for it on the bones are irregular. On the first phalanx they are on the margins: the lines of attachment then cross the capsule and lateral ligaments to reach the margins of the second phalanx, then pass on to the third where they meet, more or less fused with the rough basal ridge that marks the insertion of the Flexor profundus tendon. The markings for the different bands that compose the sheath cannot be recognised with any certainty on the bone, but it can be noticed that the upper crucial fibres cause a roughness on the first phalanx behind the line of its margin.

Between the margins the bones are covered by the synovial sheath of the tendons, and on the second phalanx the markings of insertion of Flexor sublimis are easily seen, distinct from the marginal sheath marking, opposite the middle and proximal parts of the shaft.

Where the capsules of the interphalangeal joints support the tendons they are thickened into a single sesamoid body which occupies the anterior notch between the condyles of the phalanges: this raises the tendons slightly from the bones to which they are going, and so far increases their power of action on these bones.

The last phalanges have a distal expansion that lies under the nail, but its greater thickness is on the palmar surface, where it is roughened by the attachment of the fibro-fatty pad that makes the palmar finger-tip.

The strong phalanges of the thumb are in general similar to the first and last phalanges of the other fingers, but it should be noticed that the base of the first phalanx exhibits a slight central prominence in front which resembles the inter-condylar point on the bases of the second phalanges of the other fingers. This is probably associated with the large size of the sesamoids in the joint and the fact that the intercornual notch in the first metacarpal is more horizontally placed than in the other bones, so that the phalanx fits, in its front part, for practical purposes on two badly-formed condyles.

The phalanges of the hand are much stronger than those of the foot, for they are used as opponents in grasping, whereas little weight is thrown on the foot bones, except in the case of the big toe. So the appearance of the two classes of bones, conforming with this, is enough (Fig. 85) to enable one to distinguish between them at a glance.

Development of Bones of Hand.

All the bony elements of the hand are preformed in cartilage, and in the case of the carpus the cartilaginous units are usually greater in number than the bony. Possibly this may have an atavistic significance : in lower vertebrate forms the typical carpus consists of two rows, a proximal containing three elements, *radial, ulnar,* and *intermediate,* and a distal with five " *carpalia,*" each one supporting a metacarpal and numbered from without inwards : between the two rows is placed an " *os centrale,*" round which the others might be said to be grouped (see Fig. 157).

In the human cartilaginous carpus the os radiale is represented by the scaphoid, the intermedium by the semilunar, and ulnare by cuneiform : the pisiform is usually regarded as of the nature of a sesamoid, although this view is not incontestable.

The five carpalia are represented by—(1) Trapezium, (2) Trapezoid, (3) Os Magnum, (4) and (5) Unciform : the double nature of the unciform is indicated by the facts that it supports two metacarpals, is chondrified from two cartilaginous centres, and is said to be ossified also from two centres. The os centrale is represented by a small cartilaginous nodule usually present in the embryo, which either becomes divided and disappears later or joins with one of the other elements, usually the scaphoid.

Many irregular and differing " elements " have been described from time to time in association with the carpus : of these the most constantly found are the separated styloid process of the 3rd metacarpal and (rarely) a small ossicle on the outer side of the scaphoid, the *radiale externum.* In many cases the additional so-called " elements " are ununited results of old fractures of the carpus.

The trapezium is at first smaller than the trapezoid, and the hook of the unciform is developed in cartilage after the body.

The *carpus* is cartilaginous at birth, but there is occasionally a centre in the os magnum. This centre appears as a rule a few months after birth, and the whole carpus, with the exception of the pisiform, is ossified by the sixth or seventh year : the pisiform centre does not appear until about ten to twelve years. The centres appear in a more or less definite order, and for practical purposes the times shown in the table below may be taken as approximately correct. It is to be noted that the centres appear from a few months to a year earlier in the female hand :—

Bones.	Time of Appearance of Centres.
Os magnum	1st year.
Unciform	2nd ,,
Cuneiform	3rd ,
Semilunar	4th ,,
Scaphoid	} 5-6th ,,
Trapezium	
Trapezoid	7th ,,
Pisiform	11th ,,

The *metacarpals* are ossified from two centres, one for the shaft and one for a distal epiphysis, so that at birth the heads are cartilaginous, because the shaft centres appear in the ninth week, whereas the epiphysial centres are found in the second year. The thumb metacarpal is an exception, in that its epiphysis is at the proximal end, the centre appearing in the third year : in about 6 per cent. of cases there is also a small distal epiphysis, appearing later and uniting with the shaft in a few years. The normal metacarpal epiphyses join the shafts between fifteen and twenty. Additional centres are sometimes found for the styloid process of the middle metacarpal and for the proximal end of the index metacarpal (3 per cent.).

Phalanges.—These are preformed in cartilage with the exception of the ungual tubercle on the last phalanges : at one stage the cartilaginous terminal phalanges are longer than the middle ones. Ossification starts in the shafts, except in the terminal phalanges, where the primary centre is at the distal end, and at birth the proximal ends of each phalanx are cartilaginous : here the epiphysial centres appear about the second year.

The shaft centres for the last phalanges come first, in the seventh to eighth week, then those for the first phalanges, ninth week, and lastly, about the eleventh or twelfth week, those for the middle row. The proximal epiphysial centres come first in the first row, and later in the others.

The last phalanx of the thumb is the earliest to present a centre in its shaft : this is the first ossifying nucleus in the hand.

Ossification of epiphyses appears to start about a year earlier in females in the metacarpus and phalanges, and in them the fusion with the shaft is also earlier : in male bones this occurs about eighteen to twenty.

Chapter V

THE LOWER EXTREMITY AND PELVIS

The *bony pelvis*, which, covered by soft tissues, forms the lower support of the abdominal contents, is formed by two *ossa innominata* fixed behind to the sides of the sacrum and meeting in front at the *symphysis*. It forms a firm and strong base on which the trunk can rest, and through which its weight can be transferred to the free limbs below.

The sacrum and coccyx have been described already (p. 38), but it is necessary to give a short description of the various parts of the innominate bone before considering the pelvis as a whole.

OS INNOMINATUM.

A large irregular bone that helps to form the bony pelvis in conjunction with its fellow and the sacrum. It articulates at its upper and back part with the side of the sacrum, below and in front with its fellow in the middle line, and with the femur externally. The bone is formed by the fusion of three separate main pieces: these are (Fig. 95) the **Ilium** above, the **Ischium** below and behind, and the **Pubis** below and in front. Where the three parts meet is placed the *acetabulum*, on the outer side, for articulation with the femur, so that the articular surface of this cavity is composed of portions of all three parts. The three parts are joined only by cartilage till just before puberty, so that the following description applies to the adult bone.

Looked at from the outer side (Fig. 96) the bone shows, below its middle, the large and deep *acetabulum* to receive the head of the femur, surrounded by the prominent *acetabular rim*, best marked behind and above. The wall of the cavity is deficient below, constituting the *cotyloid notch*.

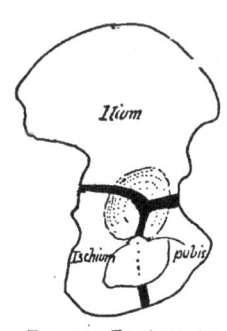

Fig. 95.—To show the three divisions of the innominate bone which fuse to form it about puberty; the thick lines are cartilage.

Above the acetabulum is the wide outer surface of the ilium (*dorsum ilii*) for the gluteal muscles, and their areas of origin are marked on the surface by the *upper, middle, and lower curved lines*, due to the presence of tendinous fibres on the surfaces of the muscles. Above, the bone terminates in the thick *crest* of the Ilium, and this ends in front and behind in the *anterior superior* and *posterior superior* spines: below these spines are notches which separate them from the *anterior* and *posterior inferior* spines of the Ilium. Below the anterior inferior spine is another ill-defined notch and the *ilio-pectineal* eminence, which marks the junction of Ilium and pubis: below the posterior inferior spine is the deep *great sciatic notch*, and near the front border of this notch, on the base of the wall of the acetabulum, is another rounded thickness which marks the junction of Ilium and Ischium.

The *body* of the Ischium is the mass of bone below and behind the acetabulum, and forming part of this cavity: this is connected by an *upper ramus* with a large *tuberosity* with *facets* on its postero-external aspect for the origin of the hamstring muscles: above this is the *lesser sciatic notch*, which is separated from the great notch by the sharply projecting *spine of the Ischium*. In front of the Ischium and between it and the *body of the pubis* is a large hole, the *obturator foramen* (or *thyroid foramen*), which is bounded below by the *Ischio-pubic ramus*, a compound of a *lower ramus* projecting forward from the Ischium, and one, the *lower ramus of the pubis*, projecting back from the pubis. The *body of the pubis* forms the front boundary of the obturator foramen and, with the Ischio-pubic ramus, gives origin to the Adductor group of

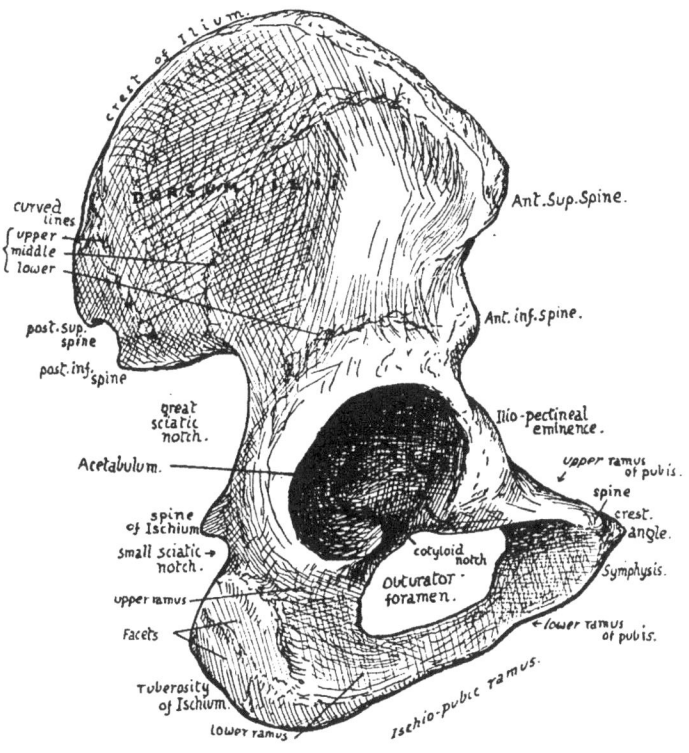

FIG. 96.—View of right innominate from the outer side.

muscles its inner border is the *symphysis*, which is connected with its fellow of the opposite side by means of an intervening plate of fibro-cartilage: the upper end of the symphysis forms the *angle* at its junction with the *crest*, which gives attachment to the Rectus abdominis and terminates externally in the prominent point of the *spine*, to which Poupart's ligament is attached. Outside this the *upper ramus* extends to the Ilio-pectineal eminence, where it joins the Ilium; its front border gives attachment to the puho-femoral band of the hip capsule, and overhangs the obturator foramen.

Thus the outer aspect of the bone shows four margins, of which the upper and lower are the crest of the Ilium and the Ischio-pubic ramus respectively ; the anterior margin is on the whole a concave line, allowing exit for muscles, etc., below Poupart's ligament, and presents, from above downwards, anterior superior spine, notch, anterior inferior spine, notch, Ilio-pectineal eminence, upper surface of superior ramus of pubis : posterior margin is more irregular and shows posterior superior spine, notch, posterior inferior spine, great sciatic notch, Ischial spine, lesser sciatic notch and tuberosity of Ischium.

Looked at from the inner side (Fig. 98) the same parts can be recognised in the margins, but three distinct main surfaces are apparent. The first of these is in the upper and back part and looks directly inward : it has an articular area shaped like an ear and hence termed the *auricular* surface : above and behind this a rough *ligamentous* area for the strong posterior sacro-iliac ligaments, and above and behind this again is a *muscular* area, which really occupies the expanded posterior part of the crest, for the Erector spinæ muscle. A strong bony buttress runs obliquely downwards and forwards from the auricular region to the pubis, called the *Ilio-pectineal line*, and forming the bony brim of the true pelvis in this region : this divides the bone into the remaining two surfaces, an upper one, which is abdominal and formed by the *venter ilii* and part of the pubis, and a lower pelvic part, consisting of Ischium with a large proportion of pubis and a very small amount of Ilium. The ventral surface of the Ilium is slightly concave, looking upwards and forwards and inwards, and forming the bony Iliac fossa giving origin to and covered by Iliacus. The wall of the true pelvis, or lower portion of the bone, is concave from before backwards, and is mainly covered by Obturator internus, which also arises from it : above this muscle, however, is a deep groove leading to the top of the obturator foramen, and below it, on the inner side of the tuberosity of the Ischium, is a rough *falciform ridge* for the attachment of the falciform process of the great sciatic ligament. The Ischio-pubic ramus is seen, from this aspect, to have its lower part everted, affording an area for attachment of the crus penis or clitoridis.

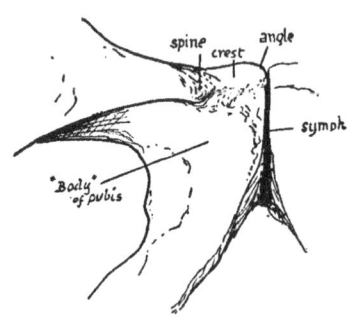

FIG. 97.—Pubic part of right bone from the front.

The Ilio-pectineal line, traced on to the upper ramus of the pubis, presents secondary ridges, and the main part of these, made by Gimbernat's ligament, turns forward to run into the spine of the pubis : other parts are directed internal to this, as will be described later.

The Levator ani arises from the back of the body of the pubis, from the inner side of the ischial spine, and from the fascia covering the Obturator internus between these points. It follows, therefore, that below a line drawn between these bony points the Ischio-pubic portion is in relation with and gives attachment to structures in the perineal and Ischio-rectal regions, while above this line its relations are with structures in the pelvic cavity.

It is convenient now, before proceeding to a more detailed examination of these bones, to consider the general structure of the pelvis. For this purpose the complete skeleton may be used, but it is advisable to have in addition an articulated pelvis with the ligaments *in situ*.

PELVIS.

Notice first of all the position of the pelvis: it is placed so that the cavity it encloses opens forwards as well as upwards, and the ilio-pectineal lines on the innominate bones are directed obliquely downwards and forwards. This, combined with the

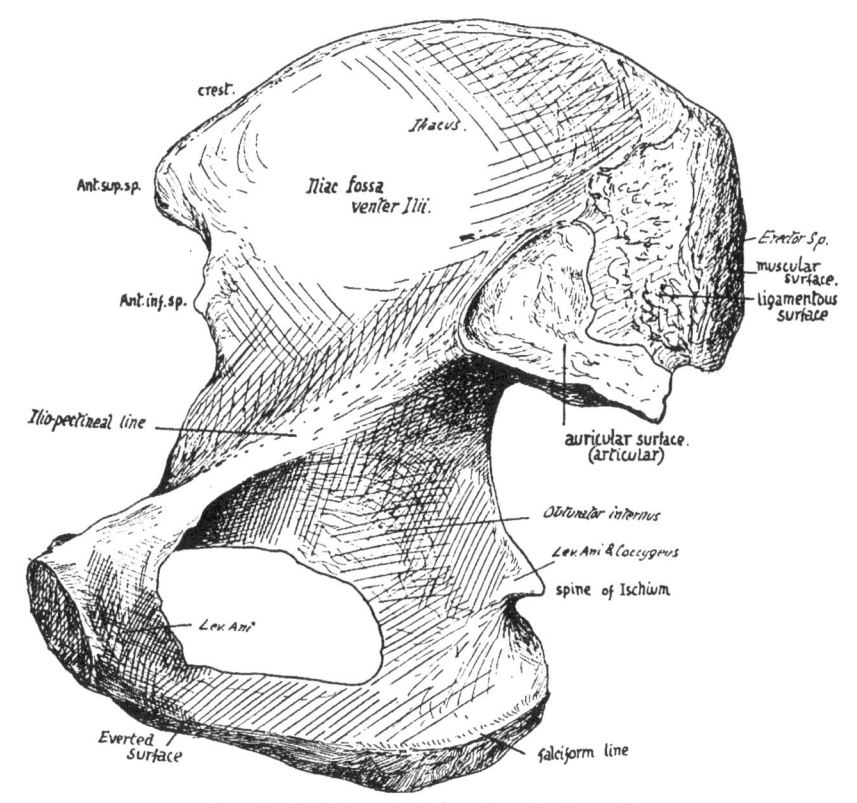

FIG. 98.—Right innominate bone from the inner side.

direction of the sacrum, produces the position of the whole pelvis that is often described as "oblique."

The cavity is divided into two parts by an oblique plane that corresponds with the level of the ilio-pectineal lines. These lines, with the promontory of the sacrum behind and the upper part of the pubes in front, constitute what is termed the *brim of the pelvis*, and the cavity below the level of the brim is that of the *true pelvis*, while the expanded and open portion above the brim forms the *abdominal* or *false pelvis*: it is really a part of the true abdominal cavity.

The passage through the level of the brim into the true pelvis is called the *inlet* or *upper aperture* of that cavity, and the lower opening is termed the *outlet* or *lower aperature*. The sacrum is fixed between the innominate bones by strong posterior sacro-iliac ligaments (see p. 125). It is also attached to these bones (Fig. 99) along the length of its borders by the great sacro-sciatic ligaments passing to the tuberosities of the ischia, and by the lesser ligaments deep to these and fastened by their apices to the ischial spines: these ligaments also reach the coccyx.

Now examine the outlet of the pelvis (Fig. 100). Its bony boundaries antero-laterally are provided by the ischial tuberosities and ischio-pubic rami, and by meeting in front at the lower part of the symphysis they form the pubic or *subpubic arch*. The coccyx limits the outlet behind, and between this and the tuberosity of the ischium stretches the great sciatic ligament on each side: in the recent state a ligamentous transverse band below the symphysis, called the subpubic ligament, rounds off the angle made by the meeting of the bones of each side.

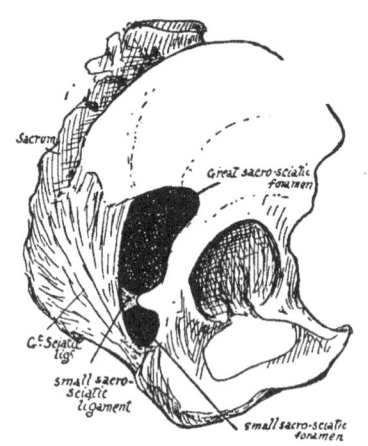

FIG. 99.—Right innominate *in situ*, with sacrum and ligaments.

It is in the width of the pubic arch that the most striking difference is seen between the male and female pelvis (Fig. 100): in the latter the lines of the rami form a right angle with each other, whereas the male arch is only about 70 degrees to 75 degrees or even less.

It is apparent at a glance that the posterior wall of the true pelvis is higher than the anterior: the vertical depth of the back wall is between 5 and 6 inches, and that

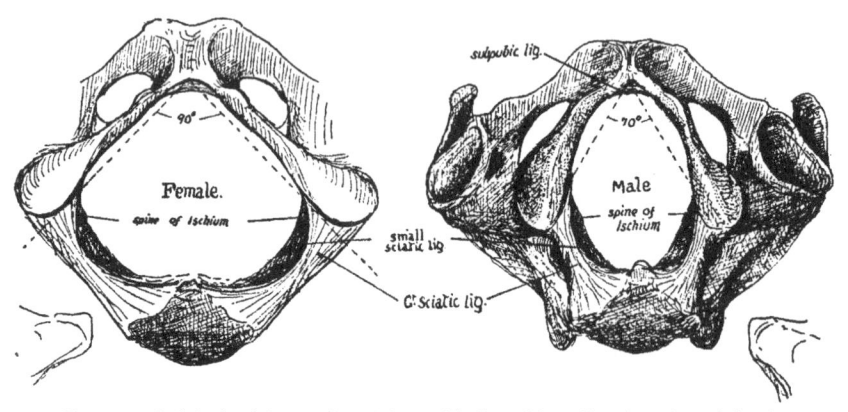

FIG. 100.—Outlet of pelvis seen from below, with the pubic outlines in male and female.

of the front wall about .2 inches. The plane of the pelvic inlet forms an angle of about 60 degrees with the horizon, whereas that of the outlet is inclined at an angle of about 15 degrees, open in the opposite direction.

Consequently (Fig. 101) the *axis of the cavity*, extending from inlet to outlet, is curved with a backward convexity.

The size of the cavity or passage of the true pelvis is naturally a matter of great importance to the obstetrician, and it is necessary to know the average measurements of the different parts of the cavity. For this purpose arbitrary points are taken for measurement, and different observers have made use of different standards ; but there are certain diameters which are recommended by their natural value, so so speak, and these are : (1) *Antero-posterior* or *conjugate*, in the middle line ; (2) *Transverse*, at the widest spot : (3) *Oblique* (valuable in considering the mechanism of labour, in which the fœtal head is obliquely placed), between the sacro-iliac joint and the situation on the opposite wall of the cavity corresponding with the level of the ilio-pectineal eminence. These measurements on the upper aperture of the male pelvis, *i.e.*, at the brim, give as an average— (1) 4 inches, (2) 5 inches, (3) 4¾ inches. In the female they are all increased to—(1) 4½ inches, (2) 5¼ inches, (3) 5 inches. (The conjugate is measured from the sacral promontory to the top of the symphysis.)

At the lower aperture or outlet the antero-posterior diameter admits of increase through mobility of the coccyx, but the averages in the male may be given as—(1) 3¾ inches, (2) 3½ inches, (3) 3½ inches ; in the female as—(1) 4½ to 5 inches, (2) 4½ inches, (3) 4½ inches. Speaking roughly, the diameters in the female show an increase of ¼ inch or more at the upper aperture and 1 inch at the lower aperture.

The corresponding diameters half-way down the cavity are also of interest. In the male they average about—(1) 4¼ inches, (2) 4¾ inches, (3) 4¾ inches ; in the female—(1) 5⅓ inches, (2) 5 inches, (3) 5½ inches.

The least increase is in the transverse diameter : possibly the less curved sacrum in the female accounts for the greater increase in the other measurements.

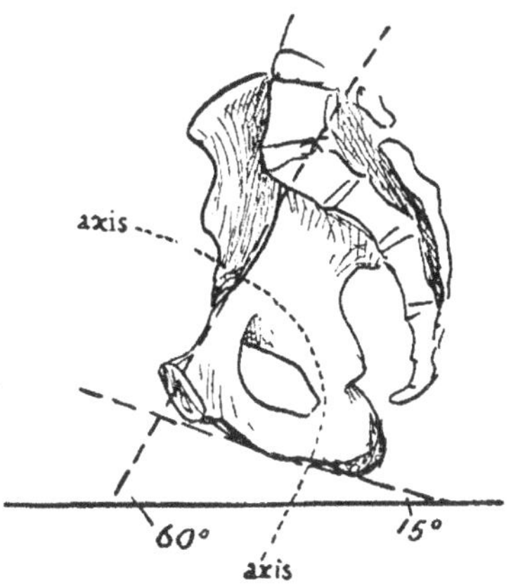

If we now compare the measurements in the female pelvis from above downwards we find that the greatest diameter is transverse above, oblique in the middle, and antero-posterior below : this change is correlated with and responsible for the rotation of the fœtal head that places its long axis in the sagittal direction.

FIG. 101.—Diagram to show the angles made with the horizontal line by the brim and ischio-pubic ramus respectively when the pelvis is in the standing position ; also to show the direction of the axis of the cavity.

Fig. 102 is an outline of male and female types drawn to the above measurements and reduced to half size. It exhibits also certain sexual features of the Ilium, for the male bone is seen to have a wider iliac fossa with an anterior superior spine somewhat incurved : the female bone has a smaller fossa, but more open, and the anterior spines are slightly wider apart, though the crest is not so prominent. The outlines are placed so that the pubes and anterior superior spines fall together in both—which is not quite a natural coincidence, for the bony spine is further forward in the female— and it can be seen that the pelvis, looked at from above, has its sexual characters, though these are not nearly so marked as they are when it is seen from below.

The general characters of female bones are usually well exemplified in the pelvis.

Looked at from the side (Fig. 99) the greater and lesser sciatic *notches* on the innominate bones are seen to be converted into *foramina* by the great and small sacro-sciatic ligaments : the former of these bands can be traced as a falciform edge along the inner side of the tuber ischii.

124　　　　　　　　Anatomy of Skeleton

Notice also that, when the pelvis is in position, the vertical plane of the anterior superior spines is only just behind that of the symphysis in men, and just in front of it in women, owing to the slightly greater obliquity of pelvis in them.

Further details of pelvic structure, including other minor sexual characters, will be better considered in connection with the innominate bones, while certain others have already been discussed under the Sacrum ; but before returning to the os innominatum it may be pointed out here that the distinction between male and female pelvis can be made as early as the third or fourth month of development, and it can

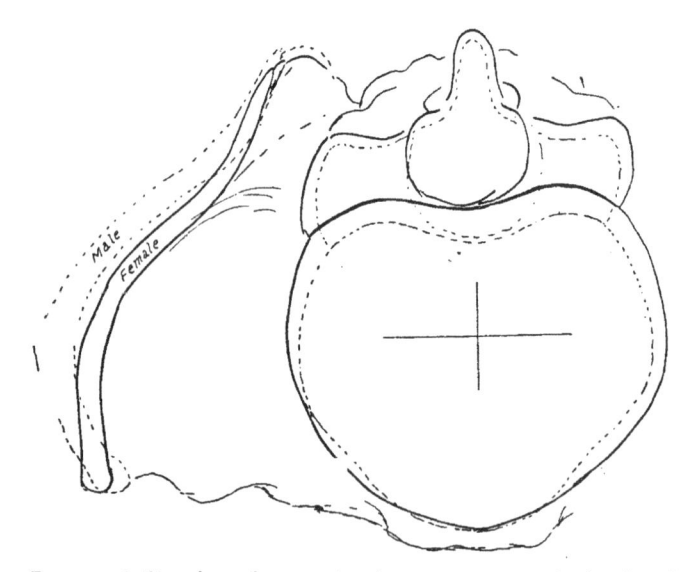

FIG. 102.—Outlines drawn from a series of average measurements of male and female bones and superimposed for comparison. The male outline is interrupted. It can be seen that the female pelvis is broader and deeper, while the false pelvis is not so wide as in the male, nor its crest so prominent ; the male crest is also more curved and turned in front, but this last character is not usually so marked as it appears in this imaginary outline. It must not be forgotten that the female false pelvis is relatively wider than it seems in the diagram, because of shorter stature.

be stated that the appearance of the pubic arch (Fig. 100) is the best and most reliable test of sex in the bony pelvis.

Detailed Consideration of Os Innominatum.

The general " build " of the bone is associated with the transference of weight from the vertebral column to the heads of the femora. The bone can be placed in the position it occupies in the complete pelvis by holding it so that the plane of its symphysial articular surface is sagittal, while the cotyloid notch is opening directly downwards : when held thus it can be seen that the os innominatum admits of division into two parts that lie in different planes. The upper and back portion looks in general

2^4

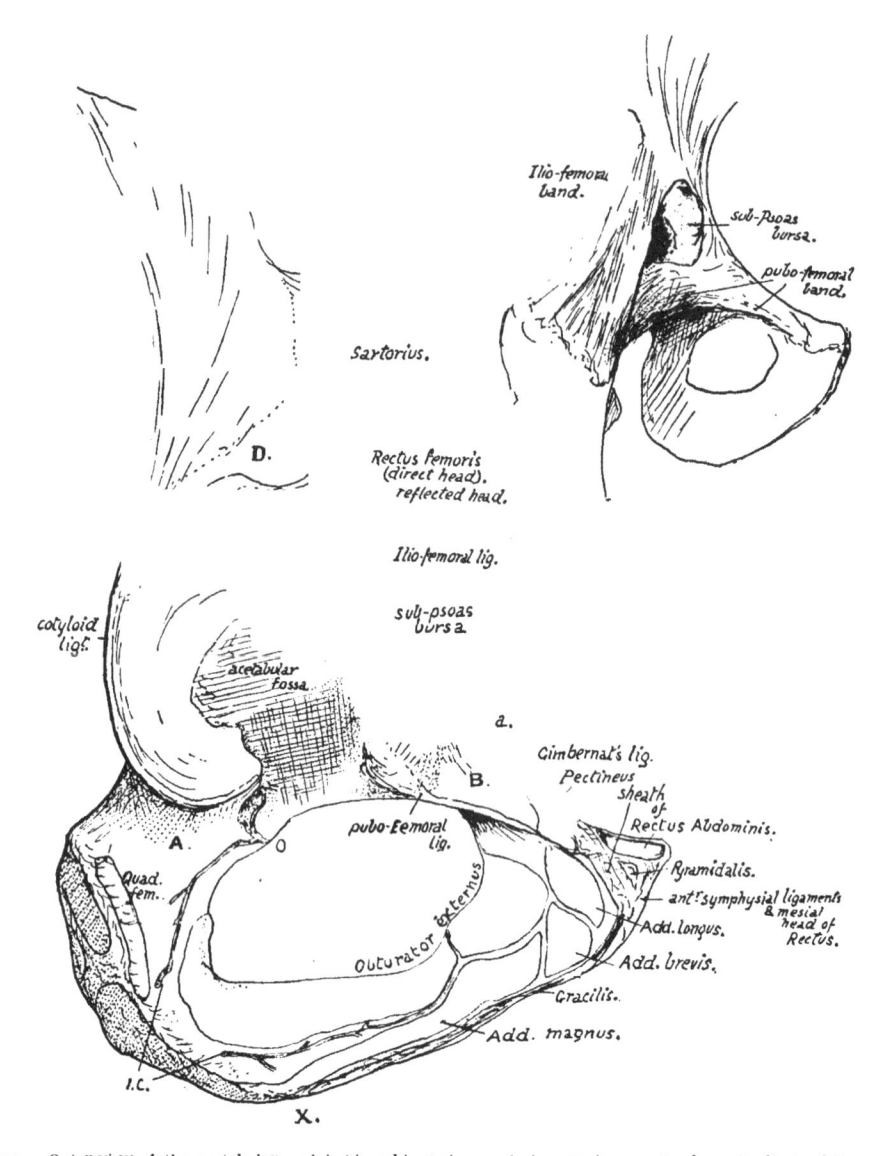

FIG. 104.—Outer view of the acetabular and ischio-pubic regions. A. is a region on the bone in front of the positi Quadratus femoris which is in relation with the tendon of Obturator externus and some loose fibro-fatty tissue that pe changes in place of the tendon with movement of the joint. B. is a sloping surface of bone which supports Pectineu does not give origin to it; the surface extends out to the ilio-pectineal eminence where the fascia covering the Pect (pubic portion of fascia lata) reaches the bone at a. The front of the eminence is roughened by fibres belonging to the femoral group of ligaments. These are separated from the ilio-femoral set by an interval, apparent on the bone and ma the limit of the secondarily added pubic area of articular surface (see Fig. 106), where the synovial cavity is protruded . sub-Psoas bursa; this is seen in the smaller figure. The Psoas lies on the bursa and the surface C. D. is covered by Gl minimus, which arises above the dotted line; below D. the muscle lies on the reflected head of the Rectus and the caps the joint. O. and I.C. are branches of the obturator and internal circumflex arteries anastomosing round Obturator ext and giving twigs through the cotyloid notch into the cotyloid fossa and so into the lig. teres, which is attached roun margins of the fossa and to the transverse ligament that extends between the lips of the notch. X. marks an ill-d depression on the margin, which indicates the spot where the origin of Adductor magnus passes from the outer side pubic ramus to the lower aspect of the ischial tuberosity; it therefore also marks the hinder limit of origin of Gracilis.

The Lower Extremity and Pelvis

backward and outward, whereas the lower and front parts looks forward and outward. The two planes cut each other nearly at a right angle, and the upper one roughly corresponds with the Ilium, the lower with the Ischium and pubis.

The scheme in Fig. 103 illustrates this : a is the upper and back plane and b the lower and front one. The projection forwards of a could be taken to represent the schematic position of the anterior superior spine, while the hinder end of b would represent the projection of the tuberosity of the Ischium. The transition from one plane to the other is comparatively abrupt on the outer side, and occurs about a line drawn from below the Ischial spine to the anterior superior spine : this places the acetabulum on the lower segment, but, as can be seen on the other side of the scheme and is apparent on the bone, the part of the wall of this cavity that rests on the head of the femur and transmits weight to it is supported by the front edge of the upper plane, which carries the weight to it from the sacrum. Accordingly we find that that portion of the Ilium between the auricular surface and the top of the acetabulum is very thick and strong, and forms the acetabular rim along the line of transition from one plane to the other.

On the inner side, however, the transition is modified by the curved bar c of the ilio-pectineal line, whose mechanical function is concerned with the mode of transmission of the weight from the sacrum to the ilium. The appearance of a section through the pelvis along the ilio-pectineal line would be in its essentials like the scheme in the figure : notice from the direction of the articular surfaces that the sacrum is not wedged down between the two hip bones, thus tending to drive them apart, but is suspended from them by posterior sacro-iliac ligaments (x), and therefore would draw them together as it throws weight on them. The bracket c counteracts the inward bending of a on b and makes the whole structure strong and rigid, dispensing with the necessity for strong anterior sacro-iliac ligaments to resist the separation of sacrum and ilium in front.

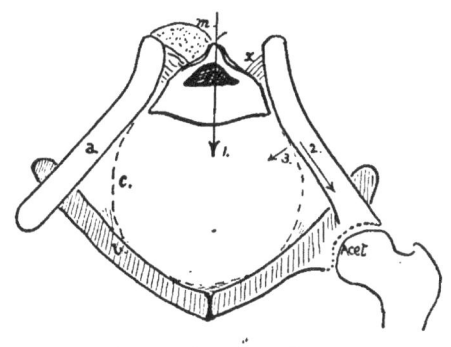

FIG. 103.—Scheme to show the mechanical build of the pelvis; it practically represents a frontal section. Explanation in text.

In this way we see why there must be such a large ligamentous area above and behind the auricular surface, separating it from the post-vertebral muscle area, with a very ill-marked ligamentous line below and in front of it, and at the same time it becomes clear that the pubic symphysis is not concerned in carrying or transmitting the weight of the trunk, and is on this account not provided with excessively strong ligaments.

The Acetabular Region (Fig. 104).—It has already been said that the three main elements of the bone are all represented in the acetabulum. The Ischium is credited with forming rather more than two-fifths, the Ilium with rather less than this, while the pubis accounts for the remaining fifth. But as a matter of fact the triradiate cartilage that separates these elements, almost till puberty, develops a variable number of small ossifying centres in the floor of the acetabulum : these may fuse to form a small separate bone, the *os acetabuli*, but in any case there is ultimate junction of the various parts, and the acetabular centres are usually described as forming a part of the pubic element.

The articular surface of the acetabulum surrounds on three sides the non-articular *cotyloid* or *acetabular fossa*, which contains the fatty tissue of the " Haversian gland," and opens below at the cotyloid notch. The notch is bridged across by the transverse ligament, a tendinous structure continuous with the fibro-cartilaginous cotyloid ligament that is attached to the whole length of the edge of the brim : vessels and nerves pass through the notch, under the transverse ligament, to enter the ligamentum teres. The fibrous basis of the ligamentum teres is attached to the ischial and iliac parts of the cotyloid fossa and to the transverse ligament : its synovial covering is attached to the whole margin of the fossa and the whole length of the transverse ligament below, covering the fossa but lying free on its surface. The round ligament is the remains of the original ventral wall of the capsule (Fig. 106). In the human embryo the Ischium and Ilium alone are concerned in the articulation with the femur and the capsule is attached round their ventral margin : the pubic cartilage is extracapsular. As the pubis extends it is still separated by the cellular capsule from the femur. Later it breaks through this capsule and becomes articular, the synovial cavity extending over it from the Ilium. In this way the original attachment of the capsule is only left on the Ischium as the fibrous basis of the Ligamentum teres. In the adult the pubic articular surface is still more or less distinct from the iliac surface, and the same distinction is usually marked on the rim by a shallow notch ; here the communication may take place of the joint with the sub-psoas bursa, between the Iliac and pubic parts of the capsule. Above this notch is the large rough area for the λ-shaped Ilio-femoral band, spreading on to the lower half or more of the anterior inferior spine, and below and internal to it is the area for the pubo-femoral band which extends inward along the front edge of the upper pubic ramus, overhanging the issuing obturator nerve.

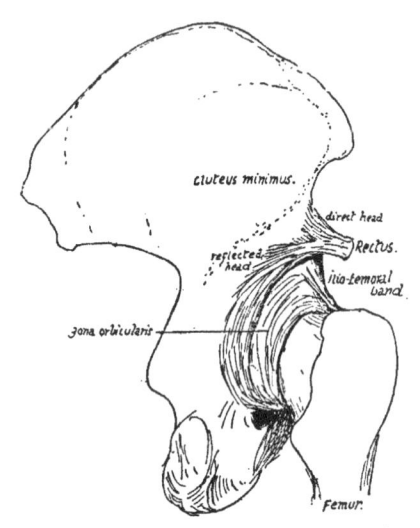

FIG. 105.—Posterior View of the capsule of hip, showing the circular fibres and reflected tendon of Rectus ; this sends some fibres to the circular band.

The direct tendon of the Rectus femoris is attached to the upper part of the anterior inferior spine, so that it rests on the ilio-femoral band at its origin : outside this the line of its attachment passes downwards and backwards, to run into the cotyloid ligament and capsule at the top of the acetabulum, thus forming the reflected head. Observe that this must be under cover of Gluteus minimus, arising above the inferior curved line (see Fig. 105).

At the back of the acetabulum synovial membrane comes over the cotyloid ligament and touches the bone : this occurs from the transverse ligament below to the reflected tendon of Rectus above. In front the synovial membrane does not transgress the cotyloid ligament, but passes nearly directly from it to the strong capsule.

There is only a slight capsular attachment behind the acetabulum, for there are no

2^6

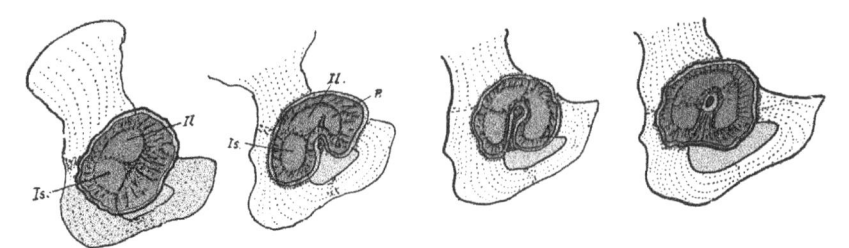

Fig. 106.

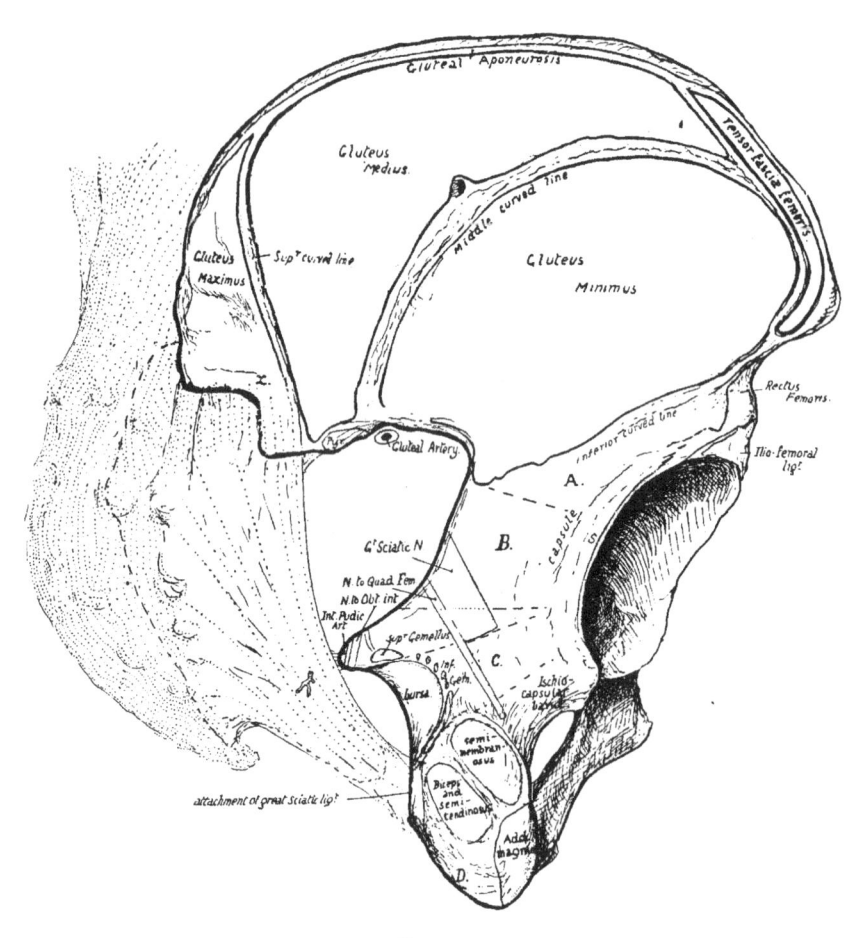

Fig. 107.

true transverse fibres on the back of the joint, and only a few of the inner marginal fibres of the circular group run to the bone in this region : a group of these below reach the upper ramus of ischium as the " ischio-capsular band."

The posterior surface of the acetabular mass is covered by Pyriformis, and may give origin to some fibres of Gluteus minimus deep to this : the sciatic nerves, etc., and nerve to Quadratus femoris pass down on it. The thin floor of the acetabulum is occasionally, like the thinnest part of the iliac fossa, found to be perforated, and the hole is then closed by membrane.

Dorsum Ilii.—The curved lines separate the gluteal planes. The upper or posterior line commences just in front of the posterior inferior spine and reaches the crest about 2½ inches in front of the upper spine. Gluteus maximus arises behind it, and can therefore extend directly on to the great sciatic ligament, which is fastened to the lower spine and the margin of the notch just in front of this (see Fig. 107), and can thus reach the sacrum. The middle line commences near the deepest part of the greater notch and ends about 2 inches behind the anterior superior spine : observe (Fig. 107) that the line does not actually reach the outer lip of the crest. Gluteus medius arises between this line and the upper one, and some of its deeper posterior fibres may be separated from it as an additional origin of Pyriformis from the top of the notch : just in front of this the margin of the notch may be slightly grooved by the Gluteal artery.

FIG. 106.—To illustrate the formation of the ligamentum teres. In its early stage the ilium (*il*) and ischium (*is*) are alone concerned in the articulation, the synovial lining passing off them on to the capsule which is attached round their surfaces. The pubis is covered by these fibres and has no articular area. In the next stage the covering fibres are destroyed and the pubis has acquired an articular surface (*p*). This extends, and the front part of the original ischial capsule is caught, so to speak, between the extending surface and the ischium ; these fibres persist and remain attached to the ischial region, but on their surface the synovial cavity has extended, as shown in the last diagram, and has joined the older cavity below as well, passing between the femur and the lower portion of the attached capsule. Thus a synovial funnel is formed, wider below where it includes the attachment of the fibres and narrowed at its femoral end, where it is fastened to the fovea.

FIG. 107.—Postero-external aspect of right os innominatum. The curved lines are somewhat diagrammatically drawn : for account see the text. Some of the fibres of the great sciatic ligament run on to the surface of the bone round *x*, and give origin here to part of G. maximus. The position of the sacrum and great ligament is indicated, with the origin of the muscle from it. *A* is the surface below the inferior curved line, covered by G. min. ; *B*, the area covered by Pyriformis, with the great sciatic nerve interposed ; *C*, covered by Obturator internus and Gemelli, which lie between the nerve and the bone, but have the nerve to Quadratus between them and the bone. The muscles mentioned are practically in a continuous curved plane, so that the areas *A*, *B*, and *C* make a convex surface, continuously curved and smooth, round the acetabulum ; the muscles pass to the raised trochanter, so do not mould the bone by pressure. The lower aspect of the tuber ischii, below the facets for the hamstring muscles, shows two sloping surfaces, of which one looks outwards and gives origin to fibres of Adductor magnus (ischio-condylar portion), while the other looks inwards (*D*) and is covered by fibro-fatty tissue which is continuous round the great sacro-sciatic ligament with that of the ischio-rectal fossa : in this tissue is a badly-defined bursa which lies under the tuberosity in sitting, the hamstrings and Adductor moving to the outer side of the prominence when the limbs are bent for that purpose.

The lower curved line is variable as to its commencement : it is directed toward the upper part of the anterior inferior spine, or toward the notch above this.

But the arrangement of the lines is not quite so simple as this description implies and by comparing several bones the following account can be verified.

The Gluteus maximus is in the same curved plane as the Tensor fasciæ femoris, which arises from the bone just below the crest for about 3 inches behind the anterior superior spine. The two muscles are connected by a strong aponeurotic sheet, the *Gluteal aponeurosis*, which splits behind to enclose G. maximus, and in front to enclose the Tensor : between these two it covers the medius and is attached to the outer edge of the crest. On the bone the line of this aponeurosis is easily followed, and the layer that passes deep to maximus is seen to form the superior curved line and so reach the great sciatic ligament. Again, in front, the line on the bone is seen to divide to enclose

Anatomy of Skeleton

the Tensor, and the middle curved line runs into the lower of the two divisions : thus the middle line never really reaches the crest, but the very front fibres of medius, at their origin, lie just under cover of the posterior fibres of the Tensor, which otherwise only covers the minimus. The area of Tensor fasciæ turns down along the outer side of the notch below the superior spine, with its enclosing lines, here lying to the outer side of and behind Sartorius.

The lines join again below the origin of the muscle, indicating the junction of the two layers of its covering aponeurosis, and this complete layer turns back above and outside the Rectus to form the inferior curved line, the deepest of the gluteal markings.

We thus have an aponeurotic sheet covering the Glutei and turning deeply round the front margin of Gluteus minimus, forming a sort of septum that divides this group from those in front : this will be placed in contact with the outer aspect of the tendon of Rectus, and is pierced by the ascending branch of the external circumflex artery coming from under that muscle to run among the Glutei. The front fibres of the gluteal muscles are attached to this "septum," which is in fact made from their degenerated fibres, and it thus passes with them to the trochanter of the femur : the aponeurotic band that covers the front of minimus is usually strong, and is sometimes termed the *ilio-trochanteric* band : this can therefore be placed on the Ilium at the lower front part of the Tensor area. The "septum" being continuous with the aponeurotic layer on the deep aspect of minimus, comes into contact with the capsule and is attached to it : this can sometimes be followed on the bone when the Gluteus minimus is large.

Below minimus the back of the acetabulum is covered by Pyriformis, and has the great sciatic nerve resting on it. This is soon separated from bone by Obturator internus and gemelli, but the small nerve to Quadratus femoris still lies with its accompanying vessels on bone deep to these, finally passing to the deep aspect of its muscle, superficial to the tendon of Obturator externus. (Fig. 108). The tendon of the internal Obturator turns round the lesser sciatic notch, which is lined by cartilage and a bursa : upper Gemellus has a rounded tendinous origin on the base of the spine above, while lower Gemellus has a linear origin along the top of the tuberosity below, therefore is covered somewhat by the tendon leaving the notch. The spine of the Ischium is crossed by the nerve to Obturator internus, and at its tip by the internal pudic vessels, these separating it from the coccygeal branch of the sciatic artery running to pierce the great sciatic ligament. The small sacro-sciatic ligament is attached to the tip of the spine.

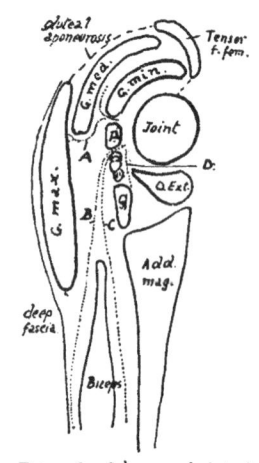

FIG. 108.—Schemes of gluteal planes arranged round the joint. *P.* Pyriformis ; *Q.* Quadratus femoris. Between these is the Obturator internus and Gemelli. *A.* plane of course of superficial division of gluteal artery ; *B.* plane of small sciatic ; *C.* plane of great sciatic ; *D.* plane of nerve to Quadratus.

The External Ischio-Pubic Region (Fig. 104).—A few words may be said about the attachments shown in the figure. The bony origin of Obturator externus can usually be made out on the bone with little trouble : observe that it comes well on to the body of the pubis, where its marginal line is fairly well marked. Its tendon, passing under the capsule, covers the surface A, and lies therefore in front of Quadratus femoris. The adductors as a whole, though they arise largely by tendon, do not generally leave recognisable secondary markings, but their origins are as on the plate, and can be placed as follows. Adductor longus is just below the spine, occupying the space between the oblique marking for the sheath of the Rectus and the Obturator externus : it extends down to just below the level of the mid-point of the symphysis, and in its lower half is separated from the Obturator by Adductor brevis.

The Lower Extremity and Pelvis

Adductor brevis comes in between the Obturator and the lower half of longus and extends down to just below the level of the lower end of the symphysis, being separated from the Obturator in its lower half or so by Adductor magnus and coming into relation, below the level of longus, with Gracilis superficially.

Adductor magnus, in its pubic origin, is up against Obturator externus, separating it in front from Adductor brevis for a little distance above the level of the lower end of the symphysis: below this it arises from the whole of the ramus left between the Obturator and the lower margin, where Gracilis arises. So the area of origin extends back on to the ischial portion of the ramus, where it soon exhibits a change in position: examine the lower edge, and a shallow depression of its prominent

FIG. 109.—Right os innominatum viewed from above, showing the iliac fossa, etc. The Iliacus passes forward over the anterior border, covering the surface D, while the Psoas magnus, lying along its inner margin, comes into relation with the bone over C. B marks the area covered by Pectineus, the origin of the muscle being shown in black. The back part of the iliac fossa is lined with ligamentous fibres, in the region x, from which Iliacus arises. The iliac fascia, which covers Iliacus and Psoas, is attached along the dotted line $i.f.$: where this is deficient in front the fascia reaches Poupart's ligament. Psoas parvus is inserted into the fascia, and a strong slip from its tendon reaches the iliopectineal line at $ps.p.$ With the fascia. Pectineus lies deep to the fascial plane, so that the iliac and pubic fasciæ blend on the muscle: an expansion from the combined fasciæ passes deep to the inner edge of Psoas and reaches the bone at a. The conjoint tendon is continuous with the front sheath of Rectus along the groove g on the inner side of the pubic spine. Compare with Figs. 104 and 110. $Art.$ and N are for iliac branch of ilio-lumbar artery and obturator nerve, separated from the bone by the Iliacus in this situation. F marks an occasional foramen for the iliac branch of the obturator artery, deep to Iliacus.

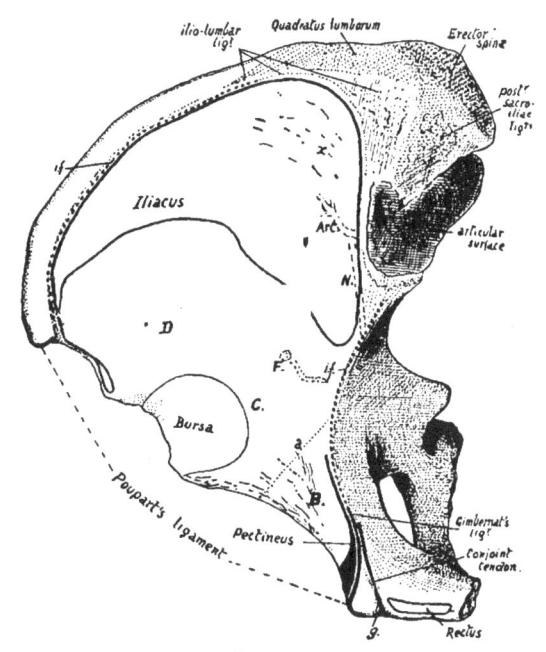

border will be noticed at X, and here the origin of the muscle passes from the outer aspect of the ramus on to its thick under surface. Here it extends on the tuberosity, occupying the outer of the two planes of this under surface, as far as the facets on the back of the tuberosity (see also Fig. 107).

Gracilis extends along the margin from half-way up the symphysis to the depression X (Fig. 104). It is a linear aponeurotic origin.

The upper ramus of the pubis has the puho-femoral band attached along its prominent front margin. Above this the bone (B) is covered by Pectineus, but this muscle only arises from a line along the back border of the upper surface and a wider area just outside the pubic spine. Gimbernat's ligament and the conjoint tendon are attached to the postero-internal margin immediately behind Pectineus (see later).

The inner edge of Psoas, lying on the area of bone C, just overlaps the outer edge of Pectineus, and the pubic portion of fascia lata that covers Pectineus sends a deep sheet under Psoas to the pubo-femoral band and to the bone above this along the line *a*.

Notice that when the bone is in its proper position the plane of the ischio-pubic adductor surface looks downwards as well as outwards and forwards, so that the group of Adductor muscles is below the Obturator externus as well as in front of it.

Arterial branches of the obturator (O.) and internal circumflex (I.C.) run on the bone among the muscle origins as shown in the plate.

Looked at from above, the bone exhibits the iliac fossa most clearly, with certain other parts. Fig. 109 is a sketch of the bone seen in this way: the origin of Iliacus is indicated in outline in the fossa. Examine the back and inner part (x) of its area, above and external to the auricular surface, and the surface of the bone will be found somewhat lined; the markings are not in the direction of the muscular fibres, and are not made by them, but by a thick ligamentous layer that covers the bone as a prolongation of the sacro-iliac ligamentous system. Iliacus arises from these, and from the rest of its area on the bone, by muscle fibre.

The muscle covers the bone between its origin and the mesial aspect of the anterior inferior spine; internal to it the Psoas lies on the bone. The brim of the pelvis in front turns inwards away from the line of Psoas, and the triangular area thus made is covered by Pectineus: observe that this has a linear origin except where it arises by muscle fibre outside and behind the pubic spine.

The line *a* marks the attachment of a fascial layer from the pubic part of fascia lata, deep to edge of Psoas, so can be taken as showing the division between the surfaces covered by Psoas and Pectineus respectively. Notice that the pectineal area can be looked on as a prolongation of the iliac fossa, and the Pectineus is considered to be a portion of Iliacus, which, from the situation of its origin, has a different functional value and therefore has separated from the main sheet: this would account for its nerve supply differing from that of the adductors with which it is descriptively grouped.

Gimbernat's ligament is attached to the front end of the ilio-pectineal line behind Pectineus, and conjoint tendon again immediately behind this: these two structures are necessarily directly continuous with the front wall of the sheath of the Rectus, the former through its connection with Poupart's ligament attached to the spine, the latter along an attachment in a groove (g) just internal to the spine and immediately external to the origin of Rectus from the crest.

(These arrangements of the structures can perhaps be understood better by reference to the plan in Fig. 110.)

Iliacus and Psoas are covered by the Iliac fascia, which is attached to the margins of the fossa In Fig. 109 the dotted line *i.f.* on the inner margin of the crest marks its attachment here: in front it passes on to the posterior aspect of Poupart's ligament, behind it reaches the ilio-lumbar ligament below the Quadratus lumborum, and from this passes over the Psoas.

Its inner attachment can be picked up on the ilio-pectineal line internal to the Psoas: it is attached along this line, reinforced along the Pectineus by the pubic part of fascia lata, and reaches Gimbernat's ligament. Its attachment is from this along the free edge of Gimbernat's to the back of Poupart's ligament, and thus to the crest of the Ilium.

The sheet of fascia can thus be regarded as a simple layer attached round the margins of the fossa, and continuous at the outer and front margins with the plane of transversalis fascia.

When the main vessels, lying on it, proceed under Poupart's ligament, they must carry with them a process of the fascia, attached in front or above to the back of

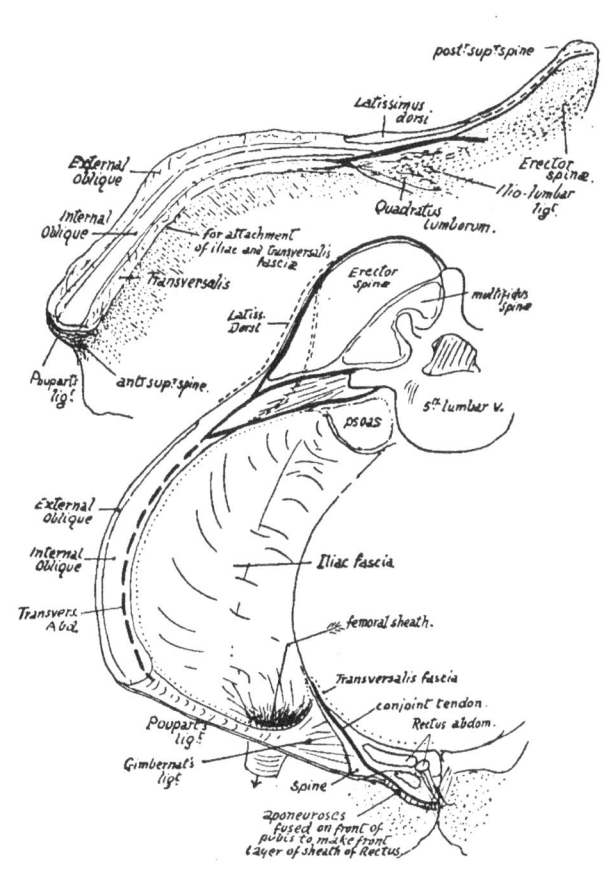

Fig. 110.—The upper figure is a schematic drawing of the crest of the right ilium to show the attachment of muscles, etc., to it. Compare with Figs. 34, 98 and 109. The lower figure is a schematic representation of the planes of muscles and fasciæ related to the bone. Poupart's ligament is stretched between anterior superior spine and pubic spine, and its inner end is expanded to form Gimbernat's ligament; the two deep muscles of the abdominal wall extend on to it and make a "conjoint" muscle arising here and inserted by a conjoint tendon. Conjoint tendon and Poupart's and Gimbernat's ligaments, being only parts of the abdominal aponeuroses, must be continuous with the front wall of the sheath of Rectus abdominis, and the lines of continuous attachment to the bone are indicated in the scheme. See Figs. 104 and 109. The dotted line indicates where the transversalis fascia of the wall becomes continuous with the iliac fascia. The figure also shows how the femoral sheath is derived from the iliac fascia, and how the three lamellæ of the Transversalis aponeurosis pass to their vertebral attachments.

FIG. III.

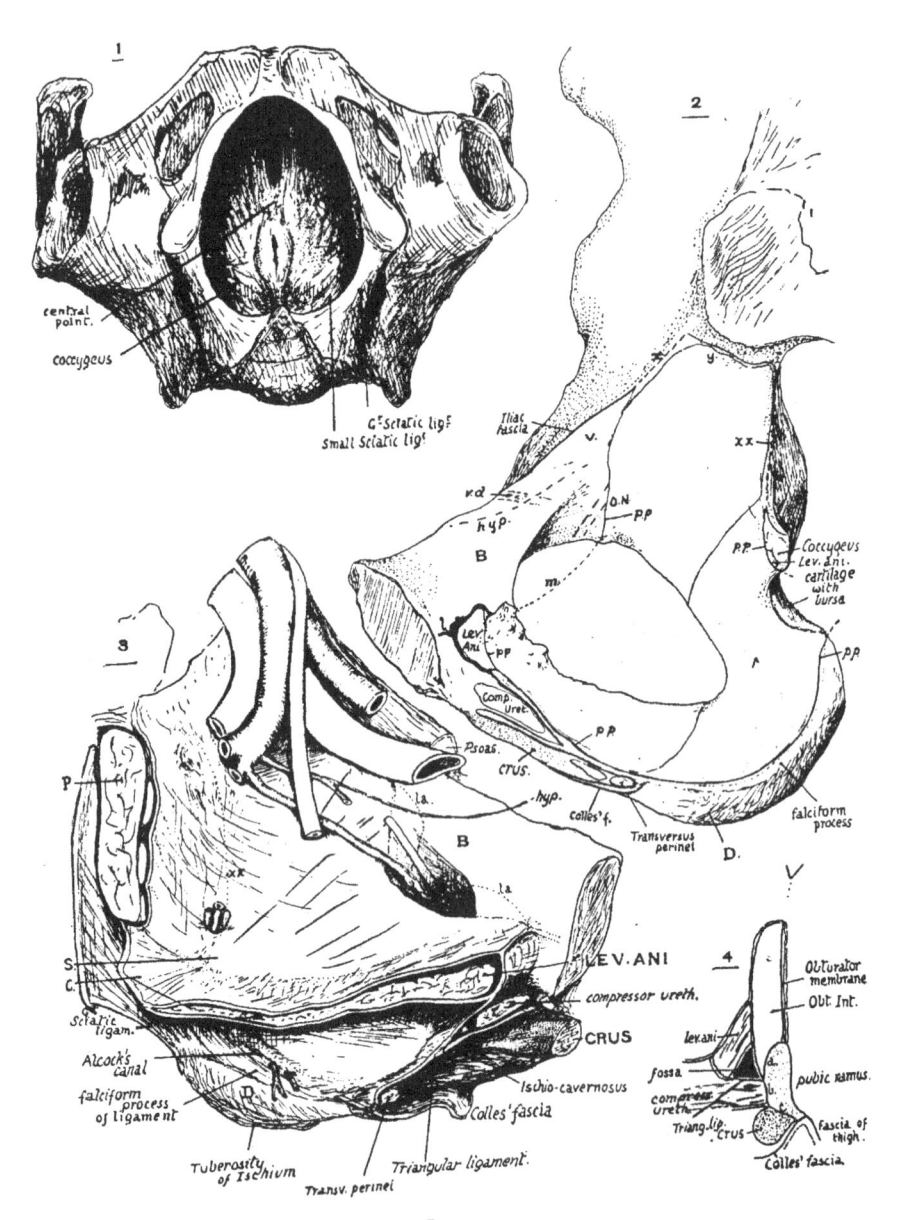

The Lower Extremity and Pelvis

Poupart's and edge of Gimbernat's ligament, and directly continuous behind with the sheet that covers the iliac fossa. The Psoas parvus is inserted into the iliac fascia, spreading out in a fan-shaped aponeurosis, of which the fibres can be traced as far out as the outer half of Poupart's ligament, into the back wall of the femoral sheath, and to the attachment of the fascia on the ilio-pectineal line. This last-mentioned place of insertion of the tendon receives a strong slip, and there is frequently a secondary ridge (*ps. p*), on the line opposite the ilio-pectineal eminence, that marks its attachment.

The **pelvic surface** can be divided into two descriptive areas by the ilio-pectineal line. The upper of these forms the bony floor of the iliac fossa, and has just been described. The iliac fascia makes a common covering for Psoas and Iliacus and is attached to the ilio-pectineal line internal to and behind the Psoas.

The lower or true pelvic surface (Fig. 111) is enlarged by the presence of the obturator membrane in the recent state, and the Obturator internus has an extensive origin (No. 2 in Fig. 111) from this membrane and from the bone as far up as the lower sacro-iliac ligaments. From this origin the fibres converge on the lesser sciatic notch, where they form a group of tendons that play round the notch on a cartilage-covered surface lined by a bursa. It is therefore evident that the area of bone (A) between the muscle origin and the notch is covered by the muscle that plays over it.

In front of the muscle and above it a surface of bone (B) is left exposed. This is covered by peritoneum, but under the peritoneum it is crossed by the obliterated hypogastric artery (*hyp.*), the vas deferens or round ligament (*v.d.*), and further back may have some relation to the external iliac vein (*V.*). Below the vein, the obturator nerve (*O.N.*) runs to the top of the foramen and is joined by the vessels.

To understand the remaining relations and attachments it is necessary to follow the disposition of the pelvic fasciæ on the bone to some extent.

The Obturator internus is covered by the parietal fascia of the pelvis which is attached all round its pelvic surface: in No. 2, it is represented by the fine black line *pp*, which is seen practically to map out the upper and front margins of the muscle origin. At *x* it is continuous with the iliac fascia with some slight attachment to the bone, and at *y* it runs into the ligaments. Along the front border of the great sciatic notch it is continuous with the fascia covering Pyriformis with some slight attachment to the bone: this is shown at *xx* in this drawing and in No. 3, where the fasciæ are represented *in situ*. The two fasciæ separate at the ischial spine: *pp* is continued along the front of the base of the spine and is carried from its lower border across the notch and tendon to the top of the tuberosity by attachment along the lower part of the lesser sciatic ligament and from this to the front border of the greater ligament. It is now attached to the falciform extension of the greater ligament, and is thus carried to the pubic ramus, where it again closely follows the muscle origin, reaching with it the upper border (*m*) of the obturator membrane.

When this parietal fascia is in position the bone is left uncovered in the area B, while the spine of the ischium projects behind and between the fascia and that covering Pyriformis. The muscular floor of the pelvis is attached to these two bony surfaces —that is, to the lower and front part of the bone above and in front of the Obturator and to the spine of the ischium, and between these points it arises from the parietal fascia in a line connecting them.

The floor is formed mainly by Levator ani, which has therefore a marking on the back of the pubic body, but the back portion of the sheet is the Coccygeus: this, then, arises from the ischial spine in addition to the hinder fibres of the Levator, and as its back and upper fibres are converted into the lesser ligament, this structure is attached to the tip of the spine and is in the plane of the pelvic floor.

Anatomy of Skeleton

The floor, cut away near its origin, is seen in No. 3 (Fig. 111) with its covering fascial sheath : its upper surface supports the viscera, so the fascia on this aspect is the *visceral layer*, that on the lower surface being the *anal fascia*. The two layers are continuous in front round the margin of the pubic origin (as seen in No. 2) and run into the lesser ligament behind.

It is evident that the structures on the bone below the level of a line drawn from the pubis to the spine of ischium must be below the origin of the pelvic floor, and therefore are concerned in forming the outer wall of the ischio-rectal fossa. These are : the lower part of the obturator internus covered by its fascia as far as the line *pp* and the falciform edge, and the inner aspect of the bone below this line.

The extent of the ischio-rectal fossa is seen in No. 1 in the figure. The muscular floor is drawn down into a sort of cone round the anal canal, so that each half is a muscular sheet forming a curved plane, but looking as a whole downwards and outwards : the fossa is seen to extend forward between the Levator and the ramus as far as the front end of the muscle, while it passes behind into the narrow cleft between the two sciatic ligaments, where it ends as these come into apposition and partial fusion. A finger placed in the fossa would have the ramus and tuberosity with their attached structures to its outer side.

The front part of the fossa as seen in No. 1 is covered in (from below) by the perineal shelf of triangular ligament, etc., thrown across from ramus to ramus. The deeper layer of the ligament runs into the anal fascia, where this turns round the free edge of the Levator in front of the central point of the perineum to join the visceral layer. Thus the front part of the fossa ends in a pyramidal *cul-de-sac* floored by the deep layer of the triangular ligament, and limited in front as well as internally by the junction of this layer with the fasciæ on the Levator. This *cul-de-sac* is seen opened up in No. 3 and on transverse section in No. 4, and it is clear that the Obturator internus and its fascia alone form the inner wall here, because the ramus below this muscle is altogether taken up by the structures that make the transverse shelf. Thus the only part of the lower margin of the pelvic aspect of the bone that is really directly concerned in forming the outer wall of the fossa is that part of the ischium that lies below the falciform ridge (D in Nos. 2 and 3), for the ramus in front of this is for the attachment of the structures lying between the deep layer of the triangular ligament above and Colles' fascia below.

Fig. 111.—1. Pelvis seen from below, with the Levator ani in position : on each side, the ischio-rectal fossa is between the muscle and the bone, which is covered in part by Obturator internus. When the perineal shelf, composed of triangular ligament, etc., is thrown across the anterior portion of the opening of the pelvis, it makes a floor for the front part of the fossa on each side ; this can be seen in No. 3. 2. Inner aspect of pelvic part of os innominatum. Right side. Account is given in the text. *A;* area covered by Obturator internus, but not giving origin to it ; hence smooth and concave. *B,* area covered by sub-peritoneal tissue, therefore smooth, and crossed under this by obliterated hypogastric artery (*hyp*) and vas deferens (*vd*) ; sometimes the Levator ani arises from part of this area (*la* in No. 3) and makes a slight roughness on the bone. *V,* situation of external iliac Vein. *D,* inner slope of lower part of ischial tuberosity, covered by fatty tissue continuous with that in ischio-rectal fossa (see Fig. 107). 3. Left side, showing in diagrammatic form the structures in position. *B* and *D* as in No. 2. Above and in front of *D* is the side wall of the ischio-rectal fossa, extending forward to the meeting of the triangular ligament and fasciæ. *C,* Coccygeus and lesser sciatic ligament. *S,* spine of ischium. *P,* Pyriformis. The figure is for comparison with No. 2. 4. Scheme of a section through the front part of the ischio-pubic ramus, to show the various surfaces in relation with the structures represented in the other figures : *a,* surface for origin of Obturator internus ; *b,* surface for Compressor urethra; *c,* everted surface for crus.

We can now place the various perineal structures on the bone. It is well to have several male bones on which the markings can be followed, for different bones vary considerably in the extent and nature of the secondary lines that are apparent on them, and it is very exceptional to find a bone that has its markings complete here.

First look at the everted surface. This carries the crus with its muscle, so that its upper and inner border must be for the superficial layer of the triangular ligament and its lower border for Colles' fascia. Trace the falciform ridge forward and it leads to the upper border, thus showing that the falciform process of the great sciatic ligament is in plane with the triangular ligament. Occasionally a line can be traced from the falciform ridge to the lower border of the everted surface : this shows the attachment of Colles' fascia to the bone.

The deep layer of the triangular ligament is attached to the bone in the line of the lower and front attachment of the parietal fascia, and this can as a rule be found on the bone without difficulty : the Compressor urethræ can now be placed on the bone, and behind this the two layers of the triangular ligament are seen to fuse, so that they form a single layer at their bony attachment, when they are joined by Colles' fascia.

The various lines of fasciæ are shown in No. 2 in Fig. 111, where the superficial layer and deep are placed as blue lines and Colles' fascia green : observe that the deep layer is continuous on the bone with the fascia covering the Levator (black).

The crus is firmly fastened by fibrous tissue to the everted surface : behind and internal to it is the origin of Ischio-cavernosus and behind this Transversus perinei, and all these must be between the blue and green lines.

The extent to which these different areas can be traced backwards varies in individual bones, and they are of course badly marked and much smaller in female bones, but the plan of arrangement remains the same. The everted surface ends below in a thick edge to the inner lip of which Colles' fascia is attached, while the outer lip is for the fascia of the inner side of the thigh.

No. 4, Fig. 111, is a plan of the areas just described on the inner aspect of the pubic ramus as shown by section : (a) surface of origin of obturator, (b) surface for compressor, (c) everted surface.

Now examine the area above the falciform ridge. The lower margin of the obturator internus comes down to the ridge, for the muscle is a fairly thick one and makes a mass that fills the slight concavity and has a convex lower margin. The parietal fascia covers this part and is attached below it to the falciform process, making a sheath (*Alcock's canal*) for the internal pudic vessels and nerve, which are thus carried along the top of the falciform process and brought by it to the back of the triangular ligament. They reach the sheath by crossing the inner or ischio-rectal aspect of the obturator tendon. In No. 3 the vessels are shown piercing * the lining fascia above the level of the ischial spine (S) ; thus, when they come in again below the spine they are deep to the fascia and below the level of the pelvic floor. The figure also shows the course of the canal.

The extent of origin of the Levator from the pubis is variable : it may be as shown in No. 2, Fig. 111, or its level may be at the dotted line *la* in No. 3, or even higher. There may be tendinous fibres causing roughnesses, but as a rule the rough area near this region is for the Obturator, as in No. 2. The covering fasciæ also reach the bone, and thickenings in the visceral layer form the anterior pubo-prostatic ligaments, which may apparently be sometimes indicated.

Borders.

Crest of Ilium.—Observe the curved line formed by this crest, and notice that it is thick and lipped in its anterior half or two-thirds. Certain areas can be found on it without much difficulty. It ends behind at the posterior superior spine, and for the first few inches in front of this it presents a blunt edge with sloping surfaces on

---

* In the figure this is shown very schematically indeed, and no other vessels are represented : as a matter of fact the sciatic and pudic vessels are very variable both in the position of their points of exit and their relations to each other as they leave the cavity of the pelvis.

It may be remarked also that the description of the pelvic fasciæ given above is naturally very incomplete—for instance, no mention is made of the double layer above the Levator ani, etc. But from the present standpoint the account given is sufficient to enable the student to place the various structures properly on the bone.

each side. The inner one of these surfaces presents two areas (Fig. 109), of which the posterior is the larger, slightly convex, and gives origin to Erector spinæ: it is separated from the auricular surface by the ligamentous area of the "tuberosity." Anterior to this the Quadratus lumborum has an area of about the same length, but not so deep. These two muscular areas should be definitely recognised on the bone, the Erector slightly overlapping the Quadrate surface, and it can then be seen that the ligamentous areas that lie below them on the inner side of the bone are distinctly in different planes: that below the Erector is for the posterior sacro-iliac ligaments, whereas that below the Quadratus is for the ilio-lumbar ligament. The muscles obtain part of their origin from the ligaments.

It is now easy to place the other attachments. The crest is described as having inner, middle, and outer lips in its front part, made by the musculo-tendinous attachments of the three large muscles of the abdominal wall. The Transversalis is on the inner lip: traced back, it becomes aponeurotic just outside Quadratus lumborum, and the anterior lamella runs in front of this muscle on to the ilio-lumbar ligament, while the main lamella passes behind it and then between it and the Erector spinæ to reach the ligament also, and thus pass to the lumbar transverse processes (Fig. 110) (whence the ridge that separates the two muscle areas), and the posterior lamella comes off the main sheet at the outer side of Erector spinæ and is attached behind this to the blunt edge, which brings it to the aponeuroses on the back of the sacral musculature (see Fig. 110 and also Fig. 34). The posterior lamella is thickened by junction with the aponeurotic origin of Latissimus dorsi, but this has a muscular origin further out, behind the Quadratus. Internal oblique takes the middle lip as far back as Quadratus, where it passes on to the Transversalis aponeurosis. External oblique is on the outer lip and ends before reaching the level of Quadratus. Thus the External oblique may be said to take more than the anterior half of the length of the outer lip of the crest, and the other two take less than the anterior two-thirds of the middle and inner lips. External oblique becomes aponeurotic at the anterior superior spine, and Poupart's ligament, the lower edge of the aponeurosis, is attached to the prominent convexity of the spine: its other attachments and the related structures are also shown in Fig. 110 and a sufficient description given of them.

The whole of the crest, including its two terminal spines, is a traction epiphysis on the bone, developed from anterior and posterior centres that fuse and make a single epiphysis.

Anterior Border.—The Sartorius arises from the notch below the anterior superior spine (Fig. 104). It has Tensor fasciæ femoris outside it, and is separated from Iliacus by a thin aponeurosis that passes to the bone from Poupart's ligament and holds up the Iliacus here: the line of this aponeurosis may be picked up on the bone occasionally. The remainder of the border has been already sufficiently considered.

Crest of Pubis: presents from without inwards the spine for Poupart's ligament, a groove for the conjoint tendon, and an area of origin for the Rectus abdominis. It ends at the angle, which has a separate ossific centre that has been homologised with the marsupial "epipubic bone."

Symphysis is coated with hyaline cartilage, which is united with its fellow by an intervening disc of fibro-cartilage: a small cavity, not lined by synovial membrane, is found in the upper and back part of this after the first eight or ten years of life.

The ligamentous markings are weaker behind than in front, where they are strengthened by the origin of the inner heads of the two Recti, which decussate on them.

The pubic arch is below the symphysis; the sub-pubic ligament stretches across below the lower and back aspect of the two bones. The penis turns down from its root about the level of the lower part of the symphysis *in the horizontal plane*, but its level in planes drawn at right angles to the front of the bones is much higher, the upper line of the curve almost reaching the plane of the top of the symphysis: this is owing, of course, to the obliquity of the bone. The clitoris, in the same way, reaches about one

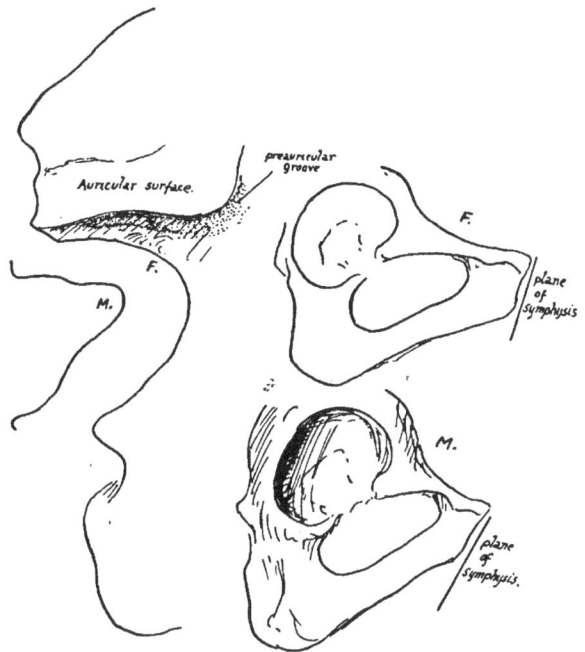

FIG. 112.—On the left somewhat exaggerated outlines of differences in the posterior borders of male and female bones, showing the more angled notch in the former. On the right the sexual differences are illustrated in the pubic region; observe that the wider female arch is correlated with a longer pubis, and this, with the relatively smaller acetabulum, makes the breadth of the acetabulum less than the distance between it and the symphysis. Notice the pubic angle formed with the vertical plane.

quarter to half the distance up, and the lower edge of the symphysis is only a little in front of the urethra in women.

Ischio-pubic Margin.—This has been already considered: observe the hinder part, where there is an outer slope occupied by the origin of Adductor magnus, and an inner one in relation with ischio-rectal fatty tissue. This region is also epiphysial, possibly a traction epiphysis, but more probably one of the atavistic variety, representing a remnant of the hypoischium of birds and reptiles.

Back of Tuberosity of Ischium: facets, upper and outer for Semimembranosus and lower and inner for Biceps and Semitendinosus, are well marked, and the latter is closely bordered by the attachment of the great sciatic ligament, from which the

Biceps obtains some origin. The ligament is considered to be a part of the flexor musculature secondarily cut off by the development of the tuberosity.

Lesser Sciatic Notch : the internal pudic vessels and nerve do not touch the bone, though they go through the foramen, but are separated by the tendon from it : the nerve to the Obturator internus runs on the bone above the tendon to reach the muscle.

Spine of Ischium.—Observe that this is incurved (Fig. 100), and the inward projection is less in the female than in the male.

Greater Notch : the foramen gives exit to Pyriformis, to seven nerves, and to three groups of vessels.* The typical shape of the notch differs in male and female bones (Fig. 112).

Posterior Inferior Spine gives attachment to great sciatic ligament (Fig. 34) and *posterior superior spine* to superficial oblique band of sacro-iliac ligament. The lower spine is at level of lower end of articular surface. The bone above and in front of it is covered by ligamentous fibres, on the surface of which the Gluteus maximus arises.

The main points of distinction between the male and female pelvis—with the exception of the breadth and curve of the sacrum and the measurements of the cavity—can be appreciated on the Os innominatum. Thus the male bone is rougher, thicker, and heavier than the female, and is higher proportionately. The difference in width between the male and female pubic arches can be appreciated on the separate bone if this is held in its proper position. Also in the female the everted surface of the ischio-pubic ramus is smaller, the symphysis is less deep, the acetabulum is shallower, and the distance from the acetabulum to the symphysis is greater than the breadth of the acetabulum, but is not so in the male.

In the typical female bone the Great Sciatic Notch is broader and more rounded than in the male (Fig. 112), and the distance between the tuberosity of the Ischium and the top of the great notch is greater.

There is a slight difference in the shape of the thyroid foramen in well-marked bones, but this depends on the breadth of the ischio-pubic region ; and it may be stated here that the width of the pubic arch is the principal point of distinction, and if this is doubtful the other and slighter sexual characters will be doubtful also : this applies with special force to the shape of the foramen and the distance between the ischial tuberosities in the complete pelvis, and with less force, perhaps, to the shape of the sciatic notch and the distance between the tuberosity and the margin of the notch.

The *Pre.auricular Groove* (Fig. 112) is as a rule only found on female bones : its *raison d'être* is not very apparent.

Other sexual differences can be better appreciated when the bones are articulated : thus the ilium in the female is more vertical, so that the false pelvis is relatively narrower, especially when compared with the greater breadth and capacity of the true pelvis.

The expansion of the Ilium, with its elongated curved crest, is a human characteristic, as are also the shortness of the symphysis and the comparative size and strength of the ischial tuberosities. The characters of the ilium and the great capacity of the true pelvic cavity give the human character of relative shallowness to the pelvis as a whole.

The ilium corresponds with the body of the scapula in being the dorsal element in the girdle, but the doubtful value of the ventral elements in the shoulder girdle precludes any definite comparison with the ischium and pubis, which are the ventral bars of the hip girdle. Definite homologies have been sought between the various parts of the

---

* Sciatic, pudic, and gluteal Vessels, N. great and small sciatic, internal pudic, N. to Obturator, Quadratus femoris, upper and lower gluteal.

scapula and innominate bone, and it has been assumed that the dorsal elements of the girdles have undergone a rotation on their long axes in opposite directions, so that the iliacus represents the infraspinatus and the ilio-pectineal line the spine of the scapula, with other parts to correspond. This seems to be a pure assumption without any evidence or ground of support, and it would probably be safer merely to look on the dorsal and ventral parts as corresponding in that relation without endeavouring to establish definite homologies (see pp. 63 and 80).

It is of course a different and justifiable thing to establish homology between the human and other animal bones, and in the case of the pelvis the general correspondence is very evident. Some details, however, require a little consideration. In reptiles and a few birds, and in some mammals, the ischia meet as well as the pubis, the "hypoischium" being a prolongation of the connecting bar between them (Fig. 113). This is fibro-cartilaginous, and as the pubic arch opens out in response to the demand for space for parturition, the cartilage is stranded, so to speak, on the ischial processes as the epiphysis there. The homology between the angular epiphysis and the marsupial bone, itself the homologue of the epipubis of lower vertebrates, has already been mentioned. This must not be confounded with the prepubic process found in many reptiles (Fig. 113), which, if it has any homologous counterpart in the human bone, is represented by the spine of the pubis or possibly by the ilio-pectineal eminence.

In structure the bone consists, like other flat bones, of a layer of cancellous tissue enclosed between two planes of compact bone. Where the weight of the trunk is transmitted through it—that is, in the ilio-pectineal line and its immediate neighbourhood, between the auricular surface and the upper and back part of the acetabulum—the strong compact layers are very thick, with corresponding decrease in the cancellous layer. The density of the cancellous tissue is increased especially in the tuberosity of the ischium.

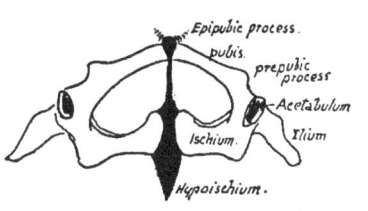

FIG. 113.—Pelvis of lizard seen from below. Cartilage black.

The iliac crest can be felt directly in its whole length, and the anterior superior spine is a very useful point for many measurements during life. Of the posterior spines, the upper can be felt easily, and a transverse line joining it with its fellow goes across the second sacral spine ; the lower is felt less readily, and gives the general level of the lower limit of the sacro-iliac joint and the upper margin of the great sciatic notch. The ischial tuberosity is an important landmark. It is sometimes difficult to place definitely in subjects with well-developed buttocks when lying on their backs, but it can be found by reaching it from below ; this is because it is covered in extension by the thick lower edge of the Gluteus maximus, and by fibro-fatty tissue round this. When the leg is flexed, the edge of the muscle is drawn up and exposes the tuberosity. The ischio-pubic ramus can be followed from the tuberosity forwards. The body of the pubis can be felt, but its parts are defined with difficulty. In women, when the bladder is empty, the bone can be examined bimanually on the surface and through the vagina. The bone in the neighbourhood of the obturator foramen can be more or less examined also through the vagina.

Development of Ossification.

The innominate bone is represented in the early embryonic stages by a mass of mesenchyme which is situated in the base of the limb bud : about the beginning of the second month chondrification commences in this, the mesenchyme comes into relation with the vertebral region in the sixth week, and the cartilage is in position in the seventh week. There are three centres of chondrification corresponding with the three primary parts of the bone, and these fuse and form a shallow acetabulum in the seventh week. By the end of the second month the two innominate plates meet at the symphysis, and the pubic chondrification extends towards this later.

Ossification commences now, that is at the end of the eighth week, as the first of the primary centres, that for the Ilium. The centre for the Ischium is found just before the fourth month, and that for the pubis about a month later. These primary centres appear in their regions near the acetabulum.

At birth each of these main centres has formed a small piece of the corresponding acetabular wall, the rest of the hollow being cartilaginous, as are also the iliac crest, the front margin of the great sciatic notch, the region of the ilio-pectineal eminence, the region of the symphysis, and the ischio-pubic ramus.

Ossification extends slowly in this cartilage, so that about the tenth or eleventh year the ischial and pubic bones have met in the lower ramus, and just after this acetabular centres appear in the triradiate cartilage, which still separates the three primary bones as they have slowly extended on to the floor of the acetabulum.

The acetabular centres appear to be very variable in their number : they may join the neighbouring bones, or may fuse to form a single separate ossicle, the *os acetabuli*, which ultimately fuses with the pubic mass. In this way the acetabular region is consolidated by the age of puberty, fusion occurring first on the pelvic aspect ; but the solid bone is still edged by cartilage on the crest, along the front and back borders in places, in the region of the symphysis, and along the back part of the margin of the lower ramus.

In this cartilaginous border secondary centres appear as soon as the bone is consolidated, at or just after puberty. They are :—

(1) For the anterior superior spine and front part of crest ;
(2) For posterior superior spine and back part of crest ;
(3) For anterior inferior spine ;
(4) For spine of Ischium ;
(5) For marginal surface of ischial tuberosity (hypoischium) ;
(6) Somewhat later, for angle of symphysis ; and
(7) For spine, apparently not constant.

All these secondary centres are usually fused with the main mass by the age of twenty or twenty-one, the last two being a few years later in consolidating.

FEMUR.

A long bone forming the skeleton of the thigh, articulating with the innominate bone above and with the tibia below, and carrying the patella in front of its lower end : consisting of a *shaft*, which is directed downwards, inwards and slightly forwards, and an *upper* and *lower* end (Fig. 114).

The upper end includes *head*, *neck*, and two *trochanters*. The head is connected with the shaft by the elongated neck, which is directed upwards and inwards and somewhat forwards,* forming an angle of about 125 degrees with the shaft. The head is about two-thirds of a sphere, and has on it a depression, the *fovea femoris*, for the attach-

* Variable in amount of rotation ; may be even directed slightly backwards.

The Lower Extremity and Pelvis

ment of the round ligament of the joint. The neck expands towards the shaft and is overhung externally by the *great trochanter*: there is a deep *digital fossa* under.

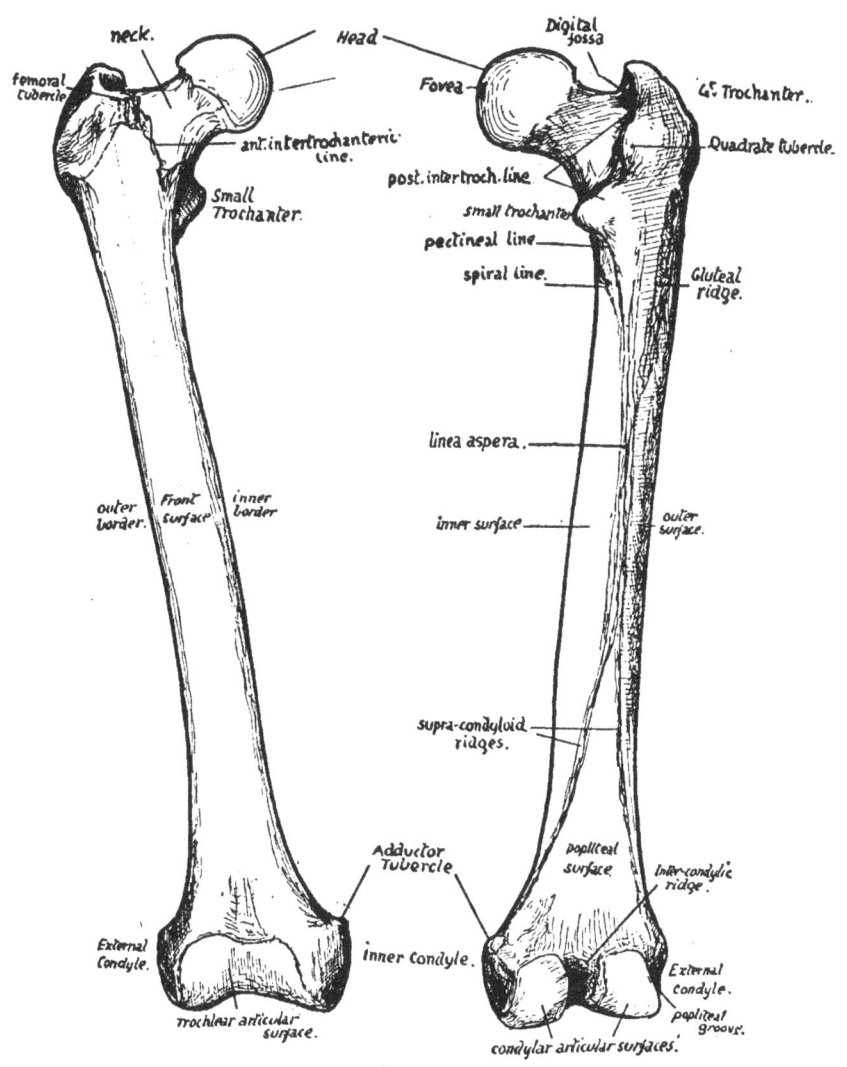

FIG. 114.—Anterior and posterior views of right femur. Left bone is usually a few millimetres longer.

cover of the back part of this, for the insertion of Obturator externus. The great trochanter is for the attachment of muscles of the gluteal group. At the back the great trochanter is connected by the *posterior intertrochanteric line* with the *small trochanter*, which is a rounded process projecting inwards from the back aspect of the bone, for the attachment of the Ilio-Psoas: the *quadrate tubercle* is a rounded mass in the intertrochanteric line to which the Quadratus femoris is attached. In front a rough line, *anterior intertrochanteric*, at the junction of the neck and shaft, marks the attachment of the ilio-femoral ligament: it does not reach the trochanter minor: its upper end, on the front of the great trochanter, presents a prominent *femoral tubercle*. The shaft is triangular on section in its centre, presenting *inner*, *outer*, and *front* surfaces, separated by *inner*, *outer*, and *posterior* borders: the first two of these borders are rounded primary edges, but the last is crowned by a longitudinally-running *linea aspera*, a rough compound secondary crest marking the attachment of adductor and other muscles. The three surfaces of the shaft are covered by the extensor muscles of the knee. The linea aspera can be traced up to the level of the small trochanter: outside its upper part is a rough broad ridge—sometimes a depression—the *gluteal ridge*, for the Gluteus maximus: leading to the small trochanter from the linea aspera is a line which may be termed the *pectineal* line, for insertion of Pectineus. About an inch below the pectineal line the *spiral line* leaves the linea aspera and crosses obliquely upwards and forwards over the inner surface. Traced down, the linea aspera is seen to divide into inner and outer *supracondylar lines*, enclosing between them the triangular *popliteal surface* of the bone: the inner supracondylar is for the attachment of Adductor magnus, and leads to the *adductor tubercle*, which is for the ischio-condylar part of the muscle and is situated on the top of the *internal condyle*, a mass of bone that is matched on the outer side by the *external condyle*. Between the condyles is the *intercondylic* notch, containing the crucial ligaments, and this notch, seen best from behind, is separated from the shaft here by the *intercondylic ridge* for the attachment of the posterior ligament covering the crucial ligaments behind. The condyles have curved articular surfaces which are carried back on the posterior projections of the condyles. The condyles are united in front, and their articular surfaces join to form a *trochlear* surface for the patella: the patellar surface is marked off by a more or less distinct transverse groove on each condyle from the lower or tibial surface. The lower part of the shaft widens gradually to carry the condylar masses, and owing to the obliquity of the shaft the inner condyle is more prominent, though it does not come lower than the outer one; but the outer condyle for the same reason is more directly concerned in weight transmission, and is therefore stronger and thicker. On the side of each condyle are tubercles for the lateral ligaments, and below this tubercle on the outer condyle is a groove that ends in front in a pit or marking for the origin of Popliteus: the tendon of the muscle is said to lie in the groove when the leg is fully flexed. The femora carry the pelvis, and the weight of the trunk is transmitted to each bone through the upper and back part of the acetabulum. The bones articulate strongly at the hip joint.

The Hip Joint.

A simple ball-and-socket joint, in which the spherical head of the femur works in a deep cavity, and the bones are held together not only by the atmospheric pressure, that is able to act on them as a result of the close fitting of the surfaces, but also by exceedingly strong transversely-arranged ligaments which are confined to the front aspect of the joint and are tight in the upright position. The bony acetabulum is further deepened by the fibro-cartilaginous cotyloid ligament on its rim, and this is continued across the cotyloid notch as the fibrous transverse ligament. The greater part of the spherical head is thus received in the cavity, and the ligamentous margin of the cavity is somewhat constricted to be applied to it. Outside the cotyloid ligament is the orbicular zone of circular fibres of the capsule, which is a broad band extending to the neck of the bone (Fig. 105). This band is not as a whole continuous with the cotyloid ligament, but is lined by synovial membrane which passes in on the superficial aspect of the cotyloid ligament (in the posterior part of the joint) and is then reflected back over it on to its deep or articular surface.

The circular band appears to derive its fibres from the deep tendons of the gluteal musculature and from the reflected head of the Rectus (Fig. 105). At the lower part of the neck many of its fibres pass into the inferior retinaculum. It covers in the synovial membrane on the neck and closely surrounds the bone, thus acting as a kind of suction pad in resisting separation between the two bones.

The transverse fibres cover the circular zone in front : they are transverse, or nearly so, to the plane of the joint, but they are more nearly vertical in their relation to outside planes. They form a broad and very thick mass of fibres extending from the front margin of the innominate bone to the anterior intertrochanteric line. The mass varies in thickness and direction of its fibres, those attached to the anterior inferior spine being exceedingly thick and diverging to reach the intertrochanteric line, while further in the fibres attached to the upper pubic ramus are thinner and converge to the lower end of the intertrochanteric line and the recurved part of it: the thick part is termed the *ilio-femoral* band or Y-shaped ligament of Bigelow, and the inner one is the *pubo-femoral* band.

The ilio-femoral ligament receives some fibres of Iliacus, which lies with the Psoas on its front surface. The pubo-femoral band is covered by Pectineus and near the femur by Psoas. The tendon of Rectus lies on the ilio-femoral band outside Iliacus, and external to the tendon of the joint is covered by Gluteus

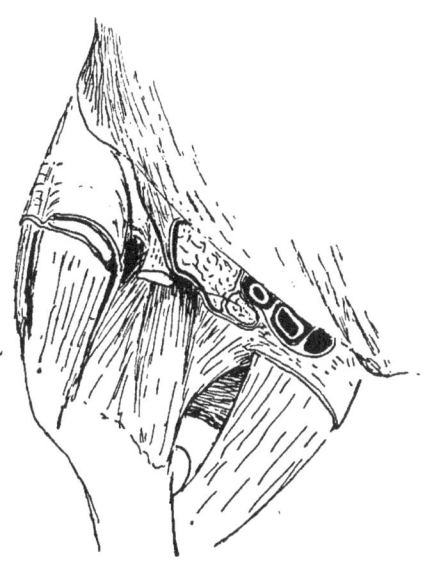

FIG. 115.—A diagrammatic sketch to illustrate the relations of the hip from the front. The capsule is seen exposed in the centre of the drawing. The femoral sheath is cut, and the artery seen resting in the sheath on the cut Psoas ; outside this is the cut Iliacus. If these muscles were not removed they would cover the front of the capsule. The tendon of Rectus appears from under the outer edge of Iliacus, cut short ; it rests on the ligaments. It emerges between a thin fascial sheet from Poupart's lig. round Iliacus and the thicker covering of the gluteal group externally. Inner part of the sheath lies on and is fused with the fascia on Pectineus ; a small area of fascia remains showing how it extends under Psoas and reaches the pubo-femoral band.

minimus. Aponeurotic fasciæ cover the Gluteus and the Iliacus and are attached to the Rectus tendon between them, passing back on each side of this to gain attachment to the capsule ; another sheet from the pubic fascia or Pectineus fascia goes under Psoas to reach the pubo-femoral band and the bone above it (Fig. 115). The circular zone behind is covered by Gluteus minimus, Pyriformis, Obturator internus and Gemelli, and Obturator externus is in relation with its lower and back aspect (see Figs. 107 and 108).

The Ligamentum teres is a weak synovial attachment of the head of the femur to the cotyloid fossa and transverse ligament : it is (p. 127) the remains of the primitive capsule isolated by the secondary taking up of the pubic surface into the joint, and has little mechanical value, but carries some small vessels and nerves to the head of the bone.

The construction of this joint can be contrasted with that of the shoulder. In both there is power of movement in any direction, although the range of movement is much less in the hips. In the shoulder the freedom of movement is obtained by sacrificing the strength of a supporting bony articulation, but security of the joint is nevertheless well attained by the muscles inserted close to the articular surface, which again is made possible by the existence of such an insecure articulation. On the other hand, the strength of the joint by bony support is a paramount necessity in the hip, and hence a deep articulation with consequent limitation of movement ; but the range of movement and power of muscles is increased by lengthening the neck, a modification that would be impossible in the shoulder.

The Ilio-femoral ligament is tight when the upright position is assumed, and thus dispenses with the necessity for muscular effort in sustaining the attitude. This would be necessary otherwise, for the line of incidence of the centre of gravity of the upright trunk passes through the back part of the joint and would tend to bend the body back on the hip. The ligament is said to be proportionately shorter in women, a fact associated with the greater obliquity of the pelvis and convexity of the lumbar column,' but it appears doubtful whether these last-named characters should be related with the shortness of the ligament as cause or as effect. An effect of the shortness is certainly seen in the inability of a woman to extend her thigh on her pelvis further back than what is practically the upright straight line, so that the pelvis has to swing back on that side when more than short strides are taken ; the effect is well seen in the present-day combination of short and loose skirts with long striding, and may be popularly expressed by saying that a man walks from his hips, a woman from her lumbar region.

Head.

This has a separate epiphysis, and the epiphysial line practically—but not absolutely—corresponds with the edge of the articular surface. A prolongation of the cartilage-covered surface on to the front of the neck is frequently found : this underlies the Ilio-psoas tendon.

The head forms about two-thirds of a sphere, and is larger in male than in female bones. The position and size of the fovea varies : one or two small foramina sometimes seen in it are for vessels derived from the obturator artery. The synovial funnel that surrounds them is attached to the depressed margin of the fovea and is lost outside this on the cartilage.

The Neck (Figs. 117 and 118).

The neck is really a prolongation of the shaft, in its development and ossification and in its structure. At birth (Fig. 116) the neck is short and thick and ossified by the extension upwards of the shaft ossification, capped by the cartilaginous head. As development proceeds the neck is gradually elongated, still carrying the epiphysis of the head as a cap on its extremity. The angle it forms with the shaft varies with the extent of its growth, being more open in the young bones and decreasing as growth proceeds, but not after full growth is attained. In the adult the angle is as a rule about 120 degrees, but may vary between 110 degrees and 140 degrees : it is smaller in women than in men and in short than in long bones. Modification of the angle is largely brought about by the extent of development of the strong ridge on the lower and front part of the neck, that acts as a bracket in strengthening the curve. The neck should be looked on as the upper part of the shaft (Fig. 116) incurved, with the muscular trochanters applied to it, but not interfering with the lines of the walls of the shaft : intersecting arcades of cancellous lamellæ spring from these walls and support the articular surface.

The front of the neck is ridged in its long axis and shows numerous vascular foramina. These occur owing to the fibrous bands with small accompanying vessels that lie under the synovial membrane : the membrane covers the whole of this face of the bone, extending to the anterior intertrochanteric line. The fibrous bands run from the capsular attachment here towards the head : they are termed retinacula, and there are three main collections of these fibres—below and behind, below and in front, and above, internal to the front part of the great trochanter. The last-mentioned is a broad band that extends in towards the head for some distance beyond the epiphysial line of the trochanter, which (Fig. 117) encroaches here considerably on the upper aspect of the neck.

The posterior aspect is usually smooth, because there are here no true transverse fibres in the capsule to gain an insertion on the bone : a few marginal fibres of the circular group are the only ones that have any insertion into the femur, and the synovial membrane is reflected from the bone to the deep surface of the circular zone directly. The line of reflection (Fig. 117) is about half-way up the neck, continued up from the turned-up lower end of the anterior intertrochanteric line.

The Obturator externus is closely applied to the lower and back part of the neck, passing to the digital fossa, and the pressure of this muscle moulds the shape of the bone slightly : an indefinite suggestion of an oblique line (Fig. 118) directed outwards and upwards shows the upper limit of the surface affected by this muscle. Below this line the capsule is in relation with the Obturator

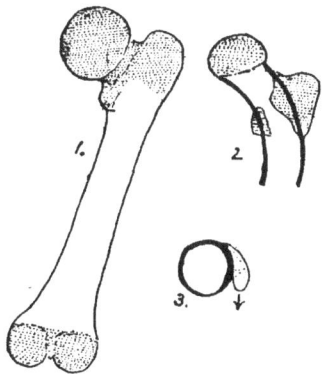

FIG. 116.—1, Diagram of femur at birth ; cartilage is dotted. Upper end shows cartilage to a lower level behind, because the small trochanter is continuous here with great. Observe the short neck. 2, A later condition, in which the shaft (thick lines) has grown in length, carrying the head away from the trochanters and so lengthening the neck. The lower part of the head is formed at first by this shaft, but this is afterwards hidden by epiphysial growth. Trochanters remain as masses applied to the wall of the shaft. 3. A scheme of a section through the small trochanter, showing it applied to the shaft ; the part of the wall which carries it is thick and forms the *calcar femorale* (see Fig. 125).

externus, and above it is covered by the Obturator internus and Gemelli, so that the Quadratus femoris has no direct relation with the capsule at all.

The existence of a long neck is a necessity in a bone which, like the femur, has a head received in a deep articular cavity for purpose of security : if such freedom of movement as the femur enjoys is to be obtained, the body of the bone must be carried at some distance from the embedded head. Necks are more vertical in long femora.

The Great Trochanter (Fig. 118).

The great trochanter is a traction epiphysis formed at the attachment of the gluteal muscles, and pulled as a sort of hood over the base of the neck. The descriptive term is applied to the whole mass of bone that stands up at the base of the neck, and does not quite correspond with the epiphysis. The epiphysial line is as shown in Fig. 118, and the epiphysis includes the femoral tubercle at the top of the anterior intertrochanteric line and a part of what is descriptively the upper aspect of the neck.

The front surface has (Fig. 118) a smooth area for a bursa under the tendon of Gluteus minimus, which is inserted into the strong ridge along its outer border. The Ilio-trochanteric aponeurosis (p. 124) reaches the ridge below and internal to the bursa. Pyriformis is inserted along the whole length of the upper crest of the trochanter. The strong, direct tendon of the muscle runs mainly into the back half of the crest, where there is usually a facetted appearance made by it and extending back to the unciform end that overhangs the digital fossa.

But this is not the whole attachment of the muscle. If the structures are looked at in an ordinary dissection the tendon of Pyriformis is seen (Fig. 117) to lie in plane superficial to that of the Obturator internus and Gemelli, but if the structures are divided and turned up the Obturator tendon is still seen to be covered by the Pyriformis; in other words, the Pyriformis partly surrounds the Obturator insertion by means of an expansion passing in front of it, and from this expansion, as shown in the drawing, another aponeurotic sheet is continued to the top of the Gluteus minimus tendon. The complete area of Pyriformis attachment to the bone therefore is as seen in the figure, and the sheet that passes in front of the Obturator tendon is also continued into the fibres of the upper retinaculum.

Obturator internus and Gemelli are inserted internal to and in front of the main tendon of Pyriformis, on the inner side of the trochanter: Gemellus inferior takes the lower and back part of this insertion, and Gemellus superior is largely attached to Pyriformis. The outer surface has a well-marked oblique ridge for the tendon of Gluteus medius, which takes the whole breadth of the ridge. Above and behind this marking may run into that for Pyriformis, when the muscles are partly fused, and in front it is continuous with the lower end of the minimus insertion: these two Glutei are usually fused along their front borders, and it is evident that the fused parts, lying in front of the level of the joint, have similar action as internal rotators.

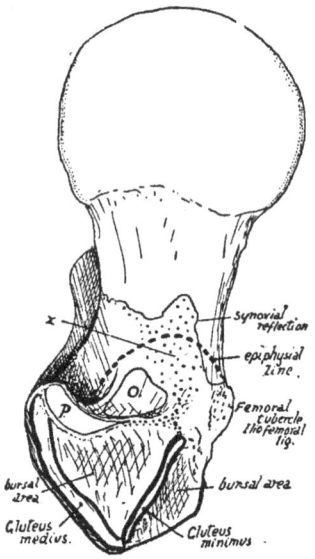

FIG. 117.—Upper aspect of head, neck and great trochanter of right femur. O. Obturator internus insertion; P. Pyriformis insertion. The area of insertion of the fibres of the superior retinaculum is marked at x, and it is seen to be continuous with that of Pyriformis round the front of O. and to extend on the neck further than the epiphysial line. Compare with Fig. 118.

Above the oblique ridge the trochanter is covered by medius with an intervening bursa, and is thus moulded to a direction differing from that of the bone below the ridge, which is covered only by maximus: this lower surface may have a bursa lying on it, but as a rule the bursa only reaches its most prominent part. The posterior surface is also directly covered by maximus, so that it forms a continuous curve with the area below the oblique ridge.

Obturator externus is not concerned in forming the overhanging trochanter, so that the epiphysial line, which goes to the bottom of the digital fossa (Fig. 118), is

144

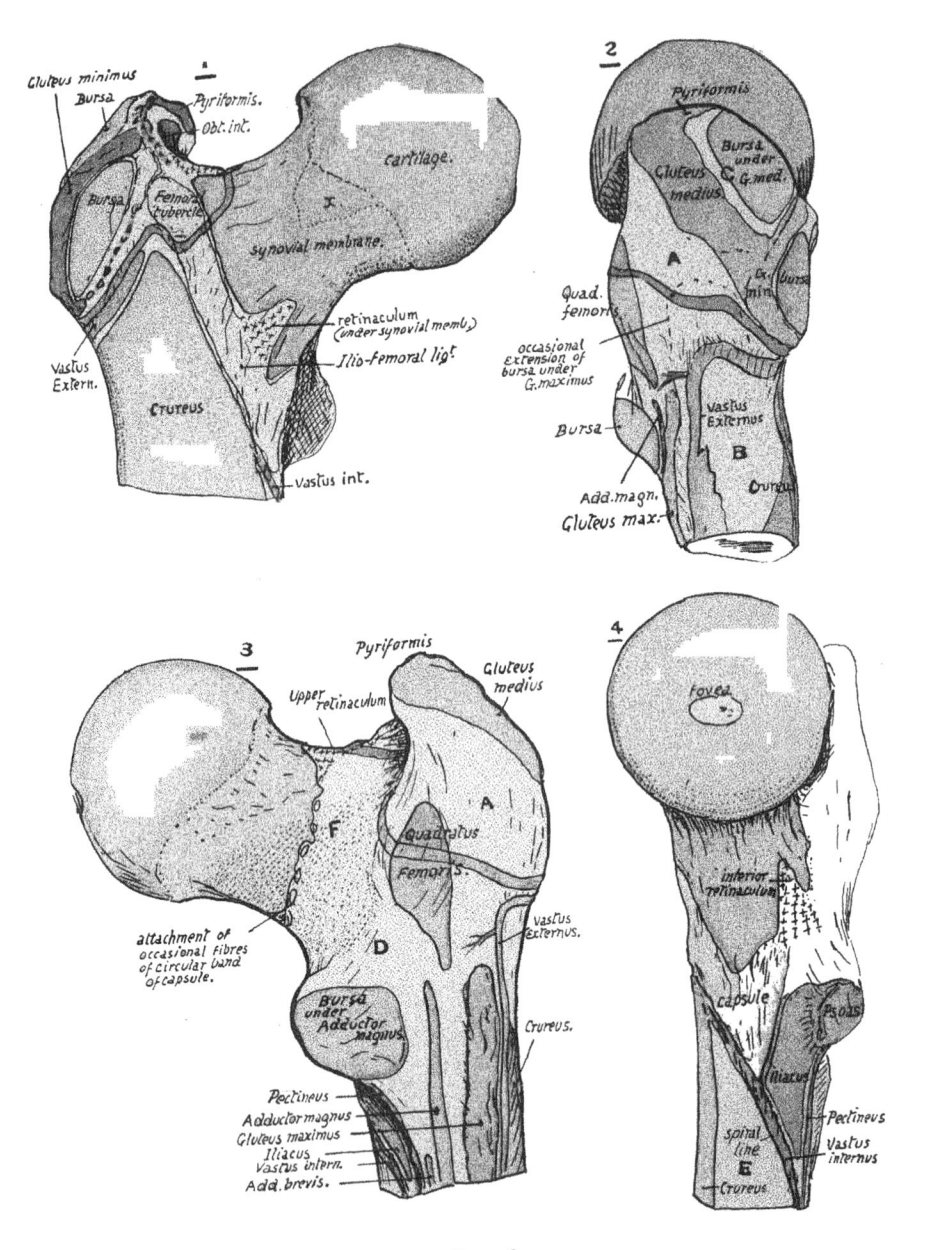

FIG. 118.

The Lower Extremity and Pelvis

further out here than the insertion of the tendon on the floor. The posterior intertrochanteric line is part of the trochanteric system and marks the pull of the Quadratus : this is inserted by muscular fibre as a rule, so that its area is not linear and has not any secondary markings : occasionally there are tendinous fibres, with corresponding secondary roughnesses. The quadrate tubercle is that part of the trochanteric mass on which the pull of the muscle is directly exerted, and it is thus of the nature of a primary marking.

FIG. 118.—Upper end of right femur. The epiphysial line for the great trochanter is marked in green round its base. The "retinacula of Weitbrecht," fibres running back toward the head under the synovial membrane, are shown only where they are congregated into their three main groups ; they are derived from the transverse capsular fibres, and the upper one obtains many fibres from Pyriformis (see Fig. 117). 1. Anterior aspect. Observe that the Gluteus minimus is attached only to the outer ridge of the trochanter, but its tendon is continuous below with an aponeurotic sheet, the ilio-trochanteric band, which covers the bursa in front and reaches the bone internal to it. The upper part of the origin of Crureus is mainly tendinous. The extension of the cartilage of the head on to the neck is shown at *x* ; this lies under the ilio-femoral band or, if the opening for the sub-Psoas bursa is large, under the tendon of the Psoas. 2. From the outer side. The oblique insertion of Gluteus medius is continuous below and in front with that of Gluteus minimus, and frequently with that of Pyriformis above and behind ; it divides this aspect of the trochanter into two areas, one, *C*, in front and above, under cover of medius and therefore bevelled off in the direction of that muscle, the other, *A*, below and behind, covered by Gluteus maximus and therefore moulded by that muscle so that it is more vertically directed and curved from before backwards. The surface *C* carries a bursa, but *A* has only occasionally an extension of the bursa situated below in relation with it. *B*, surface covered by Vastus externus and more or less flattened by it. Crureus fuses with V. externus at a lower level. 3. Posterior aspect. *D*, surface covered by Quadratus femoris ; deep to this muscle the Obturator externus lies against the bone, moulding the back and lower part of the neck in the area F as it passes to the digital fossa. 4. From the inner side. Observe the pointed area between the spiral line and pectineal line which is occupied by Iliacus. *E*, inner surface, covered by Vastus internus but not affording origin to it ; the Crureus does not transgress the inner border.

The Small Trochanter is a traction epiphysis at the insertion of the tendon of the Psoas (Fig. 118). The tendon is inserted into its inner rough aspect and, with the Iliacus, into the front surface as well : the Iliacus is also inserted by muscular fibre into the bone in front of and below its level, a triangular area bounded by the spiral and pectineal lines. The back of the small trochanter gives attachment to no muscle, but it is covered by a well-formed bursa which separates it from the top fibres of the Adductor magnus,* into the substance of which it projects when the limb is inverted. This projection can be felt in thin people when the limb is strongly inverted. The plane of the shaft wall deep to the trochanter is strengthened, and forms a bony bar or column visible on section (Fig. 116) and extending up into the neck : this is termed the *calcar femorale*, and is supposed to receive the weight transmitted through the head and from thence by the cancellous arches to the shaft (see Fig. 125). Now articulate the femur and the pelvis and see what the action of the Psoas would be in addition to flexing the joint. Its tendon passes downwards over the front of the joint and turns outwards and backwards to reach its insertion—that is, it pulls on the femur *outside* the position of the axis of rotation, which passes down through the joint nearly vertically. It is therefore an internal rotator as well as flexor, but the rotating action is probably very small.

The Anterior Intertrochanteric Line (Fig. 118).

The anterior intertrochanteric line marks the attachment of the ilio-femoral band : the outer strong limb of the ligament goes to the tubercle at the upper end of the line, the inner band to the lower end of the marking, and the intermediate thinner part to the line in between. The line turns upwards and backwards for a little distance at its lower end, this part of the marking being mostly made by pubo-femoral fibres, and

---

* Occasionally also the lowest fibres of Quadratus femoris.

this will indicate the level of the synovial reflection on the back of the neck. The interval between the line and the small trochanter affords insertion to muscular fibres of Iliacus.

The line appears at first sight continuous with the spiral line made by the aponeurotic origin of Vastus internus on the inner side of the shaft, but closer examination will show that this continuity is only along the outermost part of the intertrochanteric ridge: here the Vastus arises, while the remainder of the ridge is purely ligamentous in origin.

The Shaft.

The shaft is covered, except on the linea aspera, by the extensors. The Crureus arises from the front surface for its upper half or a little more, extending up to the front of the trochanter: its origin does not transgress the inner margin, but extends obliquely over the outer margin, an inch or more below the trochanter, to blend with the origin of Vastus externus (Fig. 118). Vastus externus arises as shown in Fig. 120: its upper fibres are from the front of the great trochanter, and, being aponeurotic, lie as an apparently free edge on the Crureus, although the two muscles are blended at their origins. Vastus internus has a linear origin, therefore, by aponeurosis, from the line shown in Fig. 119, but this only extends as far as the adductor opening: below this level the muscle has origin from the front of the tendon of Adductor magnus, and not from the bone. Thus the inner side of the shaft affords attachment to no muscle, but is covered by the Vastus internus. The upper fibres of the muscle are blended close to the anterior intertrochanteric line with those of the Crureus, wherefore these two muscles, though otherwise separated at their origins, have more or less the superficial appearance of forming one mass.

Subcrureus arises from the inner part of the front surface of the shaft about two thirds of the way down; its nerve therefore lies on the bone above this level.

A bursa is under the lower end of Crureus in young subjects, joining later with the synovial cavity and so giving it the extension upwards that characterises it: the bursa occasionally remains distinct from the joint.

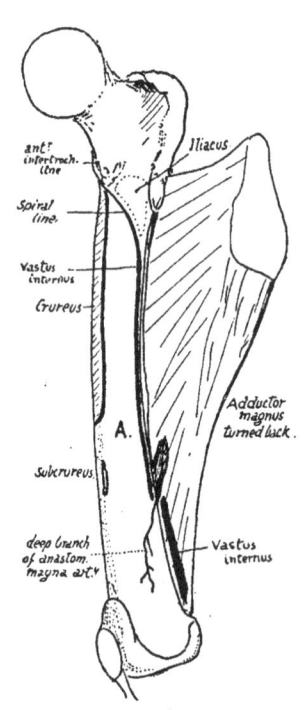

FIG. 119.—Right femur from the inner side, with Adductor magnus turned back to show the linear origin of Vastus internus from spiral line, inner lip of linea aspera and tendon of magnus. Thus the muscle lies on inner surface, A., of shaft, but does not arise from it. Notice the extent of Crureus, the origin of Subcrureus, and the plane of the deep branch of the anast. magna artery.

Region of Linea Aspera (Fig. 120).

Examine and compare the gluteal ridge in different bones: in some it is a prominent crest, in others only a broad rough area, and in others again it is represented by a rough fossa (*fossa hypo-trochanterica*), or these different aspects may be more or less combined in one specimen. Rarely there may be a traction epiphysis forming

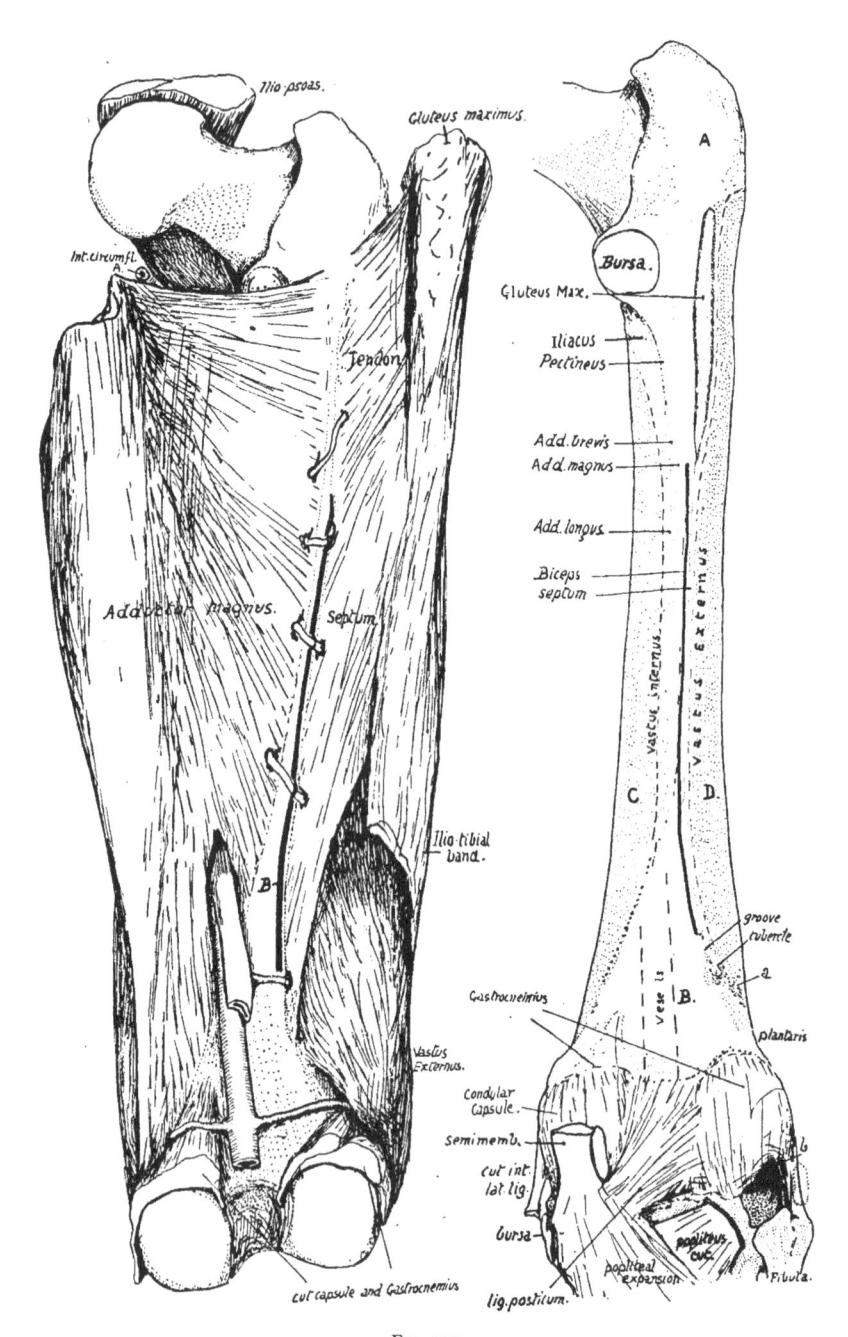

FIG. 120.

The Lower Extremity and Pelvis 147

a prominent crest, as is normally found in the femora of speedy animals such as the horse and deer. When such a centre develops in man it appears about the age of twenty-one and joins in a few years.

The gluteal ridge marks the direct tendon of Gluteus maximus, but an expansion downwards from this tendon forms the external intermuscular septum (Fig. 120), so that the line of this septum on the bone is a direct continuation of the ridge and passes between the short head of Biceps internally and origin of Vastus externus on the outer side.

The linea aspera can be more or less analysed into its constituent lines by comparison of various bones with each other. The irregularly-fused ridges that compose it are purely secondary lines situated on a primary border of the shaft, so that when a young bone is examined there is no appearance of these rough ridges, but only the rounded border of the bone, into which the aponeuroses are inserted. Later the attached aponeurotic structures have a certain amount of ossification extending into them from the periosteum, and at the same time a small artery that runs at first between the aponeuroses is by this means taken into the new-formed ridges and makes the vascular canal found in this region.

Analysis of the linea aspera places its attachments as follows (Fig. 120) : *External intermuscular septum* from gluteal ridge to a small ill-defined tubercle on the external supracondylar ridge : run the finger down this ridge and the tubercle will be felt some 2 to 3 inches above the condyle, more marked from the presence of a shallow groove just above it. The groove is for a muscular artery, and the tubercle is for the attachment of the lowest fibres of the septum, which is pierced by the artery. The tubercle is not to be confounded with a marking occasionally present at a slightly lower level and more externally, made by the fascia of Vastus externus.

FIG. 120.—Posterior aspect of right femur to show linea aspera, etc. The left figure shows how the external intermuscular septum reaches the bone, being continuous above with the tendon of Gluteus majximus, and below this with the deep aspect of the ilio-tibial band. It separates Vastus externus from the short head of Biceps, which is indicated by the thick black line (B-) between the septum and Adductor magnus : observe that the septum is pierced close to its lower end by a muscular artery, and that the Biceps is not (as a rule) so pierced. On the back of the bone (right figure) a small tubercle marks the end of the line of the septum, which runs up to the Gluteal ridge, and a groove marks the position of the piercing artery and hence the lower end of the origin of Biceps ; the little tubercle is not to be confounded with a rough spot (a) due to aponeurotic fibres in Vastus externus. The remainder of the linea aspera is seen analysed into the districts into which the tendons are inserted which are responsible for its formation, and on each side of it the origins of the two Vasti are shown as interrupted lines ; in its lower part (dotted) the Vastus internus arises from the tendon of Adductor magnus, and not from the bone. Other areas on the posterior aspect of the bone are : A, covered by Gluteus maximus ; B, popliteal surface, in relation with vessels and popliteal fat ; C and D, covered by Vastus internus and externus respectively. In the upper part of the left figure the Ilio-Psoas is seen in position, showing how the internal circumflex artery is kept away from the bone by it (compare with Fig. 121). The lower end of the bone has the origins of Gastrocnemius marked out ; observe how much of each head and of Plantaris arises from the condylar capsule. The condylar capsules cover in the condylar recesses of the joint, and the interval between them is occupied by the crucial ligaments ; in the figure, however, these have been covered in by the lig. posticum which is really an expansion from tendon of Semi-membranosus. The cut tendon is shown, lying against the condylar capsule, having first crossed part of the origin of inner head of Gastrocnemius. The inner head comes down to the inner side of the tendon, so that a bursa is necessary between them, and this must lie on the capsule, through which it sometimes reaches the joint cavity. Semimembranosus expands at its insertion and its front fibres pass under the internal lateral ligament, a bursa intervening. Tendon of Popliteus comes through the outer and back part of the outer capsule (b), deep to the ext. lat. lig. ; the latter is covered by tendon of Biceps, with a small bursa interposed.

Short head of Biceps is just internal to the line of the septum, extending from the gluteal ridge to the slight groove above the tubercle, for the artery runs below the origin of Biceps, though it pierces the septum. Sometimes the origin is prolonged a little upwards, in which case it must of course run internal to the gluteal ridge. *Adductor magnus* extends as high as the level of the top of the small trochanter, but occasionally falls short of this : it extends down to the Adductor tubercle, and as a rule no definite indication of the situation of the femoral opening can be found, for the tendinous structure is continued on the bone deep to the vessels. The upper fibres are those arising from the pubis, and by articulating the bones it is seen that they are nearly horizontal in direction and must pass behind the small trochanter, whence the necessity for a bursa on its posterior aspect.

Adductor brevis lies between the Magnus and the Longus and Pectineus, and can be placed more easily if these last two are first put into position. *Pectineus* is inserted by aponeurosis into the lower part of the small trochanter and the line leading down from it to the linea aspera for about 2 inches or more: *Adductor longus* is on the continuation of this line, occupying the middle third of the length of the shaft. *Adductor brevis* can now be placed just outside this line, extending, roughly, about half-way up the level of Pectineus and half-way down that of Longus. Notice that the upper ends of the gluteal ridge and the Adductor magnus insertion and the base of the great trochanter and upper end of small trochanter are all about on the same level: from this it can be understood that the transverse branch of the external circumflex artery, running round the base of the great trochanter, can pass just above the gluteal insertion, and is on the same level as the internal circumflex, which appears above the Adductor magnus but not close to the bone, because it is internal to the Psoas tendon (Fig. 120).

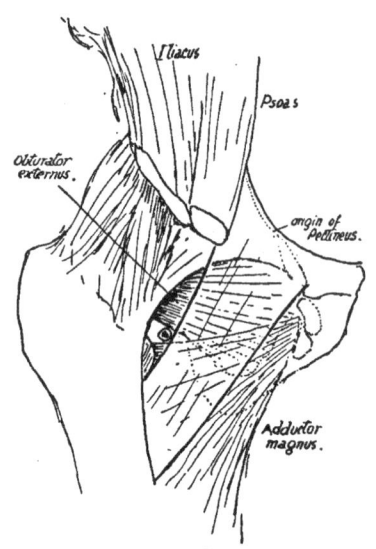

FIG. 121.—Scheme to show the course of the internal circumflex artery. Pectineus is represented as transparent, lying on Obturator externus and upper fibres of Adductor magnus. Upper part of Add. brevis comes in between them, but is too low to be in direct relation to the artery. The artery runs back in the small interval in which it is seen below the capsule and Obt. externus, above or external to Pectineus and Add. magnus; this interval is covered by Psoas, and the vessels reach it by passing back between Psoas and Pectineus, internal to the former.

The course of the internal circumflex can be understood by articulating the bones and putting the muscles into position on them, as shown in the Fig. 121. It is seen to turn back internal to the Psoas and reach the interval above the mass of the Adductors and below the capsule and Obturator externus, and therefore appears behind above the Adductor magnus and below the Quadratus femoris (which covers the Obturator externus behind).

The perforating arteries (Fig. 120) lie close to the bone, so that they run a very short course before breaking up in the substance of Vastus externus. They pierce the various aponeurotic planes attached to the linea aspera, and give branches to the muscles, etc., between these planes. The first pierces Adductor brevis and magnus and tendon of Gluteus maximus; the second goes through brevis, magnus, Biceps, external septum; the third and fourth do the same, except that they do not pierce brevis. Occasionally the first may go through the lower part of Pectineus before piercing Adductor brevis.

Lower End.

Notice the differences between the condylar masses. The outer is thicker and stronger than the inner, for it has more to do with the direct support of weight and

48

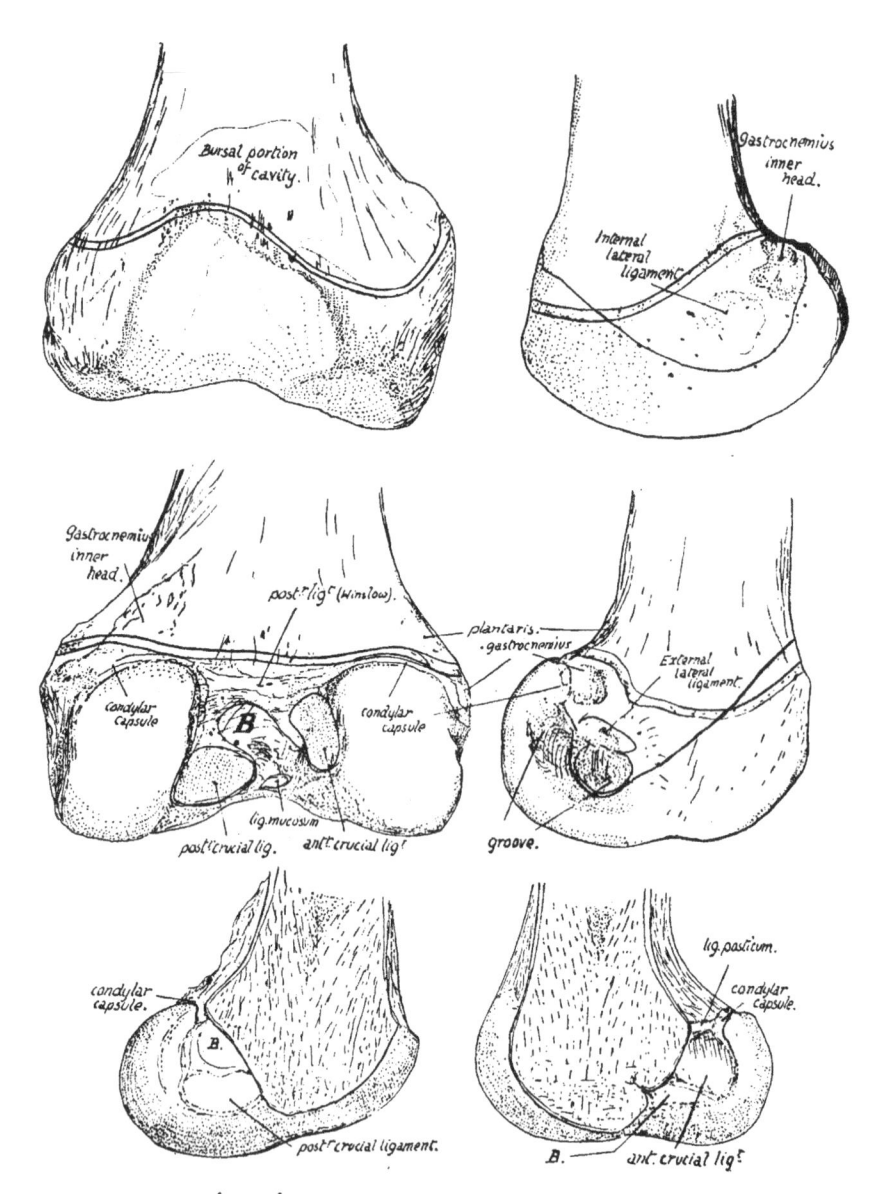

FIG. 122.—Lower end of right femur. The green line indicates the epiphyseal line. B, the bursa-like recess of the trochlear part of the cavity which extends in between the two crucial ligaments. The upper part of the trochlear cavity is formed by the addition of a bursa developed deep to the Crureus; this portion is often somewhat separated from the main cavity by fatty synovial folds. The lower figures show the two halves of the lower end divided down the centre.

transference to the tibia. On the other hand, looking at them from below, the inner is seen to be longer and curved ; this is probably in association with the terminal rotation that occurs round the attachment of the anterior crucial ligament when the leg is fully extended, when the inner part of the tibia moves forward on the femur—or the femur backward on the tibia, if the foot is on the ground—while the outer condyle is held in position by the attachment of the anterior crucial ligament to it. The attachments of the two crucial ligaments are shown in Fig. 122, and the centre of the curve made by the inner condyle is on the attachment of the anterior ligament to the outer condyle.

The two condyles are on the same horizontal level when the bone is in its natural position, so that a femur placed with its condyles on the table is in the inclined direction that it occupied in the body. It makes in this way an angle of about 9 degrees with the vertical, a little greater in woman, and at the same time sloping downward and forward at an angle of about 4 degrees or 5 degrees with the transverse vertical plane.

The popliteal surface is limited below by the intercondylic ridge to which the posterior ligament is attached. It is covered in for the most part by the belly of Semi-membranosus and externally by Biceps. Observe that the inner head of the Gastrocnemius encroaches on this space, there being a tuberculated roughness for it on the bone, so that the inner superior articular artery crosses over this head, whereas the outer vessels lie above the level of the origin of the outer head (Fig. 120).

When the inner vessels have crossed the inner head they come to the tendon of Adductor magnus on the supracondylar line : piercing this close to the bone they find themselves under cover of Vastus internus, which we have seen arises from this tendon and not from the bone. The outer vessels lie on bone all the way, passing deep to Biceps, Ilio-tibial band, and Vastus externus.

The two heads (Fig. 120) of Gastrocnemius arise largely from the vertical capsular fibres that cover the back of the condylar recesses of the joint, but they each reach the bone above : the outer tendon is attached to the well-marked facet seen in Fig. 122, and the inner tendon, in addition to the popliteal origin already mentioned, has a facet of attachment in a corresponding situation behind and below the Adductor tubercle (Fig. 122). There is usually a small sesamoid bone or cartilage to be found in the tendinous origin of the outer head, and rarely one in the inner head in old people. The " facets " may be looked on as representing the positions of small bursal prolongations from the joint deep to these sesamoids.

Below the Gastrocnemius facet on the outer condyle is a tubercle for the external lateral ligament, and below this again is a groove from the front end of which the Popliteus arises. The inner condyle is prominent below the Adductor tubercle : the internal lateral ligament is attached below the most prominent spot.

The side of each condyle is covered below and in front by synovial membrane, which is reflected on to the capsule along an oblique line directed downwards and backwards. The line passes to the under side of the attachment of each lateral ligament : behind this it turns up to become continuous with the attachment of the condylar portion of the capsule. Each condylar capsule extends up to the bone a little above the articular surface, and is lost in the intercondylar fossa by becoming continuous with the areolar tissue behind the crucial ligament and with the ligaments themselves. The synovial sac of the condylar recesses does not extend further toward the intercondylar fossa than the margin of the articular surface, but the main cavity extends back for some distance from the trochlear surface on to the crucial ligaments and comes into relation with the floor of the fossa between these ligaments (Fig. 122). The cartilage-covered surfaces rest below on the interarticular (semilunar) fibro-cartilages, and the impressions made by these are visible (Fig. 123) in the recent state, and usually also in the dry bones.

The whole continuous articular surface can be divided into three portions, two condylar and one trochlear surface, the latter being for the play of the patella; a prolongation of the trochlear surface downwards and backwards along the inner condyle, shown in Fig. 123, is in contact with the patella when the leg is flexed. These three parts were originally distinct, and the remains of the dividing walls are seen in the ligamenta alaria and mucosum. Ligamentum mucosum is attached to the front end of the intercondylic fossa, and extends from this as a synovial band to the lower end of the patella: it marks the line of the septum separating the two condylar cavities, and the ligamenta alaria are remains of the walls cutting them off from the trochlear sac (Fig. 124).

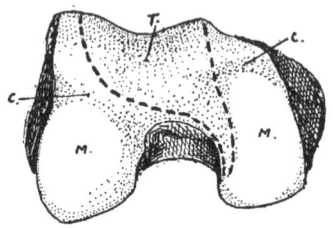

FIG. 123.—Lower end of right femur. *T*. trochlear surface, for patella, extending along inner condyle as an area with which patella articulates in flexion; *M*. menisco-tibial surfaces for resting on tibia in extension. The fibro-cartilages make marks by their front borders on the femur at *c. c.*

The whole lower end is formed from a single centre of ossification, making a pressure epiphysis which is one of the earliest of this class to appear in the body. It is formed in the middle of the cartilaginous end at or just before birth, a fact of some interest and importance in certain medico-legal inquiries. The epiphysis is that of the growing end of the bone, and the epiphysial line (Fig. 122) runs through the Adductor tubercle, above the Gastrocnemius facets, and a little distance above the articular surfaces, curving slightly downwards on the sides of the condyles. Thus there are attached to the epiphysis:—

(1) The posterior ligaments of the capsule and the lateral ligaments;
(2) Crucial ligaments and Ligamentum mucosum;
(3) Outer head and part of inner head of Gastrocnemius, with Plantaris in part;
(4) Popliteus.

The greater part of the epiphysis is in relation with the joint cavity, it being only otherwise covered by the Vasti on each side over a small area.

Fig. 125 is a scheme to illustrate the disposition of lamellæ in the construction of the bone.

FIG. 124.—A left knee at birth, showing an unusual condition of the ligam. mucosum, which is represented by a complete septum between the two condylar parts of the cavity. The ligamenta alaria are seen as wing-like folds at each side of the lower end of the patella: they are remains of septa shutting off the trochlear from the condylar cavities.

The femur can be directly examined only towards its ends during life, its shaft being indirectly palpable through the mass of extensor muscles. The great trochanter is easily felt, being only covered over its greater part by the aponeurosis of the Gluteus maximus, although the tendon of Gluteus medius somewhat obscures its upper front portion. Its breadth, etc., can be appreciated by the fingers, and comparison made with the other

The Lower Extremity and Pelvis

side. Careful estimation of its relative position and level is most useful in differential diagnosis of lesions in the neighbourhood ; its upper extremity just touches, normally, a line drawn between the tuber ischii and the anterior superior iliac spine (Nelaton's line), and a higher position on one side points to lessening of the vertical difference between the acetabular and trochanteric levels. The top of the trochanter is about on a level with the middle of the acetabulum. The rounded head of the bone lies in its joint behind the Psoas, where its mass can be felt, or its absence ascertained in states associated with such a condition. The joint lies about the centre of the line drawn from the pubic spine to the trochanter when the limb is not too much everted. The side aspects of the two condyles can be easily examined, covered by relatively thin aponeuroses. It should be remembered that the front and lower parts of these aspects of the condyles are covered by lateral recesses (Fig. 122) of the synovial sac ; this moves down with the patella when the leg is flexed, a fact utilised under certain surgical circumstances. The adductor tubercle can be felt on the upper part of the inner prominence, and the external lateral ligament is distinctly recognised on the outer side on bending the knee. The lower edges of the condyles, if not found at first, can be placed at once on flexion, and their front aspects can be examined to some extent by relaxation of the extensors and side-to-side movement of the patella.

Ossification of Femur.

The bone is represented by a short and thick cartilaginous rudiment, in which ossification commences, at the centre of the future shaft, towards the end of the second month or earlier.

At birth the shaft is bony and the neck is short, while the trochanters and the extremities are in cartilage ; in the lower end, however, there is usually a small ossific centre in the depth of the cartilage, but this may not appear till just after birth.

The centre for the head appears in the first year, that for the great trochanter at three, and for the small trochanter at twelve or thirteen. Fusion of the head epiphysis with the neck, which has become longer, occurs at about eighteen, and the trochanteric epiphyses join the shaft about the same time or a little earlier. The bony lower end remains distinct until twenty-three or twenty-four, when its epiphysial line ossifies ; so the lower is the growing end of the bone.

The occasional centre for the gluteal ridge has already been mentioned (p. 147).

The rough linea aspera becomes apparent about puberty.

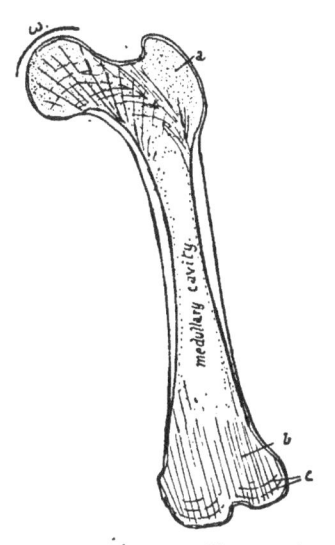

FIG. 125.—Scheme to illustrate the structure of the femur. The shaft is a hollow bony cylinder in the middle, but has cancellous tissue at the ends ; this is dense in the trochanter, *a*, but does not show any particular arrangement here, but this is not the case in the head, neck, and lower end, in the lines of weight-transmission. In the lower end there is a tendency to formation of vertical lamellæ, *b*, with cross-lamellæ near the surfaces, *c*. In the neck there are spiral lamellæ in the plane of the shaft wall, giving an appearance of arcades on section, and continued into dense cancellous bone in the head. Observe that the weight, *w*, will be transmitted from the upper part of the head through these spirals to the lower wall of the neck ; here the wall is accordingly thick and noticeable on section, and is termed the *calcar femorale*.

152   Anatomy of Skeleton

The time of appearance of the centre for the lower end seems to be somewhat variable : as a rule it appears shortly before birth, but in a small number of cases it may be found as early as the eighth month, and in others, very rarely, it may be delayed even as long as the second year after birth.

PATELLA,

A bone situated in front of the knee joint, somewhat triangular in shape, with rounded margins and its apex pointing downwards. It may be looked on as a large sesamoid developed in the tendon of the Quadriceps, receiving the insertion of these muscles on its upper margin and sides and being attached below by the Ligamentum patellæ or patellar tendon to the tubercle of the tibia. The aponeurotic fibres of the extensor group in which it exists are in fact continued over its surface into the tendon,

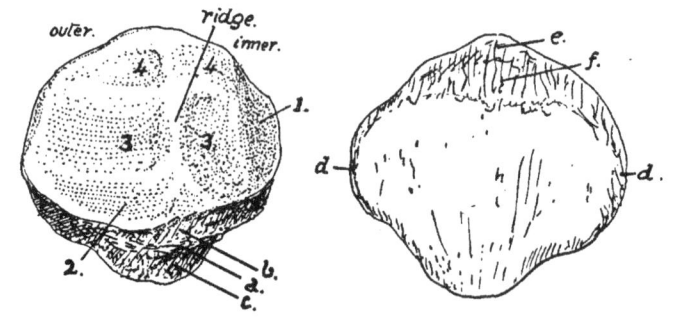

FIG. 126.—Left patella. Back and front views. 1, medial vertical facet, usually considered to rest against the condylar trochlear surface in extreme flexion, but certainly covered by fatty synovial pads in large part ; 2, surface overlaid by fatty folds below ; 3, 3, lower facets in contact with femur in extension ; 4, 4, upper facets in contact with femur in flexion. Other references given in text. To tell left from right, hold the bone so that the larger articular area is external and behind, with the " apex " down.

so that it is sheathed in front by fibrous tissue that separates it from subcutaneous tissue and skin, and, in its lower part, from a subcutaneous bursa.

But its deep or posterior surface forms part of the front wall of the joint cavity, wherefore it carries a cartilage-covered articular surface. This moves up and down on the trochlear surface of the femur as the knee is moved, being always separated from the tibial tubercle by the length of the tendon.

The *front surface* of the bone is longitudinally ridged, indicating that the ossification has involved some of the deeper fibres of the aponeurotic tissue that covers its surface.

The margins show rough markings of Vasti and Crureus, and the top margin also receives the Rectus in front of Crureus. The outer margin is occasionally the seat of a rough process, constituting an " emarginated " bone, where ossification has extended into the tendon of the Vastus externus.

The *deep surface* of the bone is shown in Fig. 126. Observe that the articular area that rests on the outer condyle is larger than the inner one : this, in conjunction with the apex pointing downward, enables one to recognise the side to which the bone belongs.

The ridge between the two surfaces lies in the hollow of the trochlear surface of

The Lower Extremity and Pelvis 153

the femur; when the knee is bent the patella is carried downwards and backwards on to the under aspect of the femur, where the trochlear surface is prolonged on to the inner condyle. Articulate the two bones and slide the patella down until it lies in the position it would assume in flexion, and it is evident that the bone becomes somewhat tilted up on the outer condyle, with the result that it is only the inner part of its articular surface which rests against the inner condyle; at the same time the shape of

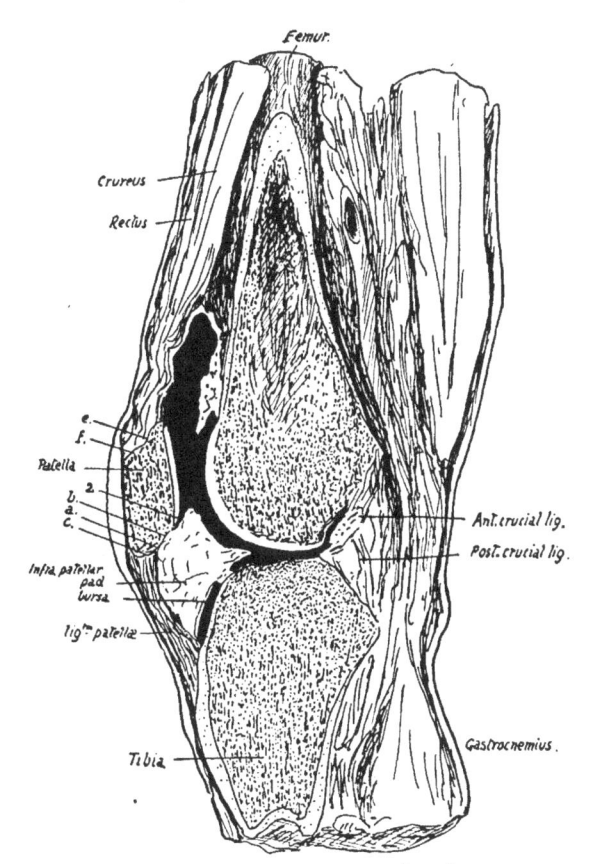

FIG. 127.—Vertical antero-posterior section through knee-joint.

the trochlear surface is such that the outer part of the patella is not in contact with the outer condyle, but the tilting is really due to the direction and level of the concavity in which the median ridge of the patella rides.

A rough sloping area extends between the lower margin of the articular surface and the point of the apex, and can be subdivided by the line a (Fig. 126) into two parts, an upper b and a lower c. The lower area c extends to the apex and is for the attachment of the patellar tendon, while the infrapatellar pad of fat, that lies deep to the tendon, comes in contact with the area *b* and is attached to it. Trace *c* round the bone : it is found to be continuous with the narrow strips *d, d* along the edges, seen on the dorsal aspect. These in their turn are continued into the wider area at the upper end of the dorsal surface, which can be subdivided as a rule without much difficulty into an upper *e* and a lower *f*. The strips *d, d* are for the Vasti, *e* for the Crureus, and *f* for the Rectus femoris : between *f* and the apex the bone is subcutaneous, although it is covered by a thick fibrous periosteum continuous with the aponeurosis, and the upper end of the bursa on the patellar tendon may lie on the lower part of this subcutaneous area.

Fig. 127 represents a vertical section through the middle of the right knee in which the cavity has been partly filled with gelatine. The infrapatellar pad is well seen : observe that it not only has a tongue-like projection backward over the tibia, but also overlaps the lower part of the articular surface of the patella. There is another similar but much smaller overlapping fold above the bone, but it is not well shown in the section. Notice also the appearance of the areas just dealt with when they are seen on section ; they are lettered to correspond with Fig. 126.

The whole circumference of the patella is covered along its articular margins by overlapping fatty synovial folds similar to that seen in Fig. 127. The folds are smallest along the upper edge, but well developed along the remainder of the margin. Along the inner and lower edges there are definite areas where the articular cartilage is in contact with these synovial folds : the areas are shown in Fig. 126. These fatty folds are continuous below and laterally with the ligamenta mucosum et alaria respectively : the former ligament extends back from the lower end of the articular surface of the patella to the intercondylic notch of the femur.

In the early condition the ligamentum mucosum forms a complete septum below the level of the patella between the two halves of the lower (condylar) parts of the joint ; the septum is generally broken through at birth, leaving the upper part only as a band connecting patella and femur. Fig. 124 is a drawing of a case in which the septum remains complete. The ligamenta alaria are remains of the former division between the condylar and trochlear parts of the articulation.

The patella is held in position partly as a result of the tension of the Quadriceps muscles retaining it in its groove, and partly owing to the attachments of the aponeuroses that lie beside it and fasten it to the tissues in the neighbourhood of the lateral ligaments : some fibres are described as lying more transversely deep to the aponeurotic sheets and attached to the femur, known as the "retinacula patellæ." When the knee is flexed the tension of the muscles and patellar tendon keep the bone firmly against the trochlear surface of the femur, and its articular surface rests on the femur by its upper areas when the knee is fully flexed, whereas its lower areas only are in contact with that bone when the joint is extended.

When the limb is extended the upper edge of the bone is about two or three fingers' breadths below the upper limit of the synovial cavity, and its lower end is a little above the upper surface of the tibia. The synovial membrane is reflected backwards and downwards from this end to the tibia. Its lower limit therefore lies somewhat below the lower end of the patella, perhaps a finger's breadth below it.

Development.

The patella is preformed in cartilage that is apparent in the third month. It is cartilaginous at birth, and a bony centre appears in the third year and extends slowly, completing the bone about puberty.

The Lower Extremity and Pelvis

TIBIA AND FIBULA.

Long bones firmly fastened together and forming, with the interosseous membrane stretched between them, the skeleton of the leg and surfaces of attachment for its muscles. The tibia is the inner and larger bone and articulates with the femur above,

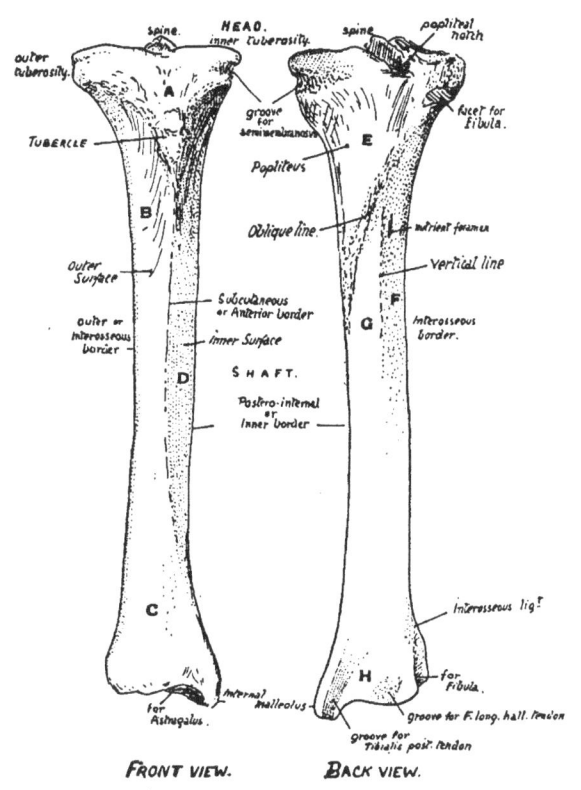

FIG. 128.—Right tibia. Anterior and posterior surfaces. *A.* area supporting fatty pad and bursa under cover of patellar tendon and capsule ; *B.* surface covered by Tibialis anticus ; *C.* lower part of outer surface looking more forward and supporting all the extensor tendons, etc. ; *D.* surface covered only by superficial fascia and skin ; *E.* covered by Popliteus ; *F.* for Tibialis posticus ; *G.* area covered by Flexor long. dig. ; *H.* area in relation with all the structures going behind inner ankle.

but both bones articulate with the astragalus to form the ankle, although the tibia alone carries the weight directly to the upper surface of the astragalus.

TIBIA. (Fig. 128).

The *upper end*, or *head*, is expanded to carry the condyles of the femur, and presents on its upper surface corresponding *inner* and *outer articular surfaces :* these are separated by an intervening non-articular area, in the middle of which is a prominent *spine*, really consisting of two tubercles placed side by side : the non-articular surface in front of the spine is wide and triangular, but is narrowed behind and falls away in a *popliteal notch*.

The two masses of bone which support the inner and outer articular surface are termed the *inner and outer tuberosities* respectively : the inner tuberosity has a *groove* on its inner and back aspect for the insertion of Semimembranosus, and the outer has the upper *fibular facet*, for the head of that bone, on its lower, outer, and back aspect.

On the front of the bone, between the two tuberosities and at a lower level, is the prominent *tubercle* for the attachment of the ligamentum patellæ : this process is included in the epiphysis of the upper end.

The *shaft* is broadened above and below, but is mainly three-sided : it has a prominent *anterior border (subcutaneous border)*, an. *inner* or *postero-internal border*, and an *outer* or *interosseous border :* the subcutaneous *inner surface* is between the front and inner margins, the slightly hollowed *outer surface* for the Tibialis anticus lies between the anterior crest and the interosseous border, and the *posterior surface* between the interosseous and inner margins. The posterior surface is wide above, where the triangular area for insertion of Popliteus is bounded below by the *oblique line* running down and in : below this the shaft is partly divided by the faint *vertical line* into an outer area for Tibialis posticus and an inner for Flexor.longus digitorum : a large *nutrient foramen*, the largest in the body, is situated a little distance below the oblique line.

Towards its lower end the shaft enlarges and the anterior sharp margin becomes less defined and turns somewhat inwards, so that the outer surface comes to look a little forwards.

The *lower end*, as a result of this thickening, presents an anterior surface covered by the tendons, etc., going to the dorsum of the foot, internal to which the ill-defined front margin is prolonged down on to the front of the (internal) *malleolus*, a projection downwards that forms the inner prominence of the ankle and rests against the side of the astragalus : it consequently has an articular surface on its outer side. The inner margin runs on to the inner and back part of the malleolus, and bounds a well-marked groove for *Tibialis posticus tendon*. The posterior surface of the lower end presents, in addition to this groove, a second broader but very badly marked groove further out for tendon of F. longus hallucis. The outer surface presents a small *lower fibular facet* or impression with a rough ligamentous area above it. The lower surface is articular for the astragalus, concave from before backwards, and slightly convex from side to side.

FIBULA. (Fig. 129).

The *upper end*, or *head*, is enlarged, presenting an *articular surface* for the tibia, looking upwards, forwards, and inwards : behind and outside this a prominent *styloid process*, for attachment of external lateral ligament and Biceps : the whole mass is continuous below with the shaft at what is termed the *neck*.

Shaft is long and slender, with a sharp *anterior margin* running up its length : immediately internal to this is the narrow *anterior surface*, for the extensor muscles, bounded by the *interosseous* or *inner border :* the flexor surface lies behind this and comprises the *inner* and the *posterior surfaces*, which are separated to a great extent by the *postero-internal border* that runs into the interosseous line in the lower part of the shaft. The posterior surface lies between the postero-internal and the *postero-external* border, and is for F. longus hallucis and, in its upper third, for Soleus, while

Tibialis posticus lies between the postero-internal and interosseous borders : between the postero-external and the anterior edges is the *external surface* that gives origin to the peroneal muscles.

The *lower end* has a triangular *subcutaneous surface* externally, ending in the (outer) *malleolus* below, a longer projection than the malleolus of the tibia : this has a triangular *articular* face internally for the astragalus, and a deep *digital fossa* behind this for the posterior part of the external lateral ligament : above the articular surface is a small facet for the tibia, and above this again is a rough ligamentous area for the lower interosseous ligament, from the upper and front part of which the interosseous border is continued up on the shaft.

The external or peroneal surface, traced down, leads to a broad groove on the back of the lower end, showing where the Peronei pass behind the ankle : Peroneus longus arises from the upper half of the outer surface, and P. brevis from the lower half.

The two bones are held together by an *interosseous membrane*, of which the fibres are mainly directed downwards and outwards and are very short below : here it is continuous with the *interosseous ligament*, consisting of several groups of fibres with fatty tissue intervening, which causes the roughened areas into which the interosseous borders are continued on the lower parts of both bones.

There are also upper and lower tibio-fibular ligaments on the upper and lower joints : these again can be divided into anterior and posterior, and in all there is the same general direction, downwards and

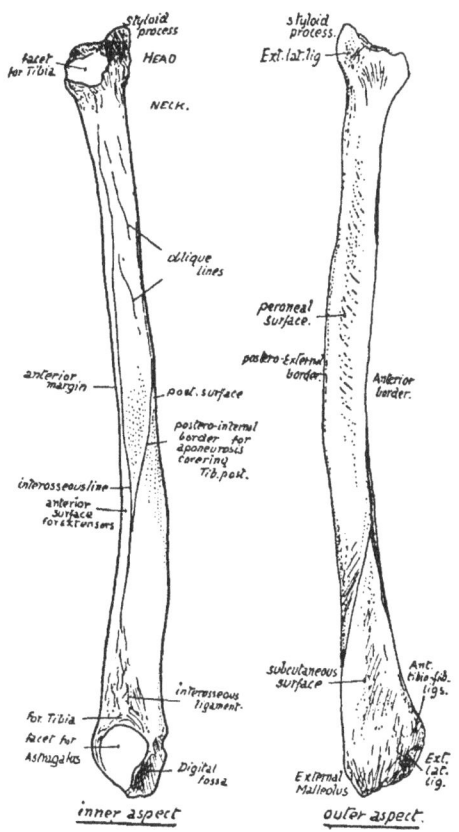

FIG. 129.—Right fibula. Inner and outer views. The bone as a whole is usually slightly concave forward, and this, with the recognition of parts at the lower end, will enable the student to place the bone on its proper side.

outwards. The lower ligaments are the strongest, and the upper anterior is stronger than the posterior. The upper joint has a weak capsule above, often perforated for a connection between its cavity and the bursa round Popliteus tendon : it has an indefinite oblique band below, directed downwards and inwards. The lower joint cavity is only a very short prolongation of synovial membrane of the ankle, and there is frequently no articular cartilage on the contiguous surfaces of the bones : the lower fibres of its posterior ligament come down below the level of the tibia (transverse lig.) and will be considered with the ankle joint (p. 174).

Fastened in this way there is still a very slight amount of movement possible between the bones, but it is so small that it may be practically disregarded and the fibula looked on as securely attached to the tibia, not carrying any weight, but completing and strengthening the articular surface of the ankle on the outer side, and affording, with the tibia and membrane, a firm ground for the origin of muscles.

The fibula is excluded from the cavity of the knee, although the outer ligaments of this joint reach it : the upper end of the tibia carries the femur and fibro-cartilages, has synovial and articular relations, and directly transmits the weight from the femur.

The knee joint is a hinge joint, in which, however, there is possible a certain small amount of rotation during flexion, and exists between femur, patella, and tibia. There is a single cavity between femur and patella, carried back as two condylar cavities under each condyle between femur and tibia. The condyles rest partly on rings of fibro-cartilage, the *semilunar cartilages*, between them and the tibia, to which the rings are fastened, and there are strong crucial ligaments connecting femur and tibia and placed between the two condylar cavities : these condylar cavities pass back on each side of the crucial ligaments, and turn up behind the condyles to reach the bone just above the articular surfaces. The whole is surrounded by ligamentous coverings.

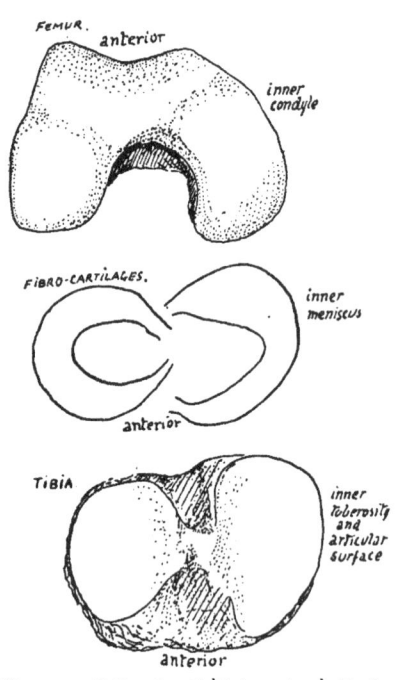

FIG. 130.—Parts of a right knee to show the correspondence between them. The inner condyle of the femur is longer, as are also the inner cartilage and tibial surface; the outer structures are shorter and rounder in shape. The inner condyle exhibits a curve, possibly related to the small amount of movement it has on the tibia rotating round the femoral attachment of the anterior crucial ligament : no definite corresponding curve is seen on the tibia, but the contiguous margins of the tibial surfaces show a comparable difference.

The upper articular aspect of the tibia naturally corresponds with the lower surface of the femur—that is, there are two surfaces on which the condyles rest, with an intervening crest for ligamentous attachments, and the inner surface is longer and rather narrower than the outer, as is the case in the femoral condyles. The semilunar fibro-cartilages which are interposed between the bones also exhibit corresponding differences, for the inner one forms a more oval figure than the outer (Fig. 130).

The outer cartilage more nearly completes a circle, so that its ends, or cornua, are closer together than those of the inner meniscus, thus placing the order of their attachment to the tibia as in Fig. 131. The two cornua of the outer cartilage are on the

The Lower Extremity and Pelvis

front and back of the outer tubercle of the spine, the anterior cornu of the inner cartilage is in front by the edge of the inner articular surface, while its posterior cornu is behind the spine also against the edge of the inner surface. The anterior crucial ligament is attached to the tibia behind the anterior cornu of the internal cartilage, but the posterior crucial ligament is on the extreme back part of the intervening non-articular area in the popliteal notch, being really placed on the posterior surface of the bone more than on its upper aspect.

When these structures are in position the arrangement of ligaments and cartilages on the top of the tibia is as seen in the figure.

The fibro-cartilages are peripheral rings, deficient centrally, wedge-shaped on section with the base of the wedge applied to the capsular structures : their projection into the cavity divides it into suprachondral and infrachondral parts widely continuous with each other round the thin edge of the wedge. The deeper fibres of the capsule fasten each cartilage (coronary ligaments) to the tibia and femur, but, as those passing to the latter are long and movable and those to the former very short, the menisci move with the tibia * on the femur, gliding forward with it in extension and thus coming to occupy by their front borders the depressions already noticed on the femoral condyles (Fig. 123).

It is evident that different parts of the menisci come under special pressure during the various ranges of movement of the joint, and that there must, in consequence, be small alterations in shape of these cartilages wedged in between the moving bony surfaces. Attention may be called in particular to the pressure on the external plate exercised by the lower part of the trochlear surface of the femur when the knee is fully extended : the front and inner margin of the cartilage, which already overlaps the edge of the tibial articular surface to some extent, is pressed further over this on to the non-articular part, and a marking may be found on the bone which indicates where this expansion takes place.

The strength of the knee joint during use depends on the breadth and practically horizontal level of the opposed femoral and tibial surfaces : when standing, for example, the femur rests directly on the tibia and is not supported by the tension of any ligament —save, perhaps, that of the Ligamentum patellæ, which is, however, tightened to keep the femur upright and not for the purpose of holding it on the tibia.

But there must be ligamentous bands whose function is to prevent slipping of the bones on one another and also to keep them in proper apposition during movement and to limit this movement. These bands are the two lateral ligaments, the patellar ligament or tendon, and the crucial ligaments. The last are mostly concerned in limiting movement and the others in holding the bones in apposition : the capsule fills up the intervals between the lateral ligaments and patellar tendon, and also covers in the posterior aspect of the condylar recesses of the joint.

The crucial ligaments are attached, as shown in Fig. 131, to the top of the tibia between the articular surfaces : the anterior one passes upwards, backwards, and outwards, to the inner aspect of the outer femoral condyle, the posterior one upwards, forwards, and inwards, to the outer aspect of the inner condyle. The area of attachment to the condyles is shown in Fig. 122.

The sides of the ligaments are covered by the synovial reflections off the condyles between which they lie, and a diverticulum of the trochlear cavity is continued back between them as a sort of bursa (Fig. 122).

The anterior crucial ligament is evidently made tight by extension of the knee

---

* This is, of course, to be expected, since the cornua are attached to the tibia.

(Fig. 127), so that it prevents over-extension,* also when it is tight there is only possible a very slight rotation outwards of the tibia on the attachment of the ligament, helped by the shape of the articular surface of the femur, so that the presence of this band is probably primarily responsible for the terminal "locking" rotation of complete extension.

The posterior crucial is made tense by movement backwards of the tibia, but it is doubtful if it limits flexion : probably it holds the bones together, while the smaller curve of the back part of the condyles enables flexion to take place without increasing its tension.

The back of the condylar cavities is covered in, between the crucial region and the lateral ligaments, by vertical capsular fibres which afford extensive origin to Gastrocnemius and Plantaris (Fig. 120). The same figure shows how the upward expansion (ligam. posticum) from Semimembranosus covers in the crucial ligaments behind, passing to the inter-condylar ridge and to the outer condylar capsule. All these fibres when tense will limit extension.

FIG. 131.—Right knee and upper end of tibia. Top left figure is a diagram of the synovial sac from the front, showing the trochlear part of the sac continued below each condyle into the condylar recesses, which are partly subdivided by the projection into them of the wedge-like semilunar cartilages. The synovial sac does not come down lower than the cleft between femur and tibia, whereas the " capsule," shown by the black lines, is brought down on the tibia in front to take in the tubercle and to be attached along the ligamentum patellæ. Thus the area A is included on the front of the bone between the sac and the capsule, behind the patellar tendon, the interval being filled by the infrapatellar pad and bursa (see Fig. 127). The front and side parts of the capsule are made by the aponeuroses of Vasti and Crureus, extending on each side to the lateral ligaments, and the deep fascia is blended with their superficial surfaces. Thus on the tibia, in the figure next below, the capsular attachment is represented by the Vasti, the lig. patellæ, and the ilio-tibial band, while the infrapatellar bursa lies in position deep to these. On the right, the back of the joint is seen ; the tendon of Semimembranosus has been cut and turned back and its expansion (see Fig. 120) removed to show the crucial ligaments which occupy the notch between the condylar recesses and capsules ; the position of the tendon and expansion are indicated by interrupted lines. The opening for Popliteus exposes the outer meniscus, and the bursa round the tendon is carried on a cartilage-covered area on the tibia ; the muscle itself is not shown, but the expansion over it from Semimembranosus is seen. Below this the areas occupied by these several structures are shown on the tibia ; notice that, internal to insertion of Popliteus, fibres of the expansion from Semimembranosus reach the bone and roughen it, the upper ones passing under the ligament and the lower ones joining it. The capsule round the fibular facet is weak above, where the bursa may perforate it. The epiphysial line is shown as a thin green line in the middle figures. The lowest figures show the upper tibial surface. On the right the crucial ligs. and semilunar cartilages are *in situ*, cut away from the femur. Each cartilage lies on the tibia, the infrachondral part of the cavity being between them and the bone ; their thick margins are attached by coronary fibres to the bone, as shown in the middle figures, such fibres being, of course, blended with those of the condylar capsule behind. The free edges of the menisci rest on the tibia, their position being indicated by the interrupted lines in the left-hand figure, in which their cornual attachments and those of the crucial ligs. are also shown. *ISC, ESC*, inner and outer semilunar cartilages. At X a part of the suprachondral cavity reaches the bone between the post. crucial and internal posterior cornu.

The Ligamentum patellæ is really the tendon of the Quadriceps, and extends from the patella to the tibial tubercle : it therefore comes down to a much lower level than the synovial membrane which passes directly from the patella to the top of the tibia. Thus between this level and the tubercle the tendon lies in front of the triangular area here on the tibia, and it is separated from this by the infrapatellar pad of fat and bursa.

The tendon has the aponeurotic expansions of the Vasti inserted into it and surrounding it, so that it and the patella are held more firmly by this means in their central position, and possibly on account of this the tendon may be considered to have some value as a means of securing the bones during movement of the joint : otherwise it is difficult to imagine that it could exercise any restraining influence on lateral movements or indeed on antero-posterior movements of an abnormal sort.

Speaking generally, the articular surface on the femur (for the tibia) shows a decreasing curve as it is followed forward : that is, there is a segment of a small circle

---

* In cases of genu recurvatum the ligament is stretched or wanting.

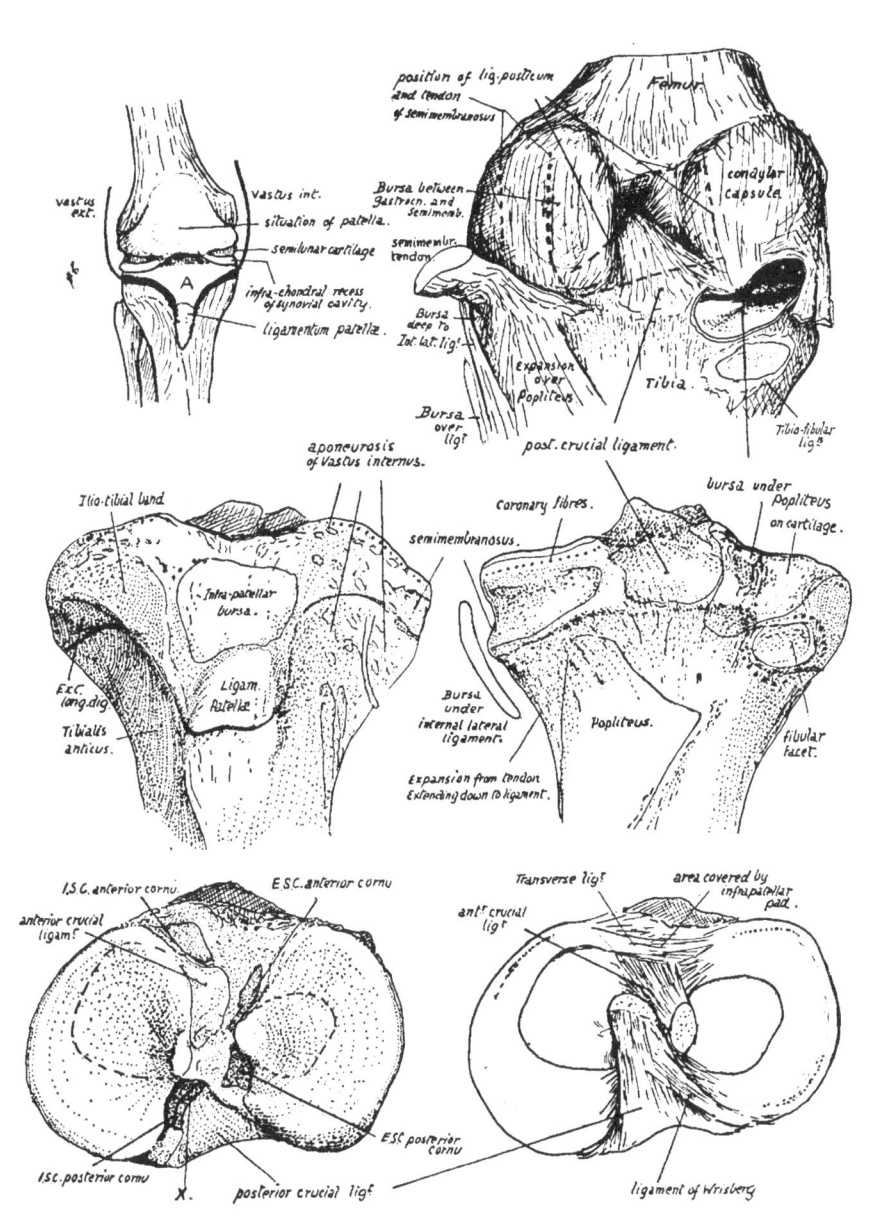

FIG. 131.

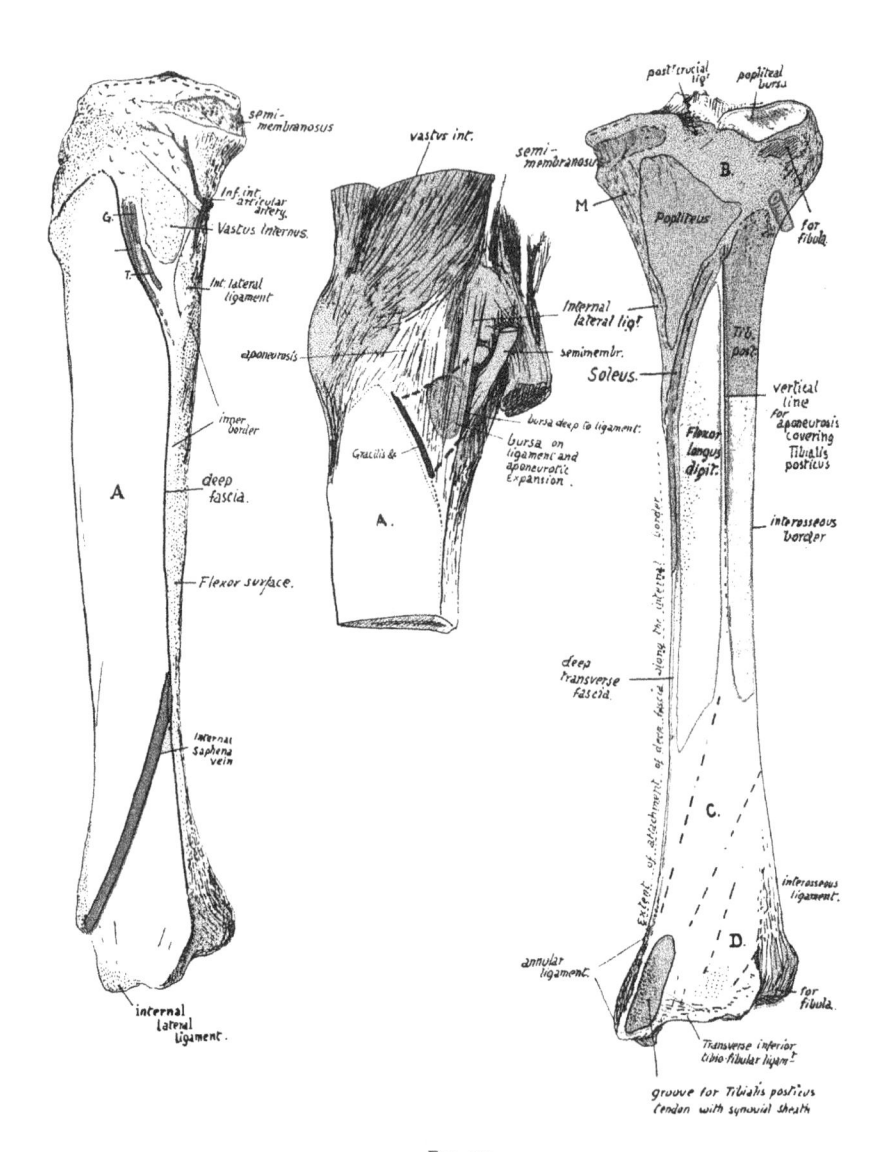

FIG. 134.

The Lower Extremity and Pelvis 161

on the back of each condyle, the segment of a larger circle becomes apparent below, and a still larger one as the surface reaches the trochlear cartilage.

The internal lateral ligament is a long and strong band from the inner condyle (Fig. 122) to the tibia not very far above the level of the inner end of the oblique line (Fig. 134). Its upper attachment is about the centre of one part of the curve in which the tibia moves in flexion, as shown in Fig. 122, and over this portion the tibia is efficiently held by the ligament : behind this the tension of the posterior crucial ligament holds the bones in apposition on the smaller curve. Observe that the internal lateral ligament comes down far below the proper level of (coronary) capsular attachment (this is indicated in Fig. 131), and below this the ligament covers the insertion of Semimembranosus, with a bursa over it, and is joined lower down by some fibres of the lower expansion from this tendon. Here it also crosses the lower articular vessels. Where it lies on the coronary fibres the ligament is attached to them, and through them has some connection with the margin of the internal semilunar cartilage.

FIG. 134.—Right tibia, inner and posterior views. *A*, in the left-hand figure, is the subcutaneous inner surface of the shaft, crossed by the long saphenous vein. This surface is thus smooth, and slightly convex from side to side in conformity with the shape of the leg ; rather concave from above down in its lower half, owing to the increasing prominence of the lower end. In its upper part, however, the inner aspect of the bone shows secondary markings for Vastus internus, Gracilis (G), Sartorius (S), and Semitendinosus (*S-T*). These are all deep to the deep fascia, which must therefore be carried in by them across the surface. Vastus internus is deep to the other three tendons, wherefore these, coming from behind and above, are inserted below and in front of the Vastus ; see the middle figure, where they are represented as a single insertion by a thick black line, and their situation over the Vastus and internal ligament is shown by interrupted lines. Vastus internus has a strong band that makes a definite rounded marking on the bone, the rest of its insertion being by more scattered fibres reaching almost to the tubercle (see Fig. 131), among which the internal inferior artery turns upwards. Internal lateral ligament is on the junction of inner and back surfaces, not very far above the inner end of the oblique line. *B*, area covered by Popliteus but not affording insertion to it ; at the extreme inner end of the triangular " popliteal surface " the bone is roughened by fibres, *N*, of Semimembranosus (expansion) reaching the bone as far down as the ligament. Below the oblique line the Tibialis posticus and Flexor longus digitorum mould the bone into surfaces in different planes, separated by the vertical line ; both arise by muscle fibre and hence show no secondary markings in their areas as a rule, but tendinous fibres in the upper part of Tib. post. may cause roughnesses near the upper part of the oblique line (see Fig. 133A). These deep flexors are covered in by the deep transverse fascia (fine green line). Tibialis posticus is directed downwards and inwards from its origin, therefore is in relation with the area *C*, while Flex. long. hallucis comes downwards and inwards from the fibula and lies on the area *D* ; between these the main vessels and nerve and tendon of F. long. dig. come into relation with the bone.

When the tibia passes on to the large front curve, the internal ligament gets taut at once, and so limits extension, like the anterior crucial ligament.

The external lateral ligament passes between the outer condyle and the head of the fibula. The tendon of Popliteus lies deep to it, between it and the tibia (Fig. 131), and makes its exit from the capsule behind it. Some of its posterior fibres make an arcuate band over the tendon, to join the condylar capsule. Its action, so far as the knee joint is concerned, is similar to that of the internal lateral.

The way in which these various structures are connected and covered in by the different tissues concerned in forming the capsule is illustrated and described in Figs. 120 and 131.

It is now possible to place these structures in position on the Tibia : they have already been considered as they lie on the upper surface of the bone.

In Fig. 131 the plan of the attachment of the capsule is shown, turning down in front to reach the insertion of the patellar tendon. This encloses the triangular surface A, on which is placed the infrapatellar pad and bursa. Observe that the capsular line behind lies near the edge of the bone and turns forward by the popliteal notch : here the proper capsular fibres are applied to the sides of the posterior crucial ligament.

F.A.

Now observe the level of the epiphysial line (Fig. 131) : it lies some distance below the proper capsule, so that the epiphysis includes the fibular facet, the insertion of Semimembranosus, and the extreme upper part of the extensor musculature, as well as all the articular attachments with the exception of the lateral ligaments and a portion of the expansion of Vastus internus, as described later

The upper epiphysis is that of the growing end. Observe that the epiphysial line cuts through the adult tubercle : this indicates that the lower part of the tubercle is really a portion of the shaft. A secondary ossification may form a traction epiphysis here, but it is more usual to find an additional centre for the upper part, joined to the upper epiphysis.

The upper end is cartilaginous at birth, but occasionally the centre, that should appear a few weeks later, is present at birth. The upper end may be somewhat tilted back, a persistence of the state normal in the infant, and a condition found regularly in some of the lower races.

Other markings, etc., to be found on the upper part of the tibia are shown in Fig. 131. A small smooth area just above the fibular facet is covered by cartilage and supports the bursa round the tendon of Popliteus : this bursa is mainly derived from the infrachondral part of the articulation and may communicate with the upper tibio-fibular joint. The Popliteus is inserted above the oblique line, over the area shown : it may have some tendinous fibres in its structures, and these may cause slight roughnesses on the bone above the oblique line. There is a clear margin of bone above the muscle, which is covered by Semimembranosus internally, but supports the popliteal artery centrally. The insertion of Semimembranosus extends well forward under the internal lateral ligament, with a bursa between them. The ligament makes a marking on the bone, on each side of its edge, a little distance above the oblique line, and inner fibres of the descending expansion from the Semimembranosus tendon reach it here, while the inferior articular vessels and nerve, which pass under the expansion, are thus brought to a deep relation with the ligament : the remainder of the expansion covers the inner part of Popliteus, extending as far out as the vessels lie, and reaches the oblique line.

The extensor area on the bone reaches up to the margin of the epiphysis. Trace the line of the covering aponeurosis up from the subcutaneous border (Fig. 132) along the outer side of the tubercle. It turns back towards the fibular facet and divides into an upper and lower line a little distance in front of this. The appearance presented varies in individual bones, but comparison between a number will enable one to establish the arrangement shown schematically in Fig. 133. The upper and lower lines mark out a triangular area which is subdivided by a vertical line, and the outer subdivision is for origin of Peroneus longus, the inner for Extensor longus digitorum. Peroneus longus extends from the head of the fibula across the anterior ligaments to the tibia : the vertical line is for the anterior peroneal septum, and the inner part of the lower line is for the septum between the long extensor and Tibialis anticus.

All these muscles are covered by an aponeurosis into which the Biceps sends an expansion, so that the upper line may be taken to represent part of the insertion of the Biceps.

Above and in front of this again, forming a rough marking on the bone between the aponeurotic and capsular lines, is the attachment of the strong ilio-tibial band.

Tibialis anticus arises from the upper half or more of the outer surface : its tendon runs downwards and slightly inwards, so that a large space is left on the widened

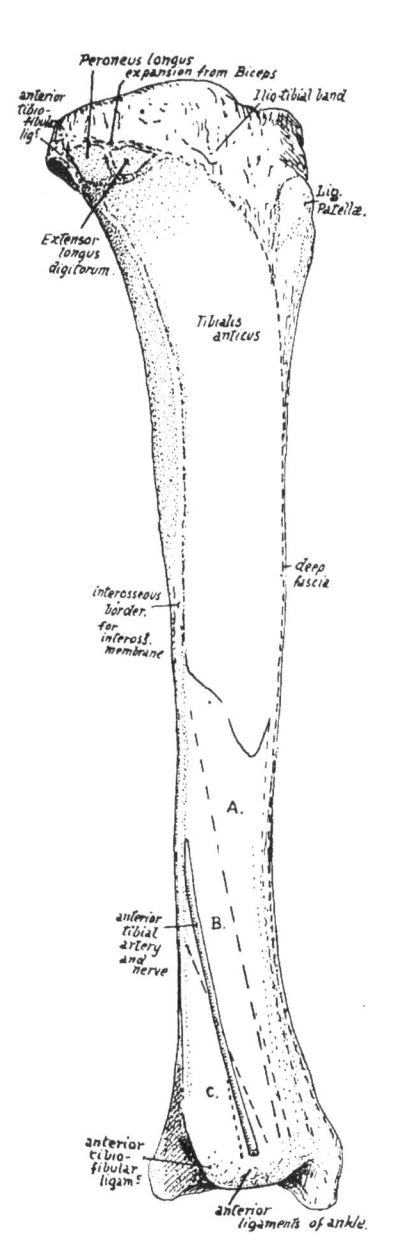

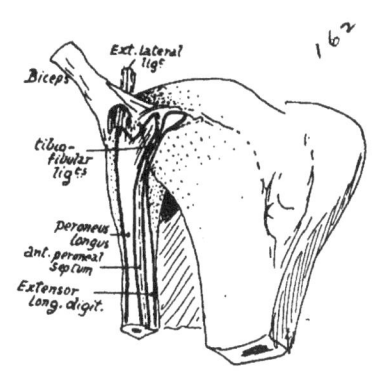

FIG. 133.—Diagram to show how the Peroneus longus and Extensor longus digitorum pass from the fibula to the tibia, and occupy the areas shown in Fig. 132. The tendon of the Biceps is seen dividing into two strong bands over the external ligament, and expansions from these bands are prolonged over the two muscles.

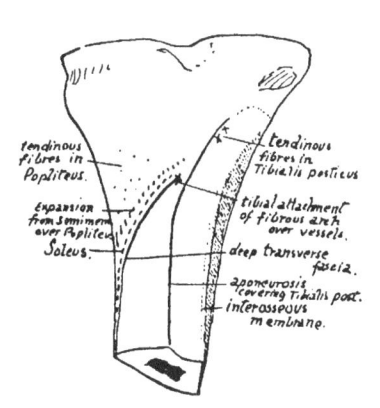

FIG. 133A.—Scheme of analysis of the oblique line on the tibia.

FIG. 132.—The outer surface of the right tibia. The tendon of Tib. anticus covers the area *A*. *B* is covered by Extensor longus hallucis. *C* is in relation with Extensor longus digitorum.

The Lower Extremity and Pelvis

bone below. Here the other extensor structures come into relation with the bone (Fig. 132). In fact, all the structures that pass down over the front of the ankle joint to reach the foot lie in front of the tibia, with the exception of the anterior peroneal artery, which usually descends in front of the fibula.

The deep fascia is attached to the front margin, thickening below to form the upper part of annular ligament (see later, p. 178). From this it passes outwards round the leg, covering it behind and passing to the postero-internal border. Thus the *inner* surface of the shaft is only covered by superficial fascia and skin, and in this the saphena vein passes upwards and backwards on the lower fourth of the bone (Fig. 134) before it comes to lie on the deep fascia. It must be noticed, however, that the deep fascia is carried on to this surface by the tendons of Sartorius, etc., in the upper part: it covers these in and is blended with them. As Sartorius gives an expansion down into the deep fascia below, it is apparent that the line of its insertion must pass backwards and downwards to reach the line of fascial attachment on the inner border.

These tendons are inserted in the order they would take when the leg is bent (Fig. 135). They pass over the internal lateral ligament, with an intervening bursa. Their markings of insertion can usually be felt if not seen on the bone: the small extension of ossification into them to form a secondary marking may be associated with their direction, for they lie on the periosteum and practically in the plane of its surface, and thus do not tend to pull it from the bone, as would be the case if they acted at an angle.

Behind the tendons, on the inner side of the upper end of the shaft, there is to be found a well-marked roughened area for attachment of a strong aponeurotic sheet of fibres derived from the expansion of Vastus internus. The area is shown in Fig. 134.

This set of fibres runs to the bone parallel with the front border of the internal lateral ligament, as seen in the central figure, and is directly continuous with more scattered fibres running to the bone above it, and above and in front of it: it is crossed by the three tendons, and the bursa between them and the ligament extends also on to the expansion. The marking for the ligament is behind and below that for the aponeurotic fibres: the inferior internal articular artery turns up above it.

FIG. 135.—When the leg is bent, in the usual position, the Sartorius, *S.*, from the front, the Gracilis, *G.*, from the inner side, and the Semitendinosus, *T.*, from the back of the thigh, reach the tibia in their proper order. When the leg is extended in the upright position, their insertions are brought forward, and the more posterior position becomes the lower, *i.e.*, Semitendinosus is inserted below Gracilis, while Sartorius is above as well as superficial owing to its fascial expansion.

The upper and more scattered fibres are attached to the epiphysis like the outer expansion, but as the whole internal structure is composed of expansion of the internal Vastus it must be considered as a functional capsule and such a term not applied only to the deeper fibres. The deep fascia covers it and is blended with it, so that the line of the fascia is as shown in the figure.

The interosseous border is very variable in its position in its upper part, depending on the size of the Tibialis posticus: this muscle varies in size independent of the general muscularity of the individual. The muscle arises from the back of the interosseous membrane, and is covered on its posterior surface by an aponeurosis, which is continued as a fascial covering over the upper end of the muscle to join the membrane. The covering aponeurosis is attached internally to the tibia and makes the vertical line on it. Therefore the area of origin of the muscle lies between the interosseous and

Anatomy of Skeleton

vertical lines (Fig. 134) and extends up to the outer part of the oblique line. Flexor longus digitorum arises from the bone internal to the vertical line and below the oblique line, which gives origin to the tibial head of Soleus : Soleus origin extends down the inner margin also for a variable distance.

We can now analyse the oblique line and see that (Fig. 133A) its value differs in its inner and outer parts : an attempt should be made to recognise the various points on the bones. The Tibialis posticus passes obliquely across the back of the bone from its origin to the groove on the back of the malleolus, and is crossed by the Flexor longus digitorum. Therefore, on the outer side of the groove, between it and the groove for the long flexor of the great toe, the back of the lower end of the bone is in relation with the long flexor of the other toes and the posterior tibial vessels and nerve.

The deep transverse fascia of the leg separates the deep flexors and main vessels lying on them from the Soleus. It is attached to the inner margin of the bone, then between the Soleus and F. longus dig. to reach the tubercle from which its attachment passes along the fibrous arch over the main vessels to reach the head of the

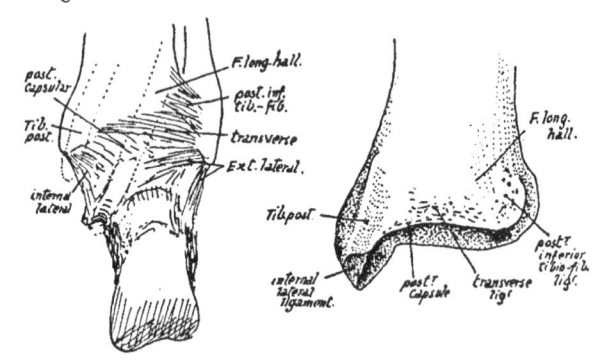

FIG. 136.—Posterior view of right ankle, and of right tibia.

fibula. It is thickened below, where it is fastened to the malleolus, to form the internal annular ligament ; just below the origin of Soleus it is joined at the inner margin by the deep fascia covering over that muscle, but lower down the deep fascia runs into it further from the margin as the Soleus narrows to its insertion, so that the annular ligament is really a compound of the two fasciæ behind the inner ankle.

The lower epiphysis includes the fibular facet : it has nothing attached to it but ligaments.

Observe that the posterior margin of the lower articular surface comes down lower than the anterior margin, but the anterior margin is wider.

The articular surface is concave sagittally and slightly convex from side to side, corresponding with the surface of astragalus : among the ligamentous markings round its margins may be found one for the transverse posterior ligament (Fig. 136). The surface is prolonged on to the malleolus.

Observe that the malleolus has its long axis directed downwards and forwards the attachments of ligaments to it are shown in the figures.

The inner surface and subcutaneous border of the bone can be directly palpated during life, practically throughout its extent. The whole upper end can be taken between thumb and fingers and examined, but its central front part, above the tubercle, is hidden by the patellar tendon and infrapatellar pad.

164

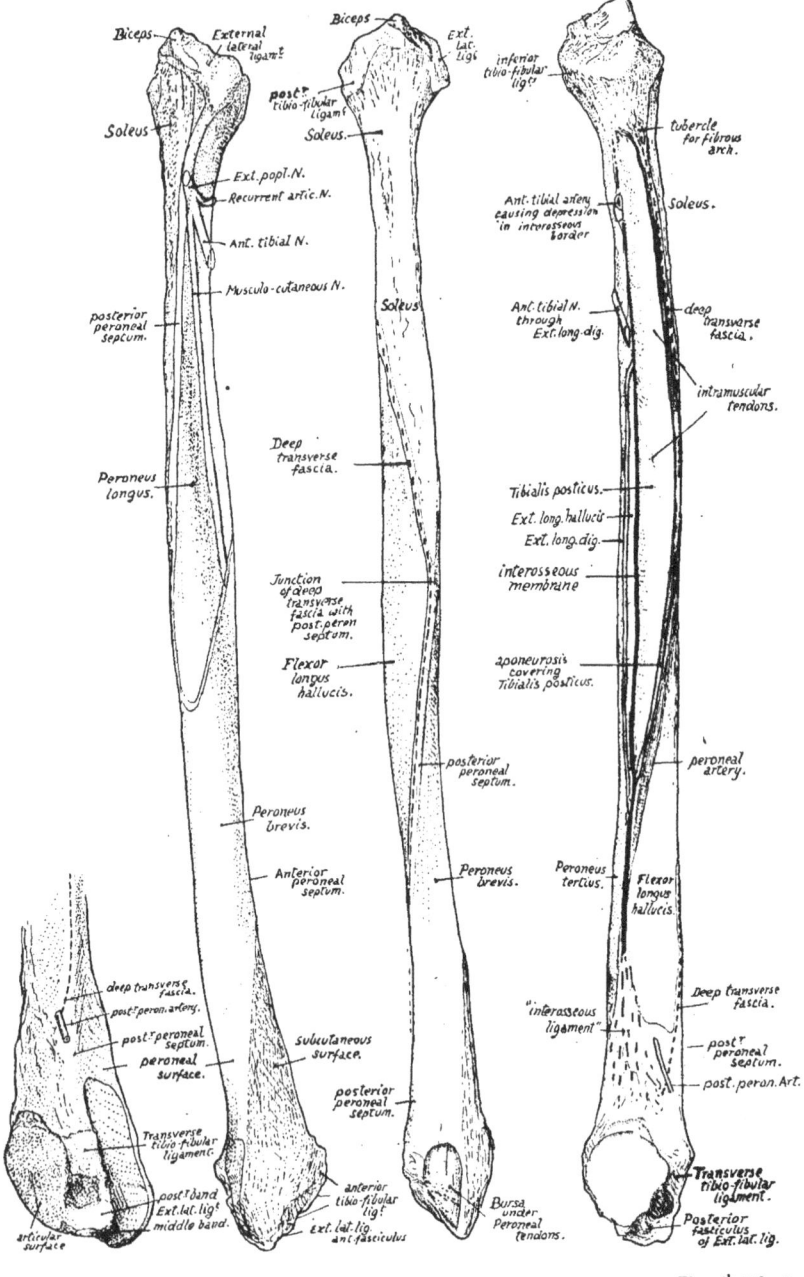

FIG. 137.—Right fibula. On the left, the lower end is seen from behind and internally. The other figures, in order, show the bone from the outside, from behind, and from the inner side. Explanations in the text.

The Lower Extremity and Pelvis 165

Fibula.

Observe that the styloid process is placed behind and outside the articular surface on the head, and that the rough markings round the articular facet show the presence of areas of ligamentous attachment; these markings are best developed in front of the styloid process, extending here on to the outer side of the head and showing the insertion of the external lateral ligament, while round the other borders of the facet they are made by the capsular ligamentous fibres of the tibio-fibular joint.

The extent of attachment of the external lateral ligament and of the Biceps tendon that covers it is seen in Fig. 137. Peroneus longus, arising from the outer side of the upper part of the shaft, extends on to the lateral ligament as well as across the anterior tibio-fibular bands to reach the tibia (Fig. 133), and the Biceps tendon gives an aponeurotic expansion over this muscle as well as over the extensors.

Soleus arises from the back surface of the upper part of the shaft and from the back of the head as far up as the base of the process: on its inner side here there is a tubercle which marks the attachment of the fibrous arch along which the origin of the muscle passes from the tibia.

On the inner side the ligamentous markings come lower down than on the other margins of the facet, and below these, on the inner side of the neck, the bone is in relation with the anterior tibial artery passing outwards, forwards and downwards above Tibialis posticus.

The Extensor longus digitorum arises from the neck in front of this arterial region, its origin running up to the ligamentous area where it passes, with the Peroneus longus, across to the tibia.

The shaft of the bone has the appearance of being twisted in its length through a quarter-turn in an outward direction: this is due to the disposition of the muscles on it, for the bone is moulded in its shape by the muscles applied to it, and its margins mark the attachment of strong fibrous septa between these muscles, so that, to understand the arrangement of surfaces and lines on it, it is necessary to consider the structures that produce them.

First find the interosseous border. The inner aspect of the bone can be recognised at once by the presence of articular facets on it, and there is a triangular rough area, above the lower facet for the astragalus, that is made by the fibres of the interosseous ligament. This ligament is really an ill-defined thickening of the lower end of the interosseous membrane, so that the line of the membrane can be immediately found as a ridge leading up from the rough area (Fig. 137). Trace this up, and it is seen to divide into two: the anterior line is the proper continuation of the interosseous border, the posterior and more salient ridge being for the aponeurosis covering Tibialis posticus.

The interosseous line can be followed up to the neck of the bone, and it divides the inner aspect of the shaft into extensor and flexor regions: the extensors of the foot lie immediately in front of it, and therefore occupy the area, variable in breadth but never very wide, which lies between the interosseous line and the front border, while the flexors of the foot lie behind it. Of these flexors the deepest is the Tibialis posticus, which arises from the membrane and from the bones on each side of the membrane, so that its area of origin on the fibula lies immediately behind the interosseous line: it is bounded behind by the prominent postero-internal border and extends up to the neck of the bone, including, within its limits, the oblique line or lines made by the intramuscular tendons in the substance of the muscle (Fig. 137). The aponeurosis covering Tibialis posticus affords origin to the next layer of muscles, the long flexors of the toes. Flexor longus hallucis is very much stronger than Flexor longus dig., so that the outer part of the aponeurosis is much thicker and stronger than the inner part, and this corresponds with the differences in the appearances of the lines of attachment of the aponeurosis to the bones : the outer line forms the prominent postero-internal border of the fibula, whereas the inner attachment makes the badly-marked vertical line on the tibia. These two lines separate the area of Tibialis posticus on each bone from the area of the corresponding long flexor, so the area for Flexor longus hallucis on the fibula is immediately behind the postero-internal border, on the posterior surface of the bone.

Tibialis posticus passes downwards and inwards from its origin (Fig. 134) across the tibia, and thus leaves the region of the interosseous membrane. In this way the aponeurosis that covers it comes into relation with the membrane as the muscle inclines away from it, and so we get the explanation of the postero-internal ridge running below into the interosseous border : it marks the lower end of origin of Tibialis posticus, and comparison of a few bones will make it evident that the extent of this origin and the consequent position of the junction of the two ridges is very variable. Moreover, the size of the muscle does not vary with the general muscularity of the individual.

Now examine the flexor surfaces of the fibula, and it is clearly apparent that the Tibialis posticus takes the whole of the inner aspect of the surface in the middle of the shaft and Flexor longus hallucis is altogether on the posterior aspect. But as we follow the surfaces down we find the Tibialis area narrowing until it disappears, and, as it narrows, the area for the long flexor, keeping close to it, gets more and more on the inner side, so that in the lower part it is altogether on the inner aspect—in other words, the long flexor, arising partly from the aponeurotic covering of the Tibialis posticus, is carried by this to the interosseous membrane when the latter muscle leaves it uncovered, and is in this way brought to the inner side of the shaft. If the Tibialis is small the long flexor comes to the inner side comparatively high up and acquires an origin from the membrane, but under opposite circumstances it may not do so.

Here we have the explanation of the twisted appearance of the bone, and at the same time we can see the " curve " of the twist, and therefore the appearance of any individual bone, depends really on the size of the Tibialis posticus. As soon as this muscle leaves the fibula the Flexor longus hallucis takes its place on the inner side of the bone, and thus affords opportunity for the peroneal tendons to come on the back of the bone so that they may pass behind the malleolus to obtain their proper action on the foot : because the bone is moulded by the muscles we therefore find that its posterior surface, traced down, becomes internal, and its outer surface can be followed down to the groove behind the malleolus.

The extensors of the foot arise from the bone immediately internal to the Peronei, that is, from the anterior surface between the anterior edge and the interosseous line. Follow the anterior surface down : when the Peronei move round towards the back of the bone, away from its outer aspect, the extensors do not follow them but remain on its front surface, and thus the lower part of the outer aspect is left uncovered and becomes the triangular " subcutaneous area."

If the foregoing account of the way in which the surfaces are disposed on the fibula is thoroughly understood there can be no difficulty in following the minor details on the bone. It is better to deal first with the muscles (Fig. 137).

Tibialis Posticus.—The area is well defined by the interosseous line and postero-internal border except at the top, where the two structures which make the lines are continuous as thin fasciæ that usually leave no marking on the bone : the area is not limited in any way by the oblique lines, which are included in it, being the attachments of intramuscular tendons.

The Lower Extremity and Pelvis 167

Flexor Longus Hallucis.—In a general way, remembering that the area of origin is modified by the extent of Tibialis posticus, this muscle may be said to arise from the posterior surface of the shaft in the middle third and from its internal aspect in the lower third. The upper limit is recognised by the existence of roughnesses, caused by its uppermost aponeurotic fibres ; the rest of its origin is by muscle fibre, as inspection of the bone demonstrates, and its lower limit is a little above the level of the lower tibio-fibular articulation.

Observe that its upper part is prolonged up for a little distance between the Soleus and Tibialis posticus, and indications of this can frequently be found on the bone, although it is often obscured by the markings of tendinous fibres in the muscles.

Soleus area extends down from the head on the back of the bone for its upper third : its lower end is outside the upper part of the origin of the last muscle and is indicated by rough markings, which may also be found in the higher part of the area. There is a slight twist in the upper part of the bone, that causes the surface for this muscle to look very slightly outwards as well as backwards and makes the muscle prominent externally in its upper part.

Peronei.—The muscle areas are as shown in Fig. 137 : the upper one reaches the tibio-fibular ligaments above and passes along them to the tibia. There are not usually any indications on the bone of the distinct areas, although there are frequently slight roughnesses for short tendinous fibres in the muscles. Notice that the divisions of the external popliteal nerve lie in the substance of the upper muscle, practically on the bone, and the musculo-cutaneous nerve comes forward between the two muscles, being separated from the bone by Peroneus brevis.

Extensor longus digitorum and *Peroneus tertius* arise from the narrow anterior surface, from the ligamentous markings at the upper end to a point easily seen about an inch above the anterior prominence of the lower end. The two muscles have a continuous origin, the division made between them being purely an artificial one. They occupy the whole of the anterior surface, except in the middle two-fourths, where *Extensor longus hallucis* arises from the surface between them and the interosseous line. Above the origin of this muscle the attachment of the long extensor of the digits is interrupted by the passage of the anterior tibial nerve.

The Extensors and Peronei are divisions of one mass of muscle tissue, the dorsal extensors of the foot. At an early stage they are included in the same cellular layer (representing enclosing fasciæ), but are separated later by the growth of the outer malleolus. Thus the enclosing layer is stranded on the surface of the subcutaneous area and covers it with a thick, felted fibrous membrane.

The anterior and posterior *peroneal septa* are attached respectively to the anterior and posterior margins of the shaft, following the peroneal muscles closely.

The *deep transverse fascia* of the leg separates the Soleus from the deep flexors and main vessels and nerves. Its attachment has already been followed on the tibia (p. 164) up to the fibrous arch over the vessels : from this it passes to the inner side of the head of the fibula and runs down (Fig. 137), separating the Soleus at first from Tibialis posticus, but then passing obliquely across the back of the bone to cover the Flexor longus hallucis, and thus separating Soleus from this muscle. This oblique line may occasionally be made out on the bone, but it is obscured as a rule by the fusion of the fascia with the aponeurotic fibres of the muscles between which it passes.

At the outer side of the long flexor the fascia blends with the posterior peroneal septum, and the two are attached together to the postero-external edge as far as the lower end, where they separate again to follow their respective muscle groups (Fig. 137).

The peroneal artery comes into relation with the fibula as it runs down on the aponeurotic covering of Tibialis posticus. There is frequently an indication of the course of this artery in the form of a slight groove in the bone (Fig. 137).

The artery follows the aponeurosis, and thus ultimately comes on to the interosseous membrane, where it divides into its anterior and posterior divisions : therefore the height at which the anterior artery appears on the front of the leg depends largely on the size of the Tibialis posticus. The posterior branch runs down deep to the long flexor to emerge from under its outer border to pass behind the malleolus : it must therefore pierce the attachment of the deep transverse fascia here, and when the line of this fascia is definite there is usually a depression in it that marks the passage of the vessels (Fig. 137, 1).

The lower end has tuberculated prominences in front and behind, at the level of the top border of its articular surface, for the anterior and posterior tibio-fibular ligaments (Fig. 136). Below the anterior mass is another marking, on the malleolus, for the anterior band of the external lateral ligament ; the other bands are attached to the margin of the digital fossa, the concavity of which is filled by a fibro-fatty pad and a synovial prolongation from the joint (see Fig. 137).

The groove on the back of the malleolus is lined by a synovial sheath for the peroneal tendons.

Notice that the articular surface has its long axis almost vertically placed, thus differing from the corresponding surface on the tibial malleolus : it is also longer than the inner malleolus, a development probably associated with the upright position.

The epiphysial line of the lower end lies above the articular surface.

This bone has the "growing end" at its upper extremity, and it is an exception to the rule that the centres for the growing end appear first, for the centre for the lower end comes as a rule some little time before that for the upper end : the latter, however, unites last—otherwise it would not be the "growing end."

Now consider the bone as a whole. The shaft is very exceptionally straight : it is nearly always bowed in a curve with the concavity forwards. It does not carry weight, so does not require the great strength of the tibia. An examination of the bone shows that there is a primary cylindrical bar running through it, covered in and largely hidden by the moulding of the muscles, so that the various surfaces and ridges may be mainly considered as secondary in nature. The primary bar is most evident in the lower half of the peroneal surface, running down to the malleolus : in the upper half of the bone it lies near its back part, and the secondary production of the bone in front of it leads to the concavities found here on the inner and outer surfaces.

The shaft of the fibula, in its middle portion, is just behind the plane-level of the posterior surface of the tibia. The head can be felt distinctly, as can also the prominent lower malleolus, but the upper two-thirds or so of the shaft is only indirectly palpable through the mass of the peroneal muscles. The end of the external popliteal nerve can be felt, in the substance of the Peroneus longus, against the neck of the bone. The line of the tibio-fibular articulation can sometimes be distinctly felt just above the ankle, immediately internal to the lower end of the fibula and external to the common mass of extensor and Peroneus tertius tendons.

Ossification of Tibia and Fibula.

Each bone is developed in cartilage and ossified from three centres, one for the shaft and one for either end : the upper epiphysis is the "growing" one in both bones.

The centres in the tibia precede those in the fibula.
Tibia :—Shaft, seventh week ;
Upper end, just after birth ;
Lower end, second year.
Fibula :—Shaft, eighth week ;
Upper end, third year ;
Lower end, second year.

The lower ends fuse with the shaft a few years after puberty in the tibia and about twenty-one in the fibula. The proximal end is fused before twenty-four : in the tibia occasionally as early as nineteen or twenty.

The upper epiphysis in the tibia includes (Fig. 131) the upper half of the tubercle : in this there is occasionally an additional centre, appearing about twelve or thirteen and joining the epiphysis as a rule.

THE FOOT.

The skeleton of the foot, like that of the hand, consists of a closely articulated number of irregular bones, the *tarsus*, carrying five long bones, the *metatarsus*, which in their turn support the *phalanges* of the free digits (Fig. 138).

THE TARSUS.—This comprises seven bones, of which one, the *astragalus* or *talus*, articulates with the bones of the leg and rests below on the *os calcis* or *calcaneum*, which makes the projection of the heel. The astragalus ends in front in a rounded *head*, which, being directed somewhat inwards as well as forwards, tends to project over the inner side of the front end of the os calcis, and the latter bone has a projection here to support it, hence termed the *sustentaculum tali*.

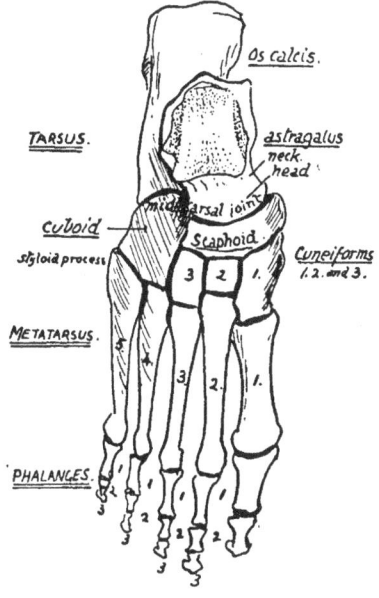

FIG. 138.—Semidiagrammatic dorsal view of right foot bones.

These two bones are the largest in the tarsus and form its hinder part : their anterior extremities, nearly on the same level, are joined at the *mid-tarsal joint* with the front part of the tarsus.

The head of the astragalus articulates with the *scaphoid* or *navicular*, and this has the three *cuneiform* bones against its anterior surface. The cuneiforms are *inner*, *middle* and *outer*, and support the inner three metatarsals.

The anterior end of the calcaneum has the *cuboid* articulating with it, and this carries on its front aspect the two outer metatarsals.

We see, then, that the astragalus is continuous with the chain of bones on the inner side of the foot, scaphoid, cuneiforms and metatarsals, while the os calcis, cuboid, and outer two metatarsals form a chain of bones that lies outside and rather below the others.

Anatomy of Skeleton

THE METATARSUS.—This can be compared with the metacarpus in the hand, when its constituent bones will be seen to be weaker and less divergent. Each bone presents a *base*, *shaft*, and *head*. Notice the obliquity of the line of the tarso-metatarsal junction, with the prominent *styloid process* that projects externally on the base of the fifth metatarsal. The first metatarsal, carrying the big toe, is very thick and strong for supporting the greater part of the weight of the body in stepping off from the foot, whereas the outer portion of the metatarsus is mainly used as a support in balancing the body, and not as a weight carrier, wherefore its comparative weakness.

THE PHALANGES.—As in the hand, these are two in number in the big toe and three in the others, numbered first, second, and third in order from behind forwards in each toe. Notice the thin shafts and thick extremities, save in the great toe, and also observe that there is frequently modification in the direction of fusion or reduction in the last phalanges of the outer toes.

When the names and general relations to each other of the bones of the foot are familiar, the complete skeleton of the region should be studied with a view to understanding the relations and significance of the various bony points, etc., in it, and after that with the object of grasping the means by which the arches of the foot are held in position to support the weight thrown on them.

Looking first at the dorsal surface (Fig. 138), we perceive that the general contour of the tarsus is markedly convex from side to side and also to a slight degree from behind forwards: there are no prominent bony points that make themselves evident above the general level. On this surface, however, there are several things to be observed. We can see that the astragalus consists of a stout body that carries the articular surfaces of the leg bones, and this is joined by a short *neck* with the rounded head. Under the outer side of the neck is the outer end of a short interosseous canal or tunnel, the *sinus tarsi*, running obliquely forward and outward between the astragalus and os calcis, and lodging an interosseous ligament connecting the bones: the os calcis has an exposed dorsal surface where this canal opens, and here the anterior annular ligament and Extensor brevis digitorum are attached.

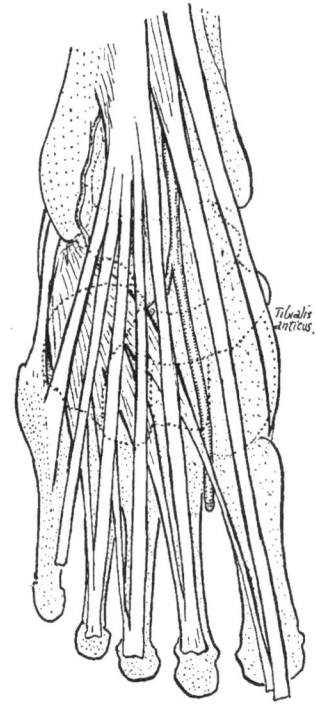

FIG. 139.—To show relation of dorsal structures to the bones.

The structures that pass on to the dorsum of the foot from the front of the tibia stream down over the head of the astragalus and mask it somewhat when the living foot is examined from the dorsum. Fig. 139 is a diagram to show the relation between these structures and the underlying bones: they can be made out, in spite of their general smoothness, by careful examination of the living foot.

Looking at the dorsal surface of the anterior tarsal region the shortness of the middle cuneiform is very evident: this gives the bone a square outline, whereas the

The Lower Extremity and Pelvis

outer cuneiform has an oblong shape when seen from the top, and this difference enables one to distinguish the two bones at once when they are loose.

Notice also how the depth of the cuboid is much less in the outer part, both from above downwards and from before backwards in conformity with the oblique line of the tarso-metatarsal joint. Finally, look at the line of this joint more closely and observe that it is irregular, for the inner and outer cuneiforms project forward beyond the line, so that the bases of the first and third metacarpals are in front of the levels of the second and fourth; or, to call attention to the most noticeable feature, the base of the second metatarsal projects beyond the line of the others in conformity with the small size of the middle cuneiform.

Now examine the lower or plantar surface of the skeleton of the foot. There is a marked concavity on this aspect that is bounded behind by (Fig. 140) the prominent *tuberosity* of the os calcis: this gives attachment to the superficial muscles of the sole, and the strong plantar fascia, so that an overhanging lip is seen on it, and inner and outer *tubercles*, the former being the larger. On its inner side, between the tuberosity and the sustentaculum tali, the bone is concave and gives origin to the inner head of Flexor accessorius.

The sustentaculum is seen to have a broad groove on its lower aspect, and this is continuous behind with a groove on the back and inner part of the astragalus, and is for the tendon of the Flexor longus hallucis.

The head of the astragalus is visible, in front of the sustentaculum, between it and the scaphoid. The inner part of the scaphoid is prolonged downwards as the *tubercle* of scaphoid. In front of this the inner cuneiform is seen to have a thick and strong base, whereas the other two cuneiform bones have their broad bases uppermost and consequently do not show very well on the plantar surface; moreover, they

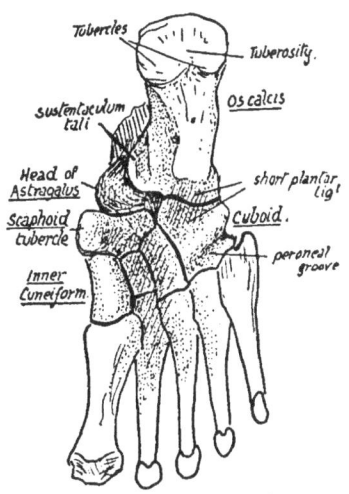

FIG. 140.—Plantar aspect of right foot. Semidiagrammatic.

are rather hidden in the deep concavity of the sole. The cuboid forms a prominent mass outside the cuneiforms, and has a deep groove crossing it obliquely just behind its metatarsal border; this is for the tendon of Peroneus longus, and is therefore directed towards the proximal end of the first metatarsal, on which bone and the internal cuneiform the tendon is inserted. This groove commences in a deep notch, on the outer border, for the tendon as it comes round from the outer side of the os calcis, and here there is a facet for a fibro-cartilaginous sesamoid thickening in the tendon. A prominent ridge crosses the cuboid behind the groove, and behind the ridge is a ligamentous hollow that is immediately in front of a corresponding hollow on the os calcis and is filled by the "short plantar ligament," a mass of fibres connecting the two bones here.

Now compare the skeleton with the living foot, and endeavour to recognise some of the bony points; these can be felt even through the boot as a rule.

On the inner side the malleolus of the tibia is, of course, recognised at once; about three-quarters of an inch below this the sustentaculum can be felt by fairly deep pressure, and about an inch in front of this the tubercle of the scaphoid is easily found. Pass the fingers carefully along the inner side in front of this, and it is possible to distinguish the bony masses of the inner cuneiform and first metatarsal from each other.

On the other side the external malleolus and the styloid prominence of the fifth metatarsal, below and in front of it, are the only points that can be found: the edge of the tarsus between these lies so deep and is so covered by soft parts that it cannot be certainly felt.

But the general level of the mid-tarsal joint can be placed on the outer side as half-way between these points, while on the inner side it lies, of course, just behind the scaphoid tubercle. This is an important joint, at which a large part of the free movements of the foot take place.

In the metatarsal region it is enough to point out at present that there is a transverse arch continuous with that of the tarsus but not so deep: attention can also be called to the presence of a couple of sesamoid bones on the plantar surface of the head of the first metatarsal, which are the largest in the body, after the patella, and which account for the marked grooves present on this surface of the articular head.

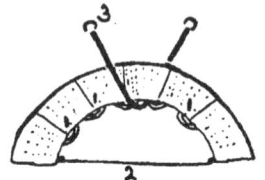

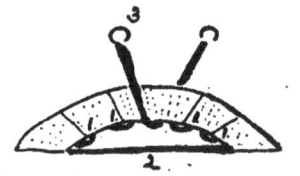

FIG. 141.—To illustrate methods by which an arch composed of segments may be held up: 1, by ties between the segments in the concavity; 2, by connecting the supporting pillars; and 3, by the direct support of a strap passing under the highest part of the arch. In the lower arch the first two methods are much less effective and the connections must therefore be proportionately stronger, but the direct action of 3 is not lessened, though more strain must be thrown on it owing to the lessened effectiveness of the other supports.

We can now examine the construction of the arches present in the foot and obtain a general idea of the mechanical factors that support them.

There is a longitudinal arch in the foot, as well as a transverse concavity, and it is convenient to deal with the former first.

If the skeleton of the foot is considered, it is evident that the bones which constitute the longitudinal arch can be divided into two groups, inner and outer, resting on a common pillar, the os calcis, at their posterior ends. The inner arch is higher, and is made by the astragalus (resting on os calcis), scaphoid, cuneiforms, and three inner metatarsals, while the outer comprises os calcis, cuboid, and two outer metatarsals. The two sets of bones are in a sense distinct, for the articulations between them permit of sliding movements only, so that one arch could be momentarily increased or lessened without affecting the other. As there is no interlocking between the arches in front of the calcaneum, each must depend on its own mechanism for support, and such mechanism will vary with the conditions of the arch.

The means used to maintain each arch can be divided into three main classes. If an arch, composed of several segments, as in Fig. 141, requires support, this can be secured in three ways: (1) The segments can be tied together on the side of the concavity; ties on the convex side would have no effect. (2) The supporting pillars can be tied together. (3) A strap can be run under the highest part of the arch, fastened to structures away from the arch.

All these varieties are represented in the supports of the longitudinal arches of the foot.

To deal first with the inner side. (1) The intersegmental ties are represented by the ligaments connecting the articulating bones, and these must therefore be strong on the concave or plantar side : no strain is thrown on them on the dorsal side. (2) Ties between the pillars are provided by the muscles and fasciæ extending between the tuberosity of the os calcis and the great toe along the inner side of the foot. (3) The expansion from the tendon of Tibialis posticus to cuboid, passing under the head of the astragalus, acts as a strap under the highest part of the arch : observe (Fig. 142) that this is not the direct tendon that goes to the scaphoid, for this can give no aid in

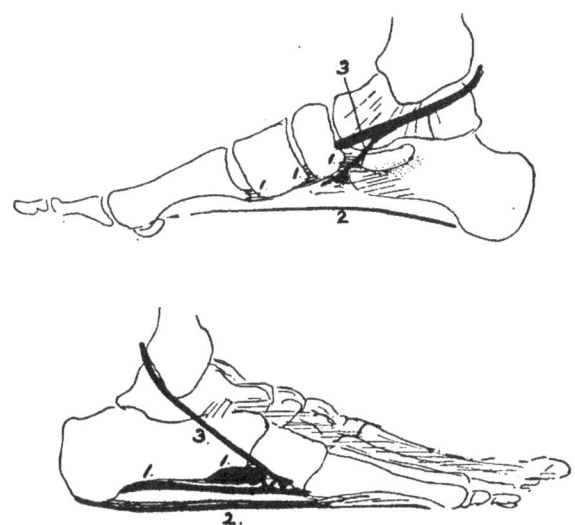

FIG. 142.—Schemes of the supporting structures in the foot, for comparison with the last figure. 1, 1, 1, are ties between the segments ; 2, between the pillars ; 3, strap-like support passing under the arches. In the low outer arch 1 and 2 are stronger.

supporting the inner arch, but its expansion crosses under the arch to be attached outside it.

The outer arch is lower than the inner, and low arches are much harder to hold up (Fig. 141) by intersegmental ties and ties between the pillars ; so we would expect to find the ligaments on the plantar side of the outer arch much stronger than those under the inner arch, and much stronger connections between the tuberosity of the os calcis and the outer toes, whereas the strap passing under such an arch would not be affected in its action, although more work would probably be thrown on it. And this is what is found in the low external arch : the long and short plantar ligaments connecting the os calcis, cuboid, and outer metatarsals are the strongest in the foot ; the fascia is very thick and the muscles largely ligamentous along the outer side of the sole, and the tendon of the Peroneus longus passes under the arch to be fastened away from it, acting under the highest part and being very strong. In this, then, we find some explanation of the occurrence of such strong ligaments and ligamentous " degeneration " of muscles in the outer moiety of the foot.

174   Anatomy of Skeleton

Turning now to the transverse arch, we find that it is highest in the front part of the tarsus. If, then, the segments could be tied together here, they would act under the best possible conditions for holding up the structure.

So we find strong transverse ligamentous fibres in this situation, fastening together the cuneiforms, the cuboid, and the scaphoid, and such fibres are necessarily on the plantar side of the bones and between them, nearer their plantar than their dorsal parts, for the nearer the ties are to the concavity the more effectively do they hold up the arch: thus we might almost expect to find that the articular surfaces between the bones arranged transversely would be nearer the dorsal than the plantar margins of their side aspects. In the same way the high arch is easily held by ties between its pillars, and the peroneal tendon passing under one pillar to be fastened to the other acts well as an extensible tie.

Thus this transverse arch is easily held up by the first and second of the two classes of supports, and the third, which would find difficulty in securing attachment away from the arch, is really not required and does not exist.

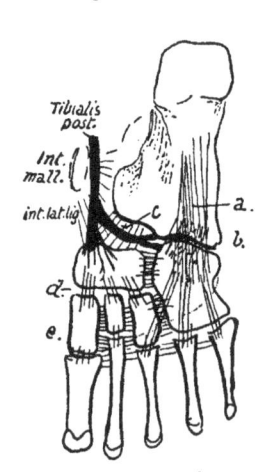

Fig. 143.—Scheme to show the general arrangement of ligaments. These can be divided into those of the transverse and longitudinal groups. The transverse fibres are found in the highest part of the transverse arch, where they can exercise most effect; thus they are found connecting the cuneiforms, scaphoid, and cuboid, and their lower fibres connect the metatarsals. The longitudinal fibres are strongest along the outer arch, where a. and b. are long and short plantar ligs., while the shorter bands on the inner side include c., calcaneo-scaphoid, d., scapho-cuneiform, and e., weaker cuneiform-metatarsal fibres. The internal lateral lig. reaches the sustentaculum by its middle fibres, and in front of this runs into the calcaneo-scaphoid set and thus into the scaphoid; observe that the tendon of Tibialis posticus sends its expansion to the cuboid under this part and thus supports the highest part of the inner arch.

If we now consider the three arches and their supports together, we obtain an idea of the meaning of the numerous ligaments which appear to run in every direction on the plantar surface of the bones (Fig. 143), and we can divide them according to the three arches they support and their consequent position, direction, and strength. But it must not be forgotten that, although the supporting bands of these arches can be analysed separately in this way, the three sets are so connected naturally in function, and in some ways in community of support, that failure in one arch will of necessity lead to consequent strain and failure in the others: in practice, one arch should never be considered to stand alone.

Observe now that the inner arch does not rest at its front end equally on the heads of the three inner metatarsals: the first metatarsal is the main weight supporter, the others, like the rest of the metatarsus, acting more as balancers and only taking weight such as is necessary in keeping the balance of the body on the "ball of the big toe." Again, on the outer side the arch, in spite of its strong ligaments, is so low that it will flatten down, under pressure, enough to bring the metatarsal edge on the ground, and the outer arch is then shortened, so that its anterior pillar becomes practically the base of the fifth metatarsal. Under these circumstances the arch of the metatarsus, not having to carry much weight, does not require strong ligamentous support: it can be easily restrained from splaying out by transverse connections at the distal end, and here we find the intermetatarsal transverse ligament connecting the plantar surfaces of the heads of the bones, with the Adductor transversus muscle.

ASTRAGALUS (Figs. 144 and 145).

Consists of *body*, *neck* and *head*. The body carries a concavo-convex *upper articular* surface for tibia, continuous with inner and outer *malleolar facets* on the sides : the outer facet, for the fibula, is longer and more vertically directed.

Lower aspect of body rests, through an oblique concave articular surface, on the os calcis : internal to this and in front of it is the *interosseous groove* which completes the sinus tarsi, separating the articular under surface of the body from the articular head. The depth of this groove causes the constriction of the neck to be more apparent below and externally : many vascular canals mark the bone in this sulcus.

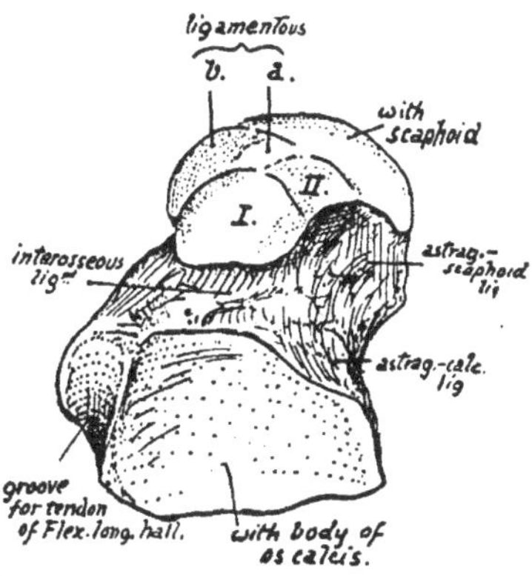

Now place the bone on the calcaneum and articulate it with the bones of the leg : the tibia rests on its upper surface and by its malleolus, with a small curved articular area directed forwards and downwards on its inner side, whereas the fibula has a larger triangular facet, concave from above downwards, on its outer side only. Evidently, therefore, the weight of the body is transmitted by the tibia, and the fibula is only a guard to strengthen the joint and prevent displacement of the foot outwards.

Every time the foot comes to the ground, in making a step forward, there would be a tendency for the leg bones to slip forward on to the dorsum of the foot, if there were no special mechanism to prevent it : this is provided by the shape of the upper articular (or trochlear) surface of the astragalus, which is broader in front than behind, so that any forward movement of the leg bones on it is checked by the increasing

FIG. 144.—Inferior surface of left astragalus. I. and II. are facets for the front part of os calcis which may be continuous or separated more or less ; *a.*, surface lying on inferior calcaneo-scaphoid ligament which extends between sustentaculum tali and scaphoid ; *b.*, the inner continuation of the same surface moulded by the internal lateral ligament which passes into the other, and by tendon of Tibialis posticus lying over it. The tendinous expansion and ligaments exhibit a fibro-cartilaginous thickening here which plays over and covers this part of the bone, and completes the capsule for its head.

breadth of the surface. The anterior and posterior ligaments of the joint are feeble, for they do not limit movement : the dorsi-flexion of the joint is limited by the broader part of the articular surface becoming engaged between the tibia and fibula, and excessive dorsi-extension is checked by tension of the posterior fibres of the lateral ligament. The lateral ligaments hold the bones in apposition. The internal lateral ligament, fastened to the margin of the malleolus, has its middle fibres attached to the sustentaculum tali, which is about three-quarters of an inch immediately below the malleolus : its most anterior and posterior fibres reach the astragalus, so that we can find markings for these (Fig. 145) on the neck and inner side of the astragalus respectively.

The external lateral ligament has three long bands, between which there are short feeble fibres attached round the articular margin. The anterior and posterior bands go to the astragalus, and the middle one to the os calcis in a downward (and slightly backward) direction. We can find the attachment of the first two bands on the astragalus (Fig. 145) : a marking on the outer and upper aspect of the neck is for the anterior band, while the posterior fasciculus reaches a groove that lies on the back and outer aspect of the bone, and a small nodule of bone that lies at the end of this and may occasionally exist as a separate ossicle, the *os trigonum*. This band, as seen in Fig. 136, lies below the level of the transverse inferior tibio-fibular band, which is covered by synovial membrane on its deep aspect, and comes into relation with the astragalus, causing the postero-external part of the trochlear surface to show a slightly bevelled-off facet for it (Fig. 145).

Now consider the action of the lateral ligaments in movement of the joint. The trochlear surface forms an arc of 120 degrees of a circle with a radius of about four-fifths of an inch. The articular surface is about half as long again as that on the tibia, so that there is not a great amount of movement possible between the bones. The centre of rotation, roughly, passes through the apex of the outer malleolus and below and behind the inner malleolus : the effect of movements on the various bands is considered with Fig. 145.

FIG. 145.—Left astragalus seen from the sides, from above, and from behind. *A*, on the upper surface of the neck of the bone, is an anterior prolongation of the synovial cavity of the ankle which is well marked in the new-born foot ; in the adult it is as a rule reduced and broken up into loculi by the presence of fibro-fatty septa which project into it. The extension of synovial cavities beyond their articular surfaces is particularly marked in this joint and in the astragalo-scaphoid articulation, and on the inner side of the neck the two synovial membranes approach each other fairly closely, being separated only by thin and indefinite capsular fibres: The articular surface for the tibia is about 4 mm. broader in front than behind ; in dorsi-flexion this broadening causes the astragalus to be locked firmly between the bones of the leg, but in dorsi-extension the narrower part of the astragalus is engaged and it is supposed that a certain small amount of lateral movement is then possible. The axis of movement of the ankle-joint passes approximately through the end of the fibular malleolus, and below and behind the inner malleolus ; thus the middle fasciculus of the external lateral ligament can be more or less tense throughout the range of movement, which is not large but rather less than 55°. But the anterior and posterior fasciculi are attached to the fibula further away from the axis, so they are of more value in limiting movement, and they become tense in dorsi-extension and dorsi-flexion respectively. The fibres of the internal ligament are strongest in the middle and posterior parts, the latter especially having a large and well-marked area of attachment on the astragalus ; these fibres limit dorsi-extension as, in that movement, the malleolus of the tibia is carried forward on the astragalus. The other ligaments on the bone are attached round the articular surfaces concerned. The line of the centre of gravity of the body falls on or in front of the front part of the ankle-joint in the standing position.

FIG. 147.—Various surfaces of left os calcis. On the outer surface, *A* is an area covered by the middle fasciculus of the external lateral ligament of the ankle passing down to the tubercle below it ; the peroneal tendons turn forward over this fasciculus, being held down by a sling-like annular band that is attached to the bone below the ligamentous tubercle. *B*, surface covered by subcutaneous tissues. *C*, on the upper aspect, is in relation with the loose fatty tissue lying between the tendo Achillis and the ankle. The upper right figure shows the back surface of the bone, on which are three areas ; upper for the bursa deep to the tendo Achillis, middle for this tendon, and lower roughened by the attachment of the thick fibro-fatty tissue of the heel.

In addition to the points already seen on the astragalus in connection with the ankle-joint, certain other details are to be noted on the bones. The posterior aspect exhibits two grooves ; one has been already examined, but the other, internal to it, is better marked and broader and is for the tendon of Flexor longus hallucis. When the astragalus and os calcis are articulated it can be seen that this groove leads directly to the groove under the sustentaculum tali. At the top of this groove is the *posterior tubercle* (or os trigonum) on the outside and a sharp point on the inner side, to which a strong band of the inferior calcaneo-astragaloid capsular ligament is fastened.

The *neck* has markings for astragalo-scaphoid dorsal ligaments in addition to lateral fibres. It also affords attachment to interosseous and anterior annular ligaments (see p. 178).

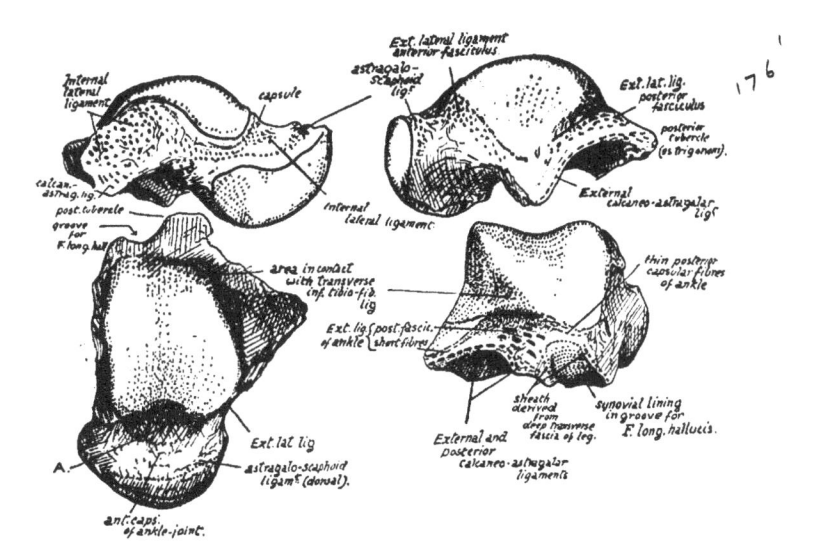

FIG. 145.

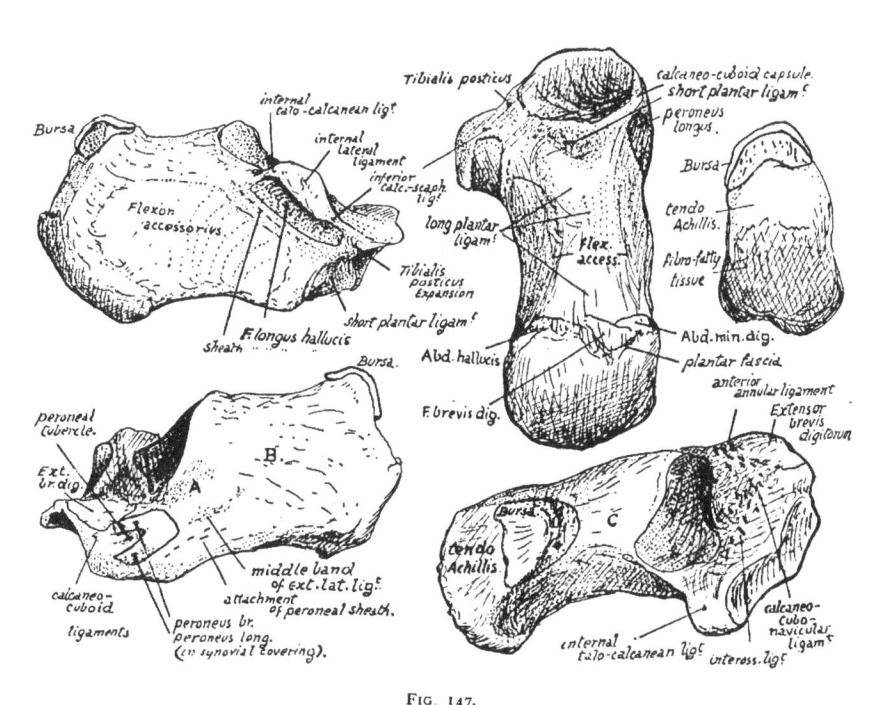

FIG. 147.

The Lower Extremity and Pelvis

The *head* has articular surfaces in front, below, and on the inner side. It rests below on the sustentaculum tali and shows a definite facet for this, which is continuous in front and externally with another inferior facet which shows where the head rests on the front part of the body of the os calcis : occasionally these two facets are more or less discontinuous. In the articulated foot it can be seen that an interval exists between the sustentaculum and scaphoid, and in this interval a part of the lower surface of the head is visible : the interval is bridged by the inferior calcaneo-navicular ligament, on which this part of the head rests, and it therefore presents another facet, between and continuous with the others, for this ligament. The inner aspect of the head lies above the level of the sustentaculum and ligament : it is covered, as shown in Fig. 143, by intermediate fibres of the internal lateral ligament, running down to decussate with those of the lower ligament, and over these lies the expansion from Tibialis posticus to the cuboid, which often presents a fibro-cartilaginous thickening.

Observe the surfaces by which the astragalus articulates with the os calcis. That under the body is concave, while the under surface of the head is convex, and their axes are practically parallel. This disposition hinders any rocking or lateral movement of the bones on one another, but the astragalus can be moved forwards and inwards on the os calcis, the two facets sliding over their opposed surfaces, and this is what occurs when weight is thrown on the astragalus : it slightly opens out the inner arch and the astragalus moves a little forward, carrying the scaphoid forward, and pressing its head on the strong inferior calcaneo-navicular ligament and the underlying expansion of Tibialis posticus.

The somewhat inward direction of the neck and head is more apparent in the fœtal and child's foot. In these the natural position of the foot is one of inversion, and the subsequent comparative straightening of the bone is associated with the change of shape of the foot that results from putting the sole on the ground.

The neck is also comparatively longer and narrower in the new-born foot and the trochlear cartilage is prolonged on to it for some distance. The angle made by the axis of the neck with that of the body in such a foot is about 148°, while in the adult astragalus it is about 152°.

The astragalus is the only bone in the tarsus that has no muscular or tendinous attachments. It is partly ossified at birth.

Os Calcis.

The *upper* aspect shows that the astragalus is carried on the anterior half of the bone : behind this the upper surface is covered by the fatty tissue that lies between the tendo Achillis and the ankle.

The astragalus facet on the body is convex, while the anterior one is concave. This last carries the head of the astragalus, and is situated on the sustentaculum and the front part of the body of the bone : it is frequently divided, more or less completely, into two, that on the sustentaculum being termed the *middle* (astragalar) facet and the remainder being the *anterior* facet. These are separated from the *posterior* facet by the interosseous sulcus, which widens out externally. The astragalar facets have their long axes put obliquely across the os calcis, so that the astragalus is directed inwards and forwards with reference to this bone and can only move in that direction, sliding on the calcaneum. It follows from this that the upper and front part of the os calcis must be exposed externally, and it is from this part, to the outer side of the astragalus, that the Extensor brevis digitorum takes origin : behind it, and fused with its origin, the lower part of the annular ligament is attached to the os calcis.

178   Anatomy of Skeleton

The anterior annular ligament is usually described as consisting of two bands, of which the upper is a simple transverse thickening of the deep fascia in front of the lower part of the leg, while the lower part, situated on the foot, is Y-shaped and on a deeper plane than the deep fascia. This lower band is not simple : the single limb of the Y is really a strong loop that passes round the extensor tendons of the digits and holds them from the outer side so that they do not slip inwards when the foot is inverted, and the diverging limbs act simply as riding lines for the loop, holding it in position as the tendons move, but not concerned in preventing any displacement of the tendons (Fig. 146).

From this description it follows that the loop must be strongly fastened externally, whereas the

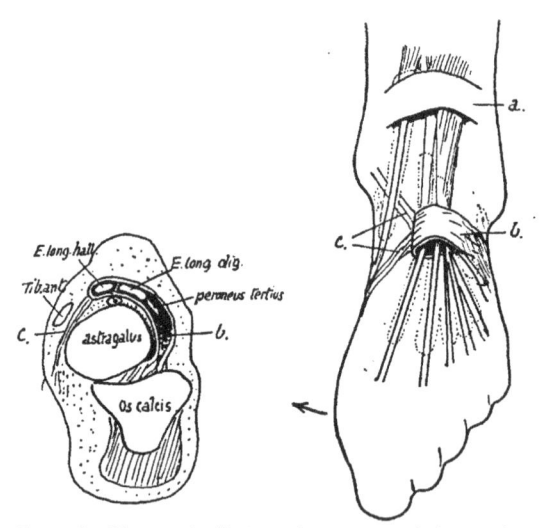

FIG. 146.—Diagrams to illustrate the structure of the anterior annular ligament. On the right the upper band is seen as a simple strap, *a.*, across the tendons, but the lower portion is a strong loop, *b.*, which holds most of the tendons so that they cannot move out of position when the foot is turned in ; thus this loop must be fastened on the outer side of the foot, and the two " stays," *c.*, which keep it from riding up and down, do not require to be strong. A scheme of the arrangement as seen on section is on the left ; observe that the loop is strongly fixed to the os calcis and astragalus in the sinus tarsi, forming an attachment continuous with the interosseous ligament. From this region the loop passes over the astragalus to surround the tendons, with the exception of the Tibialis anticus ; the vessels and nerve lie deep to it.

diverging bands need not be strong, and may go over or under the Tibialis anticus tendon ; frequently one is superficial and the other deep to the tendon. The attachment of the loop, as seen in the scheme, is to the os calcis and astragalus, and on the os calcis it is continuous with the origin of the short extensor : the two structures appear to be developed from the same continuous mass of cells in the embryo. The loop may surround the long extensor of the great toe as well as the common extensor, with a septum between them, or this may be modified in degree so that the main loop takes the common extensor and Peroneus tertius, and a secondary loop takes the extensor of the hallux : there is no essential difference in the arrangement. All the tendons running through the loop are in synovial sheaths ; the Tibialis anticus tendon is the only one that may have a sheath as it passes under the upper or leg band, because it is the only one that may be purely tendinous.

The sustentaculum tali has the internal lateral ligament of ankle fastened to its

The Lower Extremity and Pelvis

thick inner border, and the inferior calcaneo-navicular ligament further forward: these two ligaments are continuous in front of and internal to the sustentaculum, where they form part of the capsule for the head of astragalus. The groove below the process is a continuation of the postero-internal groove on the astragalus: it is lined by a synovial sheath surrounding the tendon of Flexor longus hallucis, shown in Fig. 147.

The *inner* surface of the body is practically covered by the origin of the inner head of Flexor accessorius, whose limits can as a rule be easily made out on the bone.

This limiting line is made by the thin aponeurosis covering it, which is described as part of the internal annular ligament, and thus the ligament is brought to the inner side of the calcanean tuberosity.

It would probably be more accurate to confine the description of annular ligament to the strong band covering the tendons and vessels, thus making its lower attachment on the line that runs along the outer border of the groove below the sustentaculum.

The *outer* surface presents a tuberculated point for the middle fasciculus of the external lateral ligament. The peroneal tendons turn forward superficial to this band and diverge to their insertions, so that the area in front of the ligamentous tubercle is covered by these tendons, whereas the back part of the surface is only covered by the fibro-fatty subcutaneous tissue at the side of the heel. The peroneal area may show a groove for the long tendon running to the lower corner of the surface, and above this there is frequently a ridge or tubercle that separates this tendon from that of Peroneus brevis.

The *hinder aspect* of the bone shows two surfaces, upper and lower, facing in different directions. The lower is vertically striated and ends above in an irregular ridge, affording attachment to the tendo Achillis: above this there is an area for the tendon, and above this again the bone is smooth and covered by a bursa deep to the tendon. The bursa extends a little way on to the upper surface as a rule.

Lower surface has a concave depression at its front end and internally for fibres of the short plantar ligament: behind this the broad keel of the bone is covered by, and gives attachment to, fibres of the long plantar ligament, almost as far back as the tuberosity. The pointed outer head of Flexor accessorius arises from the outer margin of this region, fused with the ligamentous fibres.

The tuberosity has a prominent lip which marks the attachment of the plantar fascia. The short muscles of the superficial layer of the sole are immediately deep to this, and their attachments to the bone are fused with the fascia on the ridge and also on the bone in front of this: they overlap one another from within outwards as seen in the figure.

Abductor hallucis, from the inner tubercle, extends its origin from this on to the fascia covering the inner head of Flexor accessorius, and thus gets an extensive origin from the annular ligament and covers the structures entering the sole here. This origin may extend as far as the tuberosity of the scaphoid, from which the muscle may even obtain a small origin.

In addition to the ligaments mentioned in connection with the bone, the markings for others can be found on the various surfaces as indicated in the figures. Attention may be called particularly to the calcaneo-cubo-navicular and the internal talo-calcanean band (Fig. 147).

The bone rests solely on its tuberosity, with its front portion raised from the ground. Its concavo-convex facet for the cuboid has its upper lip overhanging and thus resting on the cuboid: this keeps the bone from sliding up on the calcaneum, so

Anatomy of Skeleton

that any flattening of the outer arch can only occur with stretching of the long and short plantar ligaments, thus increasing their efficiency.

The bone is the only tarsal member that normally possesses an epiphysis: this is

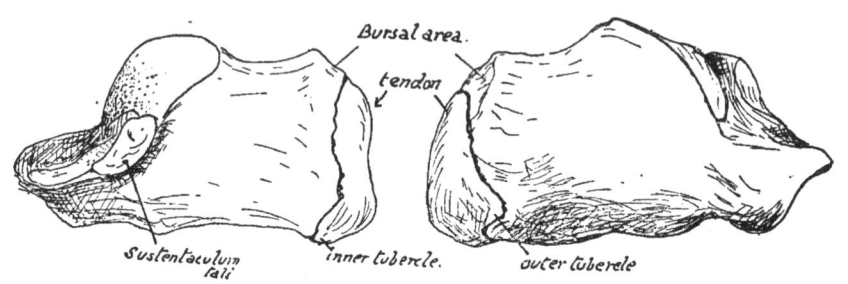

FIG. 148.—Right os calcis showing the epiphysis.

situated on its hinder part (Fig. 148) and is a traction epiphysis for the tendo Achillis. It is considered to correspond with the pisiform in the hand.

The bone is very vascular, and several small foramina can be seen on its sides ; those on the inner side are largely arterial, with venous exit on the outer side.

It is partly ossified at birth.

CUBOID.

This bone is really more pyramidal than cuboid in its general shape, for its dorsal, plantar, and anterior surfaces slope towards its narrow and short outer surface or border, with its broad base facing inwards and articulating with the outer cuneiform. The *dorsal surface* looks outwards and upwards, is roughened slightly by ligamentous fibres connecting it with the adjacent bones, and is covered by the tendons of Peroneus tertius and extensors for the outer toe. The *plantar* surface is marked by a deep groove, running obliquely across it immediately behind its front margin, for the tendon of Peroneus longus: behind this a thick ridge for the long plantar ligament, some fibres of which pass from the ridge over the groove to the metatarsus, and thus help to cover in the tendon. Behind the ridge is a hollow area for the attachment of the short plantar ligament. A facetted surface at the outer end of the ridge is for the play of the fibro-cartilaginous thickening in the tendon.

The posterior aspect has a concavo-convex articular surface for the os calcis, and the *anterior* surface has two continuous facets for the two outer metatarsals, the inner quadrilateral and the outer triangular. The *outer* surface or edge is notched for the peroneal tendon turning into the groove, is overlapped somewhat by the styloid process of the last metatarsal in front, and is covered by the muscular and ligamentous abductor of the outer toe.

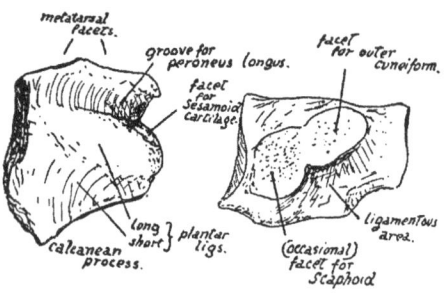

FIG. 149.—Left cuboid, from below and from the inner side.

The *inner* side has a rounded facet for the outer cuneiform, often separated from the posterior surface by a non-articular area : in many bones, however, this area bears an articular facet for the scaphoid, and when this is present it connects the outer and posterior articular surfaces. The remainder of the inner surface is ligamentous, this being the area for outer attachment of the transverse fibres that hold up the transverse arch, connecting this bone, the scaphoid, and the cuneiforms.

The lower and inner angle of the bone is prolonged somewhat as a *calcanean process :* to this and the bone in the neighbourhood is attached the expansion from Tibialis posticus that passes below the head of astragalus. The short flexor of the great toe takes origin from this expansion and so from the bone here.

The cuboid is cartilaginous at birth, but there is very frequently an ossific centre present in it.

SCAPHOID OR NAVICULAR.

Notice the position (Fig. 140) of this bone in the foot, which is such that the rounded *tuberosity* looks downwards on the inner side. The convex *dorsal* surface has markings for dorsal ligamentous connections with neighbouring bones and for fibres of the internal lateral ligament.

The *plantar* surface has the projection of the tuberosity, into which the direct tendon of Tibialis posticus is inserted, on its inner side, and is roughly concave external to this : here are attached strong ligaments of the transverse group (cubo-navicular) and longitudinal arch (plantar calcaneo-navicular). The outer end of the bone may have ligamentous attachment to the cuboid or may present an articular facet for that bone, continuous with the facet for the outer cuneiform : very rarely there may be a small surface also for the os calcis. The *posterior* surface is practically altogether articular, the long oval concavity for the astragalus having its long axis directed downwards and inwards. The *anterior* surface is covered by the confluent triangular facets for the cuneiforms : observe that the two outer ones have their bases upwards, but the inner one has its apex upwards.

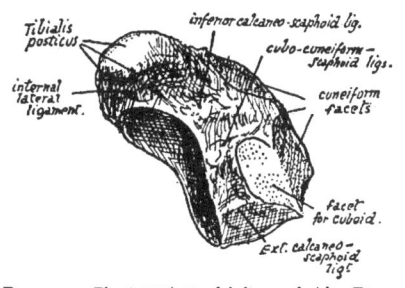

FIG. 150.—Plantar view of left scaphoid. From a foot which exhibited articulation between scaphoid and cuboid.

The scaphoid and the cuneiforms which it supports are cartilaginous at birth.

CUNEIFORM BONES.

The *first* or *internal* cuneiform is the largest, with a heavy thick base below and a thin edge above. Its general appearance distinguishes it at once from the other cuneiforms, of which the *second or middle* bone can be recognised immediately by the square cut of its dorsal surface or base, which contrasts with the oblong dorsal surface of the *third* or *outer* bone (Fig. 138). This figure also shows that the second metatarsal fits in between the inner and outer cuneiforms, so that these bones will have side facets for this metatarsal in addition to the anterior surfaces for their proper metatarsals : also the outer bone comes a little further forward than the cuboid, and consequently has another facet on its outer side for the fourth metatarsal.

Inner Cuneiform.—A triangular facet *behind*, with base below, for scaphoid, and a larger kidney-shaped surface in *front* for the first metatarsal. *Inner* surface convex from above downwards, conforming with the rounded inner border of the foot, marked by ligaments, covered by skin and superficial tissues, and crossed by tendon of Tibialis anticus, which is partly inserted into it : an oval facet below and in front is partly for its insertion and partly for a bursa under a sesamoid cartilage in the fibres that pass over it to reach the first metatarsal.

The *outer* surface is on the whole concave. It has an elongated facet along its back and upper margins for the second cuneiform and a small surface for the second metatarsal in its upper and front corner, continuous with the last ; otherwise this aspect is rough for ligamentous fibres (interosseous) of the transverse set joining it to the second cuneiform and metatarsal, but an impression in its lower and front part frequently shows insertion of Peroneus longus. The upper and front ligamentous fibres constitute " Lisfranc's ligament."

The thick base is covered by ligamentous fibres and expansions from the tendon of Tibialis posticus : superficial to these lie the fleshy mass of the short flexor and tendon

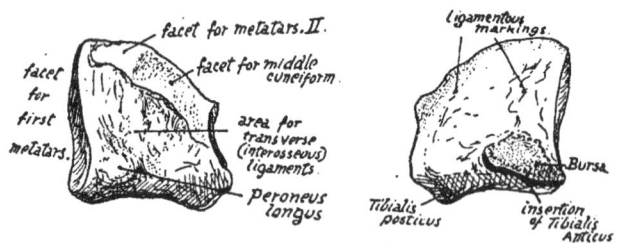

FIG. 151.—Internal cuneiform, left side. Inner and outer views.

of long flexor of the big toe and the more superficial structures, and the bone rests through these on the ground when sufficient weight is thrown on the foot to depress the inner arch to a slight extent.

Middle Cuneiform.—Triangular facet in front for second metatarsal and behind for scaphoid. *Dorsally* it has slight ligamentous markings. *Internally* a long Γ-shaped facet along the back and upper margins for the inner cuneiform : the rest of the surface for attachment of intercuneiform ligaments of the transverse set. *Externally* a long facet, slightly narrowed in the middle, along its posterior margin, for the outer cuneiform : the remaining surface ligamentous.

Outer Cuneiform.—Dorsal surface has slight ligamentous markings. Posterior facet for scaphoid does not take up the whole of this aspect, as a blunt point of bone projects slightly below it. *Internally* a long facet behind for middle cuneiform, and another, usually divided, along the front border for second metatarsal : the remaining surface ligamentous. *Externally* a rounded facet for cuboid behind and above, a small one for fourth metatarsal * in front and above, and the rest of the surface rough for ligaments of the transverse groups. It carries the third metatarsal alone on its front aspect.

The cuneiforms show a slight general convexity from side to side on their dorsal

* The connection is frequently ligamentous, when no facet is present.

The Lower Extremity and Pelvis

aspect, where they are covered by the extensor tendons, under which the middle bone is crossed by the dorsal vessels and nerve. The plantar edges of the two outer bones are covered by ligamentous fibres, from which arise fibres of the Adductor obliquus and some Interosseous muscles.

METATARSUS.

With the exception of the first, the metatarsal bones are long and slender, with heavy proximal bases and small heads compressed laterally. Though in their general appearance like the metacarpals, they can be distinguished from them at once by this lateral compression, for (Fig. 85) whereas the metacarpals are strong bones, broadening towards their solid heads when viewed from the dorsum, the metatarsals, seen in this way, exhibit narrowing when the shafts are followed down to their narrow heads. These outer four metatarsals are not directly concerned in carrying the weight of the body: this is borne mainly by the first metatarsal supported by the inner cuneiform, while the remainder of the series have their chief function in preserving the balance of the body. Thus the first metatarsal is massive and strong, and the remainder are comparatively slight, contrasting with the metacarpals, which require strength as opponents in the act of grasping.

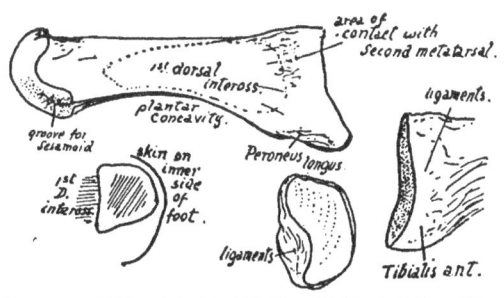

FIG. 152.—First metatarsal, left side. External view, and base from inner side and from behind. Also a scheme of section showing how the inner side of shaft is subcutaneous and hence convex.

First Metatarsal.—Shaft markedly concave below, rounded internally and rather flattened externally where the first dorsal interosseous muscle arises. The proximal

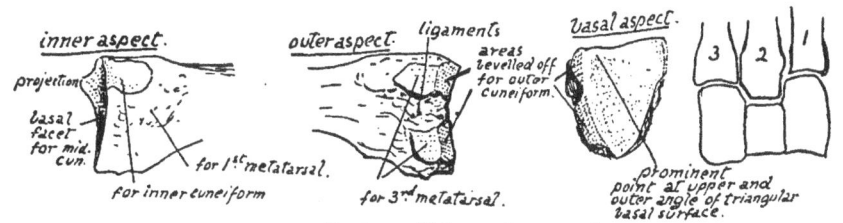

FIG. 153.—Basal end of left second metatarsal.

end or base is enlarged, especially prominent below, presenting a kidney-shaped, slightly hollowed facet for the inner cuneiform, whose borders correspond with the general shape of the shaft in being convex internally and concave externally. On the inner side of the base, nearer the plantar aspect, is a secondary marking for Tibialis anticus, and on the other side a hollowed mark for Peroneus longus. A marking—not often a definite facet—is to be found on the outer side of the proximal enlargement showing the area of contact with the second metatarsal, and in its neighbourhood

Anatomy of Skeleton

are impressions for interosseous ligaments. The head has practically no articular surface on its dorsal aspect, only on its plantar side and extremity, that in the former situation being scored by two grooves in which the two sesamoid bones slide that are in the capsule.

These two bones are the largest sesamoids in the body after the patella, and keep the long flexor tendon away from the head of the bone: they, with the soft tissues covering them, are responsible for the plantar prominence of the "ball of the toe," and receive the insertion of the Abductor, Flexor brevis, and Adductors. The inner one is slightly the larger, and there is usually a corresponding difference in the size of the grooves on the head. Sesamoids are occasionally found in the other metatarso-phalangeal joints as in the hand.

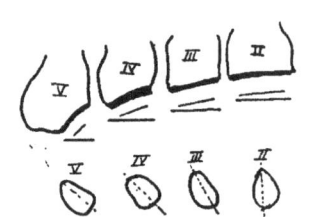

FIG. 154.—Scheme to show the obliquity of the bases of the metatarsals. Left side. The increasing obliquity is seen compared with the coronal line. The lower figure illustrates the increasing sectional obliquity of the shafts from within outwards.

The bone is superficial and can be felt on its dorsal and inner sides, but in the sole it is deeply placed and covered immediately by the short flexor of the hallux, and externally gives an extensive origin to the first Dorsal Interosseous. The communicating artery runs down in relation with the proximal part of its shaft on this side.

Second Metatarsal.—The longest of the series. Base fits in between first and third cuneiforms, so carries facets for these on each side, interrupted by interosseous (tarso-metatarsal) ligaments, in addition to an interrupted articulation on the outer side for third metatarsal, and a non-articular area on inner side for first metatarsal. The area just below the facet for the first cuneiform marks the anterior attachment of "Lisfranc's ligament." The bevelled-off cuneiform

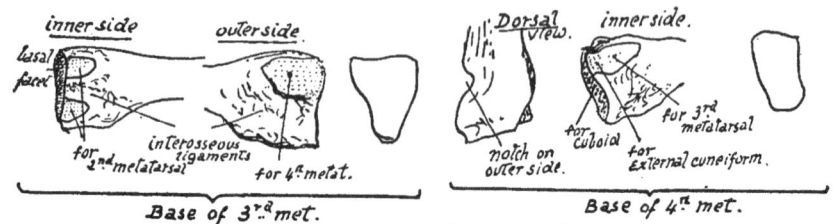

FIG. 155.—Basal ends of 3rd and 4th metatarsals, with outlines of their respective tarsal surfaces.

area on the outer side is one of the characteristics of the bone (Fig. 153). The wedge-shaped basal facet is very slightly concave, owing to a small projection of the upper and outer angle. Markings of interosseous ligaments are well developed. The *shaft* gives origin to no palmar interossei, but to the first two dorsal muscles.

Third Metatarsal.—This articulates with one of the cuneiforms, the external, which supports it, and for which it has a flat triangular basal facet: at the sides are facets for the second and fourth metatarsals respectively, that for the former interrupted and converted into double surfaces by an interosseous ligament, that for the latter single.

The obliquity of the line of its base is greater than that of the second (see Fig. 154), but not so great as in the fourth metatarsal.

The Lower Extremity and Pelvis

Fourth Metatarsal.—Notice the very oblique plane of the base, the heaviness of this end of the bone, with the appearance of a notch and twist in it; this, with the fact that its basal facet is not triangular but rather four-sided, is enough to distinguish the bone. There is a small marginal facet internally for the outer cuneiform, because this bone projects very slightly beyond the cuboid, and this facet is continuous with that for the third metatarsal. A single facet externally for the fifth metatarsal: just in front of this is the notch or pit that gives the crooked appearance to the proximal part of the bone, for the strong interosseous ligament which connects it with the last bone of the series.

Fifth Metatarsal.—The proximal end has a very oblique tarsal facet, a slightly concave metatarsal facet, and a prominent styloid process: to this are fastened the Peroneus brevis at the point and, on the dorsal side of its base, the Peroneus tertius. This latter muscle has usually a fairly extensive aponeurotic insertion spreading along the dorsal border of the bone, where there is a sharp ridge that marks its attachment and that of the dorsal fascia with which it is blended.

The Abductor metatarsi quinti, deep fibres of the abductor of the little toe, is attached, with much ligamentous tissue, to the outer and under part of the process, and the Flexor brevis has a part of its origin from the plantar surface further in.

The strong base and styloid process may support some weight when the outer arch is sufficiently flattened to bring them to the ground; this probably occurs when any pressure is made on the arch beyond that supported in balancing the body, and thus there is no extra strain thrown on the shaft or small head. The whole of the dorsal and outer surface of the bone can be palpated, but its lower aspect is covered by the marginal musculature.

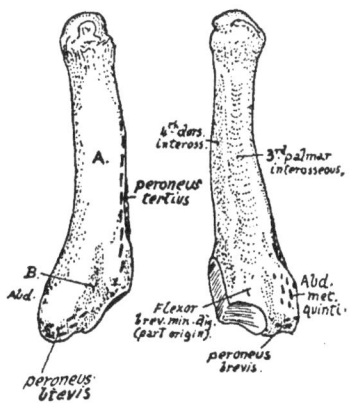

FIG. 156.—Left fifth metatarsal. Dorsal and plantar aspects. *A.* is the subcutaneous surface outside the dorsal ridge for Peroneus tertius; a tendinous slip from Peroneus brevis passes on to this surface along the groove *B.*, covered by ligamentous fibres attached to the margins of the groove.

The *heads* of the four metatarsals are comparable with those of the metacarpals (see pp. 105 *et seq.*), and the same may be said about the structure of the metatarsophalangeal joints, save that the transverse metatarsal ligament extends to the capsule of the great toe and has an Adductor muscle arising from it.

The shafts, however, differ from those in the hand in that the markings for interosseous muscles, being arranged on a slightly different plan, do not render each shaft distinguishable.

PHALANGES.

Compared with those in the hand, the phalanges of the foot in general are recognised at once (Fig. 85) by their thin and rounded shafts and large extremities: in the big toe they are broad and strong, but short for their size, and the first shows an obliquity of its base (Fig. 138), which enables one to place it at once on its proper side.

Anatomy of Skeleton

The oblique direction of the first phalanx in the great toe does not appear to be the result of wearing boots : possibly it may be correlated with the outward turn of the foot as an adaptation to such acts as running or even walking, in which the body is lifted forward on the ball of the great toe, and such a direction of the phalanx keeps it out of the way and avoids over-extension of the joint. This does not, however, seem a very satisfactory explanation of the condition.

The account already given of the phalanges of the hand can be applied, *mutatis mutandis*, to those of the foot, but a few points must be noticed first. The middle phalanges get quickly smaller from within outwards, the fifth being usually an irregular nodule of bone ; the same in a less marked degree may be said of the last phalanges, and there is frequently a fusion between the middle and last bones in the little toe, and even occasionally in the other toes.

The bones of the tarsus are composed of cancellous tissue enclosed in very thin

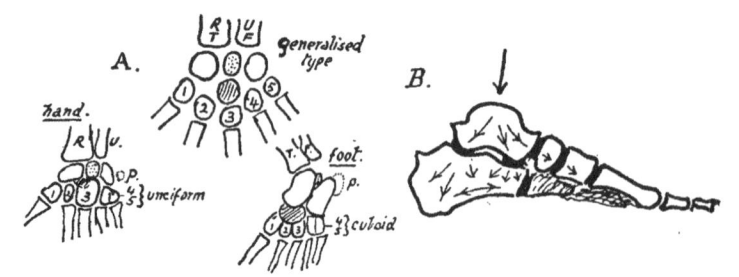

FIG. 157.—A., scheme to show the relations between the skeletons of hand, foot, and a general hypothetical type. The last has a preaxial bone, $R.T.$, radius or tibia, and a postaxial, $U.F.$, ulna or fibula ; these carry corresponding carpal or tarsal units which are separated by an "intermedium" (stippled). The distal row are the *carpalia* or *tarsalia* (numbered from preaxial side). Between the two rows an "os centrale" (shaded). This becomes the scaphoid in the foot but is fused with the os magnum or scaphoid in the hand. Os intermedium becomes semilunar in hand and os trigonum in foot. Pisiform (p) is represented by calcaneal epiphysis ; it has no place in the type form, and may stand for a rudimentary digit. B., scheme of transmission of weight and consequent direction of main lamellæ in cancellous tarsal bones ; supporting lamellæ run at right angles to the main ones.

shells of compact bone. The cancellous lamellæ are mainly directed in the lines of pressure transmitted from the tibia through the astragalus to the other bones : the arrangement is shown schematically in Fig. 157.

The correspondence of the foot with the generalised type of carpus and tarsus (p. 116) is seen in Fig. 157. In the same figure the hand and foot are compared to show the different components that "correspond" with each other.

Development of Bones of Foot.

The bones are preformed in cartilage, which at birth shows centres for the shafts of the metatarsals and nearly all the phalanges, and for two or three of the tarsal bones.

The cartilage is first apparent in the sixth week, and the form of the various bones is distinct by the middle of the third month. The metatarsals are at first rather spread out and resemble the skeleton of the hand, but by the latter part of the third month they have come to lie closer together, and the foot exhibits a definite tarsal arch.

The Lower Extremity and Pelvis

Ossification of *tarsus:* one centre for each bone except os calcis, which has an epiphysis for its back part. The centre for astragalus and the main calcanean centre appear during the sixth month and that for the cuboid about birth, so that it may or may not present a nucleus at birth. The cuneiforms follow, the external first (one to two), then the inner (two to three), and lastly the middle one (three to four). The navicular commences to ossify about a year later.

The epiphysial centre for the os calcis appears between eight and ten, and joins the main bone about the age of twenty in males and about sixteen in females.

There are occasionally separate small nuclei for the posterior tubercle of the astragalus, and for the tuberosity of the scaphoid.

Metatarsus: centres for the shaft eighth to tenth week, and epiphyses in the third to eighth year, that for the first metatarsal appearing before the others. The epiphysis for this bone is at the proximal end, while those of the remainder are distal: occasionally there is also a proximal (styloid) epiphysis found on the fifth metatarsal.

The epiphyses unite about twenty, those of women generally two or three years earlier.

Phalanges: these have proximal epiphyses. The shaft centres come first in the last phalanges (end of second month), then in the first row (fourth to fifth month), and lastly in the middle row, in which centres begin to appear in the fourth month (second toe) and show slowly, that for the little toe not being generally present at birth. Epiphyses appear about third to fourth year.

The epiphyses join the shafts between sixteen and twenty, in women some years earlier, the epiphyses of the first row being the first to join.

The foot of the infant has certain characteristics of that of the ape. It is strongly inverted, with the head of the astragalus directed more inwards and the neck longer than in the adult: the first metatarsal is also somewhat directed inwards, with an oblique articulation with the cuneiform, and is comparatively short. The extensor muscles are apparently relatively short. As the foot comes into the adult position these are lengthened, and the head of the astragalus comes to rest more on the sustentaculum, while the outer (fibular) facet on the astragalus is increased in size.

The anthropoid foot differs from that of man in the weaker development of the great toe and its greater abduction and mobility, and also in the smaller size of the tarsal components. The human embryonic foot shows at first marked abduction and mobility of the toe, but this begins to be modified early in the third month, and the modification slowly progresses into the earlier years of life.

Many races show considerable mobility of the great toe, and in some low races (*e.g.*, the Veddahs) the great toe not only stands apart from the others, but the whole tarsus is shorter and narrower, compared with the metatarsus, than in Europeans. The relative and absolute length of the great toe is greater in men than in women.

Chapter VI

SKULL AND HYOID

The skull can be divided for descriptive purposes into a *facial* and a *cranial* part. The cranial bones form the walls of the cavity that contains the brain, and the face bones are situated below the front portion of this " brain-box."

Looking at the skull from the side, or in sagittal section, we obtain a general idea

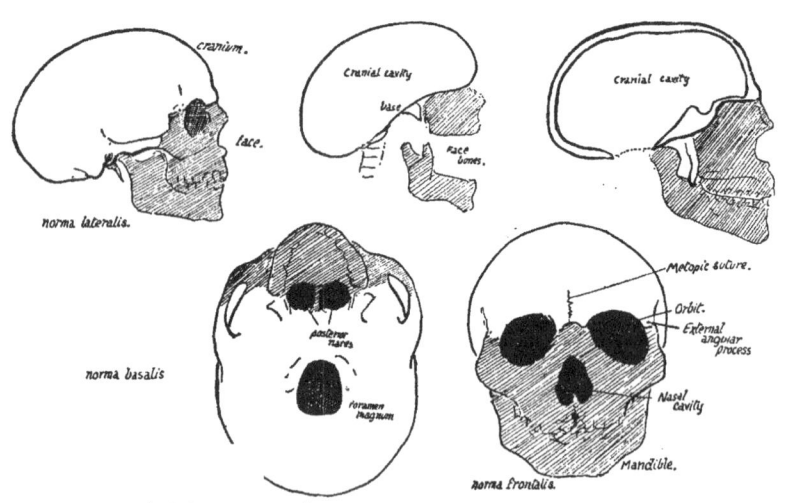

Fig. 158.—The shaded portion of each figure indicates the facial skeleton. In the upper central figure a scheme of the general construction of the skull is seen; the cranial part has a floor sloping up in front and the face bones are placed below the raised front portion. The whole is supported on the spine by articulation on the cranium. Pterygoid processes hang down from the cranial base behind the face bones; these are seen in the upper right figure of sagittal section also, but here the lower jaw is in place showing that the face bones extend round on each side of the region of pterygoid plates, while the remainder of the face lies in front of them. Other drawings are from the side, base and front.

of this relationship (Fig. 158): the cranial cavity is seen to be deepest behind, with its floor or *base* sloping upwards and forwards on the whole, and the space below it obtained in this way is occupied by the bones of the face. Looking at it from below, the face bones hide the front part of the cranial region, which is only visible behind them. From the front the cranial portion of the skull is above, the facial part below, and the cavities of the *orbits* are seen to lie between the two parts: the *nasal cavities* are placed between the two halves of the face, and the opening of the mouth passes transversely across it, separating the upper from the lower portion or *mandible*.

For practical descriptive purposes the division into cranial and facial bones is useful, but, as we will see later, it is not effected on morphological grounds, and, like

Skull and Hyoid

most purely descriptive classifications, it therefore cannot be pushed to include every detail, or even every skull bone, in its scope.

A cursory examination of the skull reveals the fact that it is composed of several bones which, with the exception of the lower jaw, are immovably articulated with each other. This is most apparent when the complete skull is viewed from above (*norma verticalis*) or from the side (*norma lateralis*), when the several bones are seen at once articulating with each other along irregular lines, known as *sutures*.*

Before proceeding to a detailed examination of these bones and of the complete skull it is advisable to become acquainted with the relative positions of the bones in the skull.

CRANIAL BONES.

Examine the skull from above (Fig. 159). We perceive that the front part is formed by a single bone, the *frontal*, behind which are the *parietal* bones, right and left, placed therefore laterally and above, and also markedly on the posterior aspect of the curved skull. The *fronto-parietal suture* is also termed *coronal*, while the *interparietal suture*,

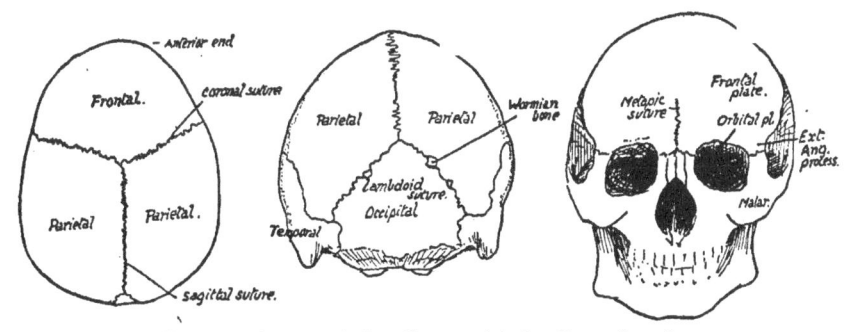

FIG. 159.—Norma verticalis. Norma occipitalis. Norma frontalis.

being in the middle line and running from before backwards, is usually termed the *sagittal suture*: observe that these sutures are very serrated.

Now turn to the posterior aspect of the skull (*norma occipitalis*) (Fig. 159), and see that another bone, the *occipital*, comes up in a point between the two parietals, so that the sagittal suture becomes continuous with two *parieto-occipital sutures*: these two together constitute the *lambdoid suture*, so termed on account of the figure formed by them. Small additional ossicles, known as *Wormian bones*, are frequently found in this suture, as also in many others in the skull, and may in exceptional cases number some hundreds : they are due to accessory centres of ossification occurring where the normal centres have failed to meet and form a normal suture.

Now look at the front of the skull (*norma frontalis*), and observe that the frontal bone alone forms this wall of the cranial cavity. It comes down to make the upper margins of the openings of the orbits, and turns back here to form the greater part of each roof of these cavities, so that the bone may be said to possess two *orbital plates* in the roof of the orbit, in addition to the large *frontal plate*. There is often a suggestion of a suture, more or less complete, running up the frontal plate in the middle line,

* In skulls from subjects of advanced years the sutures are frequently obliterated in part or altogether.

starting at the root of the nose ; this is termed the *metopic suture*, and is an indication of the fact that the bone develops in two halves which subsequently fuse. The facial bones, which form the floor of the orbits, come up to articulate with the frontal at each side of the openings of these cavities, so that the frontal bone affords support to the bones of the face : the external column of support is the prominent *external angular process* of the frontal.

We can now appreciate the position of the bones seen on the lateral aspect of the skull (Fig. 160).

Looking at it from the side, we see that the lower border of the parietal bone articulates along a curved suture with a plate of bone that is part of the *temporal* bone, known as the *squama*, or squamous portion of the temporal, and from the lower part of this an arched process extends forward to join a corresponding backward extension from one of the face bones. The arch made in this way is the *zygomatic arch*, and the

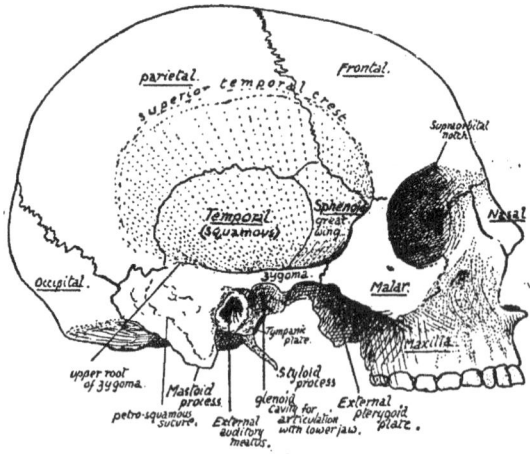

FIG. 160.—Skull from the side without lower jaw. The stippled part is the temporal fossa. The inferior temporal crest is under cover of the zygoma.

temporal process is the *zygomatic process*, or *zygoma*. The temporal articulates with the whole extent of the lower border of the parietal, tracing it backwards, thus reaching the end of the lambdoid suture, where it comes into relation with the occipital bone, and the suture between the two turns on to the base of the skull. The back and lower part of the temporal, articulating with the back part of the lower border of the parietal, is easily distinguished by its appearance from the squama, and is continued down as a blunt projection, the *mastoid process*.

In front of the temporal another bone is seen separating the squama from the frontal and articulating above with the lower front angle of the parietal : this is the outer surface of the *great wing* of the *sphenoid*.

The sphenoid is a bone that properly belongs to the base of the skull, where it has a body that lies centrally, but two pairs of " wings " project out from this body, and of these the greater wings form part of the skull base out to the sides, and also help in forming part of the side walls : here each wing articulates with the squama behind,

Skull and Hyoid

the parietal above, the frontal above and in front, and below this with one of the face bones, the *malar*. The malar is the bone at the outer side of the orbit which we have seen articulating with the external angular process of the frontal above and with the zygomatic process of the temporal externally and behind. If we look inside the orbit we can see the malar articulating with the frontal above, and, below this, by its posterior border, with the great wing of the sphenoid, but this is the *orbital surface* of the great wing.

Looking at the side of the skull as a whole, we can see that a large part of it is depressed, deep to the level of the zygomatic arch, forming the *temporal fossa*. The floor of this fossa is made by the squama, great wing of sphenoid, lower part of parietal, and, in front, a small part of the frontal and malar.

The fossa contains the temporal muscle, which arises from its floor, and it is limited above by the *superior temporal crest* which forms an indefinite curved roughness on the parietal, passes from this on to the frontal, where it becomes a prominent ridge leading to the external angular process, and from this passes on to the sharp margin of the malar.

At about the level of the zygoma the floor of the temporal fossa is found to become abruptly continuous with the base of the skull, forming a sharp border, the *inferior temporal crest:* the tendon of the temporal muscle passes between this crest and the zygoma to reach the lower jaw.

If we trace the zygoma back we find that a slightly curved ridge, *the upper root of the zygoma*, can be followed along the lower and back part of the squama, where it forms a lower boundary of the temporal fossa. Below the root of the zygoma, and just in front of the mastoid process, is the opening of the ear, the *external auditory meatus*, and the lower jaw articulates with the *under surface* of the bone just in front of this.

All the cranial bones that have been mentioned so far, with the exception of the parietals, are concerned also by their other parts in the formation of the base of the cranial cavity—in fact, there are no other bones there, with the exception of the ethmoid, which fills in the interval between the orbital plates of the frontal.

The basal view (*norma basalis*) of the skull shows us the constituent bones with a little examination : they are not so clearly defined as on the other aspects, because the basal surface, being that looking towards the neck, is covered by processes and roughnesses for attachments of various structures and pierced by numerous holes for the passage of vessels and nerves (Fig. 161).

Starting from the back, we notice that the occipital bone turns forward below, presents the large *foramen magnum* for the passage of the central nervous axis and its appendages, and narrows in front of this to a thick bar of bone. In the adult skull this bone is continuous with the central portion of the sphenoid and has to be separated by the saw, but the termination of the occipital can be taken as about an inch or less in front of the margin of the foramen magnum.

The different parts of the occipital bone can now be recognised : the portion in front of the foramen is the *basi-occipital ;* the part that projects outwards on each side of the foramen is the *ex-occipital*, frequently termed the *jugular process*, because it forms the back wall of the *jugular foramen*, from which the internal jugular vein and certain nerves leave the skull. The part behind the foramen is the *post-occipital*, and this is seen to be roughened by aponeurotic attachments, showing *upper and lower curved lines* and an *external occipital crest* running down from an external *tuberosity*.

This is the basal part of the bone, but behind the superior curved line the bone turns up on the back of the skull between the parietals, and this part is often termed the *supra-occipital*. At the junction of basi- and ex-occipitals the two articular *condyles* for the Atlas are seen on the margins of the foramen magnum.

To the outer side of the ex-occipitals the temporal bone comes into the base on each side, showing the prominence of the mastoid process externally, and then a narrow

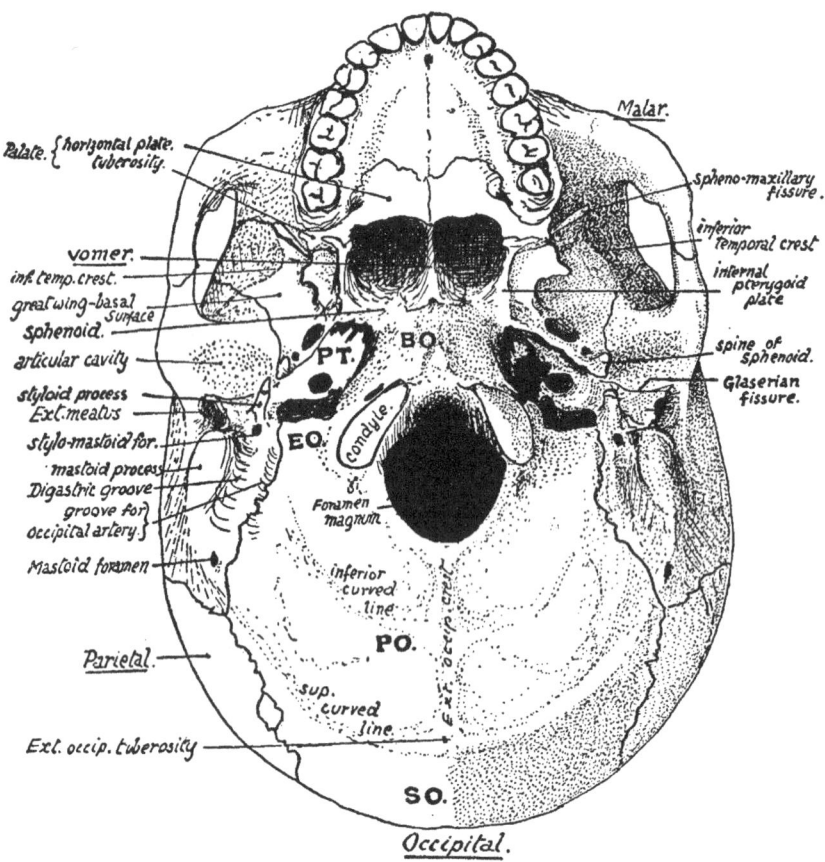

FIG. 161.—Lower aspect of base of skull. *EO*, ex-occipital; *PT*, inferior surface of petrous part of temporal; between *EO* and *PT* are carotid and jugular foramina. *BO*, basi-occipital which is continuous with basi-sphenoid. Foramen lacerum (medium) is in front of *PT* and outside *BO*. *PO*, post-occipital; *SO*, supra-occipital.

strip between the process and the ex-occipital, having a deep groove, the *digastric groove*, for the origin of the posterior belly of the digastric muscle.

Stretching forward and inward from this, in front of the ex-occipital, there is a part of the temporal bone applied loosely to the side of the basi-occiput. This is the *petrous* part of the temporal, the part that contains the organ of hearing, and named from the extreme hardness of its bone.

Skull and Hyoid

The *basi-sphenoid* is a direct continuation forward of the basi-occiput in the adult, and forms the roof of the two bony openings into the nasal cavities : the openings are separated by the *vomer*, one of the face bones, which articulates by expanded *alæ* with the under surface of the body of the sphenoid. The outer wall of each opening is made by a down-growing process that is descriptively part of the sphenoid, the *internal pterygoid plate*, ending in a hooked *hamular process* turning outwards. Outside each internal plate is an *external pterygoid plate*, placed more obliquely. The pterygoid plates are to be looked on, from a descriptive point of view, as processes from the sphenoid—the internal one from the under surface of the side of its body and the external one from the under surface of the great wing, which projects out from the body on each side. On the outer side of the base of the external plate the great wing can be seen projecting outward : this wing, if followed outward, is seen to widen rapidly antero-external to the petrous bone and then to narrow again, so that a sharply-pointed angle (Fig 161) of the wing projects back and fits in between the petrous and squamous parts of the temporal : the squama turns round on to the base in front of the external meatus to form the articular cavity for the lower jaw and the base just in front of this. The angle of the great wing can be recognised from the presence of the *spine* of the sphenoid on it.

We can therefore speak of the petrous as extending forward and inward between the basi-occiput and margin of the great wing of sphenoid, so that its apex tends to fit in between the great wing and the basi-sphenoid, but does not reach these bones : the interval between them is the *foramen lacerum*.

The under or *basal* surface of the great wing can be followed back, as just seen, to the angle between petrous and squamous, but can also be followed out to the inferior temporal ridge. The ridge is a crest that runs across this great wing and also across the squama behind it, and at the ridge the basal surfaces of these bones become continuous with their outer or temporal surfaces already described.

The *jugular foramen* is bounded internally and behind by the occipital, externally and in front by the temporal. Immediately to its outer side a long pointed *styloid process* projects from the temporal in a downward, forward, and inward direction.

Outside this the lower and front wall of the bony outer meatus is formed by the *tympanic plate*, and this plate is separated in front by a narrow but deep *Glaserian fissure* from the articular surface : the whole concavity is termed the *glenoid cavity*. It is bounded in front by the *eminentia articularis*, a root of the zygoma which forms the back limit of the passage under that arch.

The temporal is a compound bone whose various parts will be dealt with subsequently, but its general divisions can be easily and advantageously made out now. The petrous is all that part in contact with the occipital, and includes the basis of the mastoid process. The squama forms the articular area, and base just in front of this, as well as practically all the bone which appears on the lateral aspect of the skull, including the zygoma ; but in the lower and back part of this surface the squama only partly covers in the petrous basis of the mastoid, and thus a *petro-squamous suture* is visible, behind and below, while the petrous bone is seen laterally (Fig. 160). The styloid process and tympanic plate are morphologically separate portions of the bone, which may therefore be said to be compounded of four parts—not counting the ossicles of the middle ear that are enclosed in it.

A large round foramen in the basal surface of the petrous, in front of the jugular foramen, is the opening of the *carotid canal* in that bone for the passage of the internal carotid artery.

Various foramina in the basal surface of the great wing will be studied later.

In front of the line of pterygoid processes are placed the bones of the face : the pterygoid processes form a part of the support of the facial skeleton. Owing to the presence of the face it is not easy to follow the proper cranial base, but with a little trouble a fair notion of its arrangement can be obtained even with the face bones *in situ*, and this can be supplemented by an examination of a skull in which they have been removed.

Thus, looking into the orbits, the roof is seen to be considerably higher than the base of the skull we have so far examined, yet it is formed by the orbital plates of the cranial frontal bone ; at the back part, however, it may be possible to see that each orbital plate articulates with a narrow-pointed process of bone, the lesser wing of the sphenoid, which is separated below by the *sphenoidal fissure* from the orbital surface of the great wing.

At its lower border the orbital surface of the great wing becomes continuous with the under or basal surface which we have already examined, so that the base is completed in this region ; examination of this part from the side will show the continuity of the cranial base. But in the middle there is an interval between the orbital plates of the frontal ; a finger in one orbit and thumb in the other will grasp between them a delicate bony mass, the *ethmoid*, concerned in the formation of the upper part of the nasal cavities, and the upper plate of this mass fills up the interval between the orbital plates in the floor of the cranial cavity. The back surface of the ethmoid comes up against the front surface of the body of the sphenoid, and thus the base is completed centrally.

The relations to each other of these bones in the base of the cranium can be grasped more completely if this part of the skull is examined *from above*, the top of the skull being removed (Fig. 162). The upper or cerebral aspect of the base is divisible into three large (primary) *fossæ, anterior, middle* and *posterior*, each extending from one side to the other.

The posterior fossa is seen to be floored by the occipital, and the three basal parts of this bone are easily recognised arranged round the foramen magnum. At the sides in front there are obliquely set prominent bones which are directed forwards and inwards and divide the posterior from the middle fossa. Examine these bones and compare their relations with the under surface of the base, and it becomes apparent that these are simply the petrous bones standing up in the cavity of the skull : their inner and posterior aspects form part of the wall of the posterior fossa and have the jugular foramina between them and the occipital, while their upper and front aspects form part of the floor of the middle fossa. External to each petrous bone the squama can be recognised as forming a part of the side wall of the middle fossa.

The sphenoid can be easily examined from this aspect ; its central portion or body is continuous through the basi-sphenoid with the basi-occiput, and the floor of the posterior fossa can be traced forward on to the sloping back of the sphenoid. The body presents a deep *pituitary fossa* for the pituitary body, which connects the two lateral portions of the middle fossa across the mid-line, and is overhung from behind by the edge of the back part of the bone. The upper aspect of the body bears a fancied resemblance to a Turkish saddle,* and the back portion, which constitutes the front end of the posterior fossa, is hence frequently termed the *dorsum sellæ*.

On each side of the body the great wings are seen to form a large part of the floor of the middle fossa. Each great wing extends back, as seen on the lower surface, to fit into the angle between the petrous and squamous parts of the temporal : thus the floor of each half of the middle fossa is made up of sphenoid, upper, outer, and front

* The upper hollow surface is often termed the *sella turcica*.

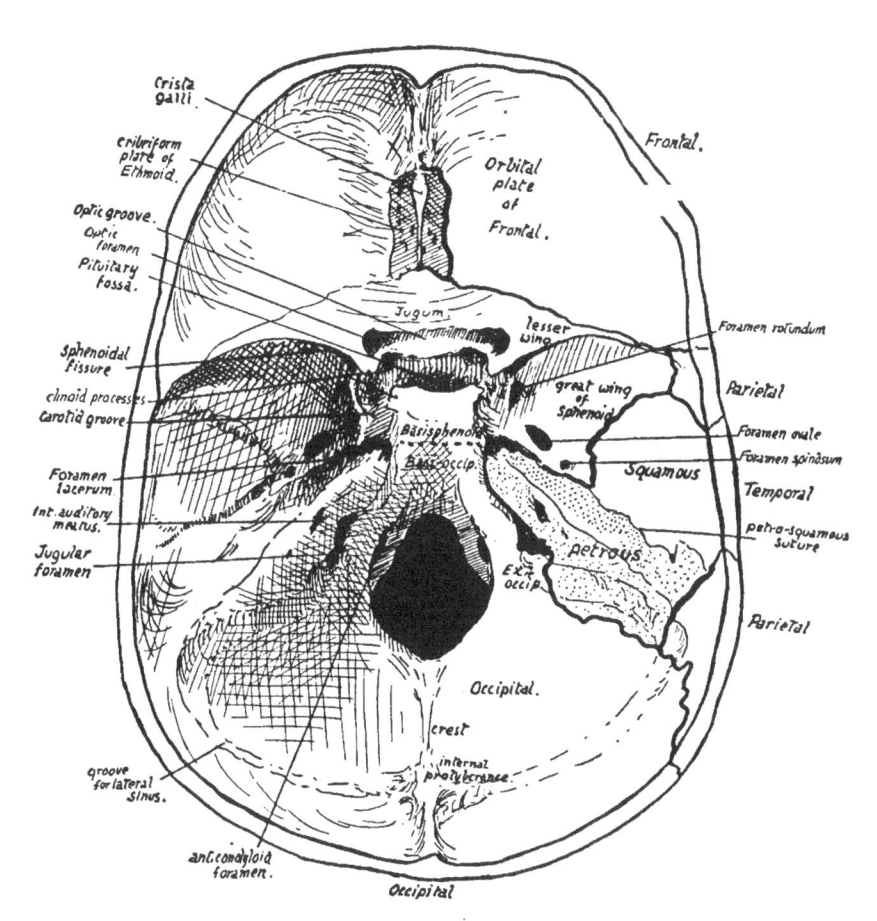

FIG. 162.—Cranial surface of base of skull. The middle fossa is coloured, and the posterior and anterior fossae lie behind and in front of this coloured part respectively. On the right side, the positions of the sutures are indicated, so that the bones which form each fossa can be seen. Compare with Fig. 181, in which the dura mater is in place.

surface of petrous, and lower part of squama. A petro-squamous suture runs back from the angle between the petrous and squama.

Where the great wing forms part of the floor of the fossa it is in a nearly horizontal plane, and its under surface is the inferior, basal, or pterygoid surface already seen below : in front it turns upward and is separated from the *lesser wing* by the *sphenoidal fissure*, and this turned-up part carries on its front aspect the *orbital surface* of the wing. Laterally it turns up on the side wall of the skull, and here its outer surface constitutes the temporal surface of the wing.

These three surfaces, which we have already inspected, are more or less flat areas, looking in different directions, and therefore separated from each other by sharply-marked borders, but on the cerebral aspect the angles between the corresponding districts are raised and rounded off, making the wing a thick bony mass hollowed to receive the temporo-sphenoidal lobe of the brain.

The back border of the small wing overhangs the greater wing to a small extent, so that one must look under the former to see the extent of the sphenoidal fissure : the fissure opens into the cavity of the orbit, being between the two wings, which are both concerned in forming part of the orbital walls.

The anterior fossa is floored mainly by the orbital plates of the frontal, but behind these are the lesser wings with which they articulate, and between them is the *cribriform plate* of the ethmoid—so termed because it is pierced by many foramina for the passage of the olfactory nerve fibres. This upper surface of the ethmoid articulates behind with a raised part of the sphenoid connecting the two lesser wings and frequently termed the *jugum sphenoidale*, which is also in the floor of the anterior fossa.

As soon as the arrangement of the bones in the fossa is seen and mastered, the student can proceed to the recognition of some of the more important points visible (see Fig. 162) on this aspect of the base. The *crista galli* is a blunt process of bone on the ethmoid for attachment of the " falx cerebri " : this is the sagittal sheet of dura mater that lies between the two cerebral hemispheres. Immediately behind the jugum sphenoidale is the *optic groove*, leading on each side to the *optic foramen*, through which the optic nerve and ophthalmic artery pass to the orbit. The *olivary eminence* is between this groove and the pituitary fossa. The groove, foramina, eminence, and fossa are all in the middle primary fossa.

The upper opening of the foramen lacerum is seen just in front of the apex of the petrous : here the internal carotid artery emerges from the apex and runs forward on to the side of the body of the sphenoid, which accordingly shows a *carotid groove* here.

The *internal auditory meatus*, for the passage of the facial and auditory nerves, is easily recognised on the inner side of the petrous, and the jugular foramen is situated below it, between the petrous bone and the ex-occipital : large grooves are seen leading to this foramen, which lodge the venous sinuses that form the internal jugular vein. Internal to this again, and on a lower level, near the margin of the foramen magnum, is another canal that ends below in front of and external to the condyle, and is therefore termed the *anterior condylar foramen* (Fig. 162).

Behind the foramen magnum is a median *internal occipital crest*, ending above in the *internal occipital protuberance*.

Grooves for the lateral sinuses run horizontally forward on each side from the protuberance and turn down on the petro-mastoid to reach the jugular foramina. The tentorium cerebelli, the partial sloping diaphragm of dura mater that separates the cerebrum and cerebellum, is attached along the horizontal parts of the grooves and then along the prominent upper edges of the petrous bones to the projecting bony processes on the sphenoid: thus the cerebellum is altogether confined to the posterior fossa. The bony sphenoidal points to which this sheet of dura mater is attached are termed *clinoid processes:* the *ante*rior process is a well-marked projection from the root of the small wing on each side, the *posterior*, also prominent, project at the upper angles of the dorsum sellæ, while the *middle*, often absent, is a small tubercle internal to each carotid groove. These processes are connected by *interclinoid ligaments*.

The position and connections of the great venous sinuses of the dura mater should now be followed in the skull (Fig. 163). They are all in relation with the bones with the exception of those that are enclosed in the free parts of the falx and tentorium, *i.e.*, the inferior longitudinal and straight sinuses.

The cranial bones are for the most part plainly of the "tabular" variety, and are built in two compact layers—the *inner* and *outer tables* of the skull—. with an intervening stratum composed of bony pillars connecting the tables and forming a coarse cancellous layer containing red marrow: this is termed the *diploe*, and large anastomosing channels run through it for the diploic veins that drain the tissue.

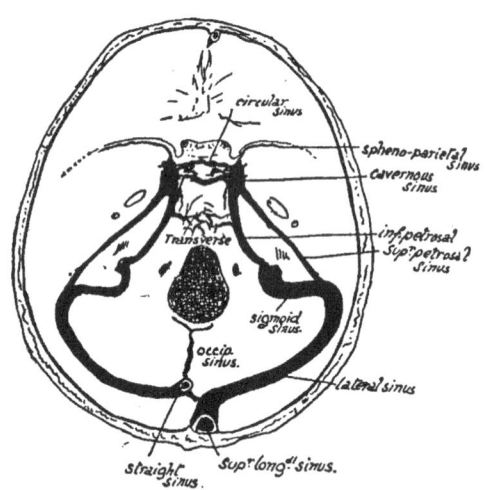

FIG. 163.—Position of cranial sinuses. Cavernous sinus is on the side of body of sphenoid; sphenoparietal sinus reaches it along the edge of lesser wing, and superior and inferior petrosal connect it with sinuses in posterior fossa. Transverse and circular sinuses are plexiform connections across the middle line. Observe that the right lateral receives the blood from the superior longitudinal sinus and is therefore larger than the left, which as a rule only drains the straight sinus. The "torcular Herophili" is the small connection between the right and left systems near the internal occipital protuberance. The right lateral sinus usually comes further forward in the petromastoid region than the left.

The diploe appears first about the age of ten; it becomes most developed in bones and regions that are thick.

The cranial bones, with the exception of the occipital and parietals, are connected with the bones of the face, so that it is advisable to gain some general notion of the arrangement of the facial skeleton from a scrutiny of the complete or halved skull before entering upon a detailed examination of the individual bones of the cranium.

FACIAL SKELETON.

This is situated altogether in front of the level of the pterygoid processes, with the exception of the mandible.

Looking at the skull from the front (Fig. 164), the two upper jaws or *maxillæ* are seen to constitute the side margins of the nasal opening and to meet each other below it. Each maxilla forms, by its upper surface, the greater part of the floor of the orbit and is prolonged up the inner margin of the orbital opening as a *nasal process* which meets the frontal. The two *nasal* bones also articulate with the frontal and separate the nasal processes of the two maxillæ. The maxillæ meet each other, not only on the front, but also below, forming a large part of the hard palate ; this can be seen on looking at the skull from below. Now look into the nasal cavity from the front : its floor is evidently formed by the bony *palatine processes* of the maxillæ, and a bony *septum* is seen dividing the cavity into right and left.

The outer margin of the orbital opening is not formed by the maxilla, but by the *malar* which rests by a broad base on the maxilla and reaches to the external angular process of the frontal above. Looked at from the side the malar is seen to throw a process backwards and outwards to complete the zygomatic arch by joining the zygomatic process of the temporal : another part of the bone turns inwards and backwards, forms the front wall of the temporal fossa, and articulates with the great wing of the sphenoid.

The orbital and nasal cavities can be more conveniently examined at a later stage, but it is well to notice here that the mass of the ethmoid separates the two orbits, the inner walls of which are largely made by the smooth *os planum* or side-plate of the ethmoid. Being in this situation the ethmoid forms the upper part of the nasal cavities, and its central *vertical plate* is in the upper part of the bony septum : the lower and back part of the septum is made by the *vomer* which can be seen between the posterior nasal openings, articulating with the inferior surface of the body of the sphenoid. The mass of the ethmoid is placed against the front of the body of the sphenoid.

The nasal cavities, then, lie not only between the maxillæ but at a higher level also, *i.e.*,

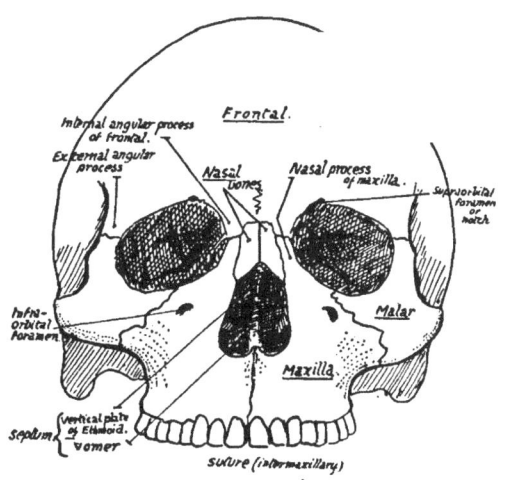

FIG. 164.—Facial aspect, without lower jaw.

between the orbits. This higher part is in the ethmoid, and a central *vertical plate* of this bone divides the two cavities here, and articulates with the vomer below.

Now turn the skull up and examine its under surface (Fig. 161), where we have seen that the maxillæ are joined below by *palatine processes*, which form the greater part of the *hard palate* and floor of the nasal cavities. But behind these processes two other smaller bones are seen completing the hard palate : these are portions of the *palate bones*. It can be seen that the palatine process of each maxilla does not extend back for the full depth of the bone, and the interval is filled by the palate. The palate bone in its greater part is placed vertically against the internal aspect of the maxilla in its back part, reaching down to the interval behind the palatine process ; here the palate bone bends in at right angles to complete the bony palate ; it is thus divisible into a *vertical portion* applied to the maxilla, and a *horizontal part* that helps to form the palate. These general relations can be comprehended from a short study of the skull.

But the palate bone also projects back further than the level of the maxilla, and the posterior border of its vertical plate is applied to the front edge of the internal pterygoid plate, so that the upper jaw is held away from the pterygoid plate by this projection of the palate. Look at the skull from the side and see that the external pterygoid palate is separated by an interval (*spheno-maxillary fossa*) from the maxilla, and this fossa has its inner wall made by the vertical plate of palate passing back from the inner aspect of the maxilla to articulate with the internal pterygoid plate : thus the vertical plate separates the spheno-maxillary fossa from the nasal cavity.

. The fossa is closed below by a mass of bone, which, in the complete skull, appears to be part of the maxilla, articulating with the lower parts of the pterygoid plates ; but this is not really so, for the maxilla does not come at all into contact with the plates, but is separated from them by this bone, which is the *tuberosity* of the palate protruding downwards and outwards behind the maxilla and thus firmly welding it to the pterygoid plates (see Fig. 161). So it is evident that the bones of the face are not only attached above to the cranial bones, but are also buttressed behind by the pterygoid plates, and, through the zygoma, by the temporal.

For a complete list of the names applied to different points on the skull as used in craniometry the reader must consult works dealing with that subject, but as some of these terms are in common use in description it is convenient to give a list of these here, with an indication of their meaning.

In the middle line :—

Nasion. Junction of nasal and frontal bones.
Glabella. Mid-point at level of superciliary ridges.
Bregma. Junction of coronal and sagittal sutures.
Lambda. Junction of sagittal and lambdoid sutures.
Inion. External occipital protuberance.

At the side of the skull :—

Auricular Point. Centre of external meatus.
Asterion. Region where occipital, parietal, and temporal meet.
Pterion. Region where frontal, parietal, sphenoid, and temporal meet.
Reid's base-line is used for certain surgical surface markings : it is drawn back (Fig. 168) from the lower margin of the orbital opening through the auricular point, and when prolonged back passes a little distance below the inion.

SEPARATE BONES OF THE SKULL.

When examining the individual bones the student should have beside him a complete or half-skull in addition to the disarticulated bones. Every point observed on the separate bone ought to be verified and extended on the whole skull, and the general position of the particular bone in the skull must be continuously borne in mind. In this way the individual bone and its related structures are followed and comprehended as parts of a connected whole, and the meaning of its attachments, etc., becomes apparent, while at the same time an extended knowledge of the whole skull is gradually built up that enables one to understand it much more fully on going over it again.

FRONTAL BONE.

A tabular bone forming the front of the cranial region and a small portion of the temporal fossa laterally, but also sending back from its lower border two orbital plates which make the greater part of the roof of each bony orbit.

Skull and Hyoid

The *frontal plate*, on its front surface, is smooth and convex, covered by the anterior part of the Occipito-frontalis sheet and, above the orbital margins, by the Orbicularis palpebrarum on each side. The supraorbital vessels and nerves run up to the scalp

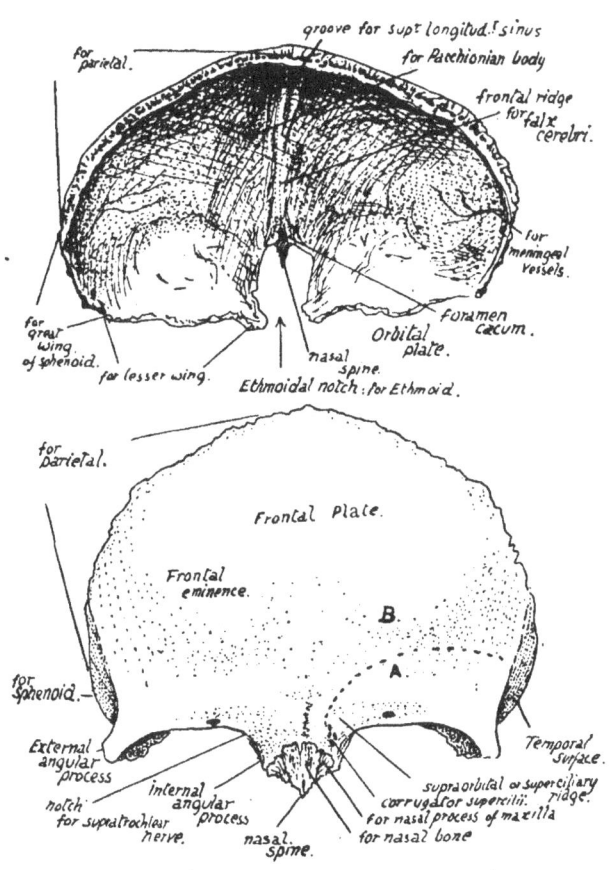

FIG. 165.—Frontal bone. Upper figure, from above and behind. Lower figure, from the front: *A.*, surface (below interrupted line) covered by Orbicularis palp.; *B.*, surface covered by Occipito-frontalis.

deep to the last-named muscle, and the bone frequently shows grooves marked on it by filaments of the nerve.

The surface is bounded above by the line of the coronal suture and below by the *orbital margins*: between these it is prolonged down for a little distance as a *nasal surface*, on which are two rough articular areas on each side, the inner for the corresponding nasal bone and the outer for the nasal process of the maxilla. This surface ends below and centrally in the *nasal spine*.

The front surface presents :—

(a) Two *frontal eminences*, marking the site of commencing ossification and due to the fixation of the earlier curves as a result of this.

(b) *Supraorbital* ridges meeting centrally and fading away above the middle of the orbital margins—due for the most part to the existence of the *frontal sinus*, a large air sinus situated here between the two tables of the bone. Covered by Orbicularis laterally and Occipito-frontalis centrally, and giving origin to fibres of Corrugator supercilii that run out to join the deep surface of Orbicularis.

(c) On the orbital margin the *supraorbital notch* for these vessels and nerve : sometimes, as in the figure, this is closed and converted into a foramen by ossification of a " supraorbital ligament " that crosses the notch. In the floor of the notch may be seen a small canal for minute vessels and nerve to the frontal sinus. The orbital margin ends externally in the *external angular process* : it is less sharply marked at its inner end, presents an ill-defined groove here for the supratrochlear nerve, and turns down as the *internal angular process* at the side of the nasal surface. It has the superior palpebral ligament attached to its outer part.

(d) The *metopic suture*, always to be found between the supraorbital ridges, and occasionally (7 or 8 per cent. Europeans) completely separating the two halves of the bone : the complete fissure occurs in only 1 per cent. of African skulls, and 4 or 5 per cent. of skulls of yellow races.

The *temporal surface* (Fig. 166) is separated from the frontal surface by the anterior part of the superior temporal crest, running down to the external angular process it gives origin to fibres of Temporal muscle. A large part of the superficial aspect of the bone here is articular, being overlapped by the parietal above and the great wing of sphenoid below.

The *orbital surface* of each orbital plate (Fig. 166) is deeply concave, and triangular in shape with the blunted apex pointing backwards and inwards ; this accords with the pyramidal shape of the orbital cavity, which forms an apex behind and internally, at the optic foramen. As the orbital plate makes the greater part of the roof of the cavity its general shape is therefore triangular, but its apex is cut off because the lesser wing of sphenoid forms the roof of the apical part of the cavity. The surface is in contact with the upper surface of the Levator palpebræ and Superior oblique muscles and the orbital fat and the supraorbital vessels and nerve and supratrochlear nerve, and presents :—

(a) Hollow for lodging the lachrymal gland.

(b) A depression, sometimes a small tubercle, that marks the attachment of the pulley of the Superior oblique—the *trochlear fossette*.

(c) Notches on its inner margin that mark the position of the anterior and posterior orbital (ethmoidal) foramina, completed by the articulation with the ethmoid.

Notice that the inner borders of the surfaces are practically parallel, while the outer ones, if prolonged, would meet at a right angle : compare this with the orbit in the skull.

The inner margins are connected by the nasal part of the bone in front, but they are separated in their posterior half or more by the *ethmoidal notch* ; this is filled by the ethmoid in the complete skull. Observe that the ethmoid lies below the level of the orbital plates, which rest on it, so that there is a rough articular surface running along the margins of the notch, for the ethmoid. This ethmoidal surface comes forward to

Skull and Hyoid

the level of the nasal spine, where it shows the *openings of the frontal sinuses* on each side. Between each of these and the spine is a smooth concave strip of bone that forms the roof of the nasal fossa here and has the nasal nerve lying on it deep to the mucous membrane.

The deeply concave *cerebral surface* (Fig. 165) of the bone covers the frontal lobe of the brain and is marked by its convolutions: notice how these are particularly apparent on the convex surface of each orbital plate. In addition to these cerebral

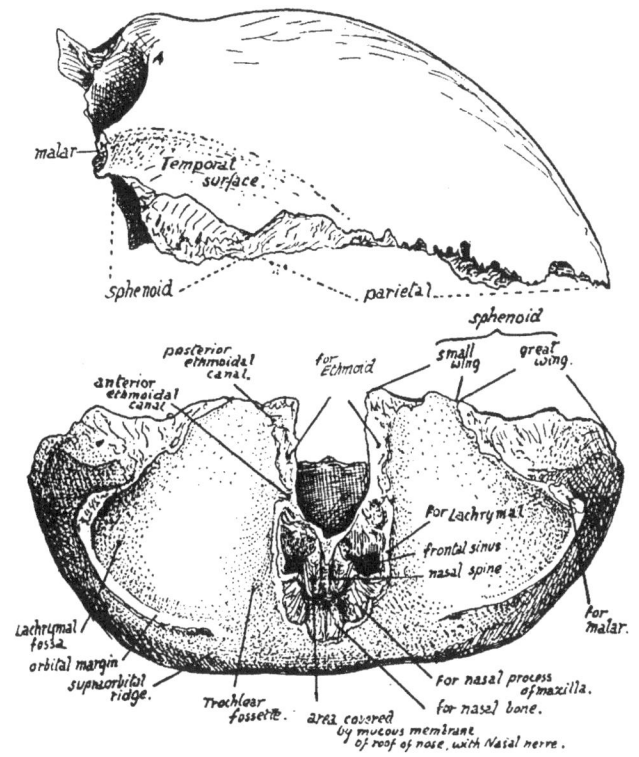

FIG. 166.—Frontal from the side and from below.

markings and those for meningeal vessels between it and the dura mater the cerebral surface shows :—

(a) A *median* ridge for the attachment of the falx cerebri, passing above into a *groove for the superior longitudinal sinus* which lies between the two layers of the falx. The ridge ends below at the margin of a small pit, the *foramen cæcum*, just in front of the ethmoidal notch : the foramen usually transmits a minute vein connecting the sinus with the veins of the frontal sinus.

(b) Depressions near the attachment of the falx for Pacchionian bodies.

The *margins* of the bone are articular. The parietal articulates along the serrated coronal border to about half-way down the temporal surface. Below this the sphenoid articulates by its greater and lesser wings as far as the ethmoidal notch, taking all the articular surface except that on the external angular process, which articulates with the malar. The ethmoid lies against the inferior aspect of the margins of the notch and also articulates with the lower border of the nasal spine, between the two nasal roof areas. Arranged round the spine antero-laterally are, from within and in front outwards and backwards, articular surfaces for the nasal, maxilla, and lachrymal. The nasal bones lie together along the front and upper border of the spine.

Development.
The bone is developed in membrane, in which two main centres appear in the eighth week, in the region of the future frontal eminences. Small additional centres, appearing later and fusing with the main halves, have been described for the nasal spine, the trochlear spine, the external angular process, and the back border of the orbital plate : it seems doubtful whether these are to be considered as of usual occurrence. The bone is in two separated halves at birth, but these come into contact in the first year, fuse in the middle region of the metopic suture in the second year, and are completely joined by the seventh or eighth year. There may be total absence of fusion, as already mentioned, or the metopic suture may remain unclosed below, forming a *metopic fontanelle*, and Wormian bones may appear here or rarely in the upper part.*

The frontal sinuses develop as extensions from the nasal cavity, and are first indicated in the bone in the first or second year, but grow slowly up to puberty. After this they increase rapidly, and continue to grow until a late period (see later, under Nasal Cavities).

PARIETAL.

A four-sided bone situated in the back and side of the head and articulating with its fellow in the middle line above (sagittal suture). The bone is a little broader above than below.

The outer surface, convex, is crossed by the *temporal ridge* about a third or more of its height up : the ridge is curved and usually shows two lines, the lower for the Temporal muscle and the upper for its covering aponeurosis. Above the ridge the bone is covered by Occipito-frontalis ; the lower part is covered by and gives origin to the Temporal muscle. Observe the differences in the surface texture in these two areas.

The *parietal eminence* is the most prominent part of the convexity, presenting a smaller curve than elsewhere : it is situated above the temporal ridge behind the centre of the bone, and marks the site of earliest ossification.

The *parietal foramen*, which may be absent, unilateral, or very large, is near the upper edge about an inch or less in front of the upper and posterior angle. At birth the position of the foramen is marked sometimes by a transverse slit in the bone (*subsagittal suture* of Pozzi), which enables one part of the bone to ride slightly over the other during parturition, thus aiding the moulding of the head. The slit is obliterated afterwards by extending ossification, and the foramen is only kept open by the persistence of small vessels that may run through it. Some phylogenetic interest is attached, moreover, to the foramen in that it has been supposed to represent in the human skull the place where, in lower vertebrates, the median eye reaches the surface.

* The writer has only once seen such a bone in the upper part of the suture, at the bottom of a gummatous ulcer in a hospital patient.

Skull and Hyoid

The concave inner or cerebral surface has slight depressions corresponding with the convolutions, particularly in its lower part. It otherwise presents:—

(a) Grooves for vessels of the *middle meningeal* series: usually two main branching grooves, one running up near the front, the other further back.

(b) A partial *groove for superior longitudinal sinus*, better marked on the right

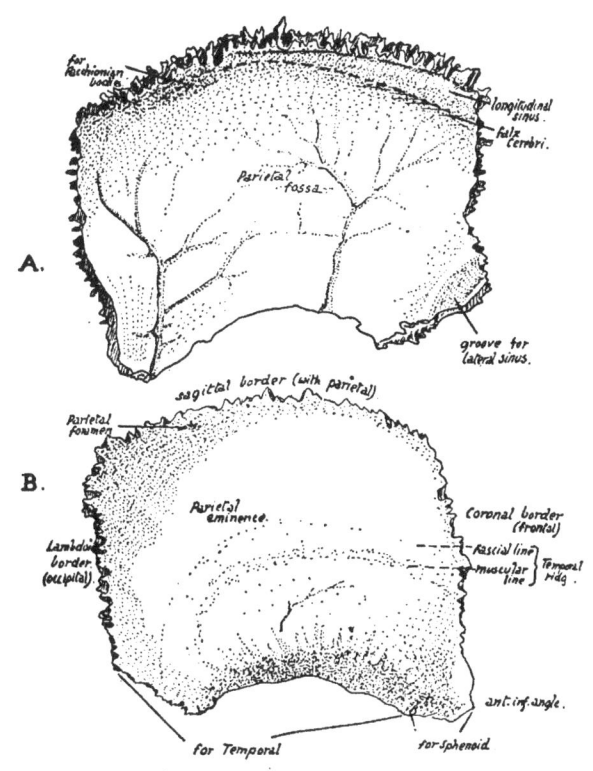

FIG. 167.—Right parietal. *A.*, inner surface; *B.*, outer surface.

bone as a rule (see Fig. 167). The margins of the groove afford *attachment to the falx cerebri*.

(c) Depressions for *Pacchionian bodies*.

(d) A groove on postero-inferior angle, where the *lateral sinus* reaches the bone from the occipital and at once turns down and passes off it on to the temporal.

(e) A *parietal fossa*, corresponding with the deepest part of the concavity of the bone.

MARGINS.—(1) Upper or *sagittal*, much serrated, but usually rather straighter opposite the foramen; articulates with the opposite parietal.

(2) Anterior or *coronal*, smaller serrations, articulating in its whole length with frontal, which it overlaps below; but in the upper part the frontal overlaps

204 Anatomy of Skeleton

the parietal to a less degree. Observe that this suture makes almost a right angle with the sagittal.

(3) Posterior or *lambdoid* ; makes an obtuse angle with the upper border, is much serrated, and articulates in its whole length with occipital.

(4) Lower border ; hinder part, starting at the angle, overlaps and articulates with the petro-mastoid. In front of this the curved margin is bevelled-off on the surface and shows fluted markings for the squama. A little distance from the antero-inferior angle this surface changes abruptly into a smoother surface, slightly concave, overlapped by the sphenoid.

Observe that the lower and front part of the bone is drawn out in a narrow angle : this proclaims the side to which the bone belongs when only seen at a distance.

The parietal covers the parietal lobe, but in addition has lying deep to it portions of all the other lobes that are seen on the surface of the brain. It must be understood

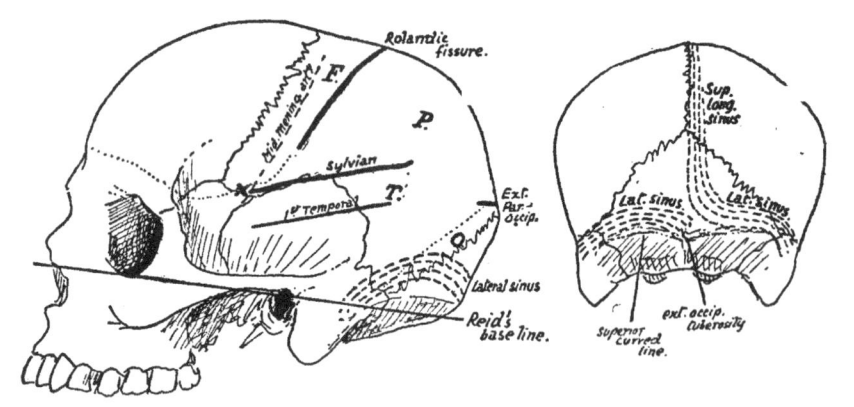

FIG. 168.—Approximate positions of main sulci and sinuses on the surface. On the parietal, *F.* is frontal lobe, *P.* parietal lobe, *O.* occipital lobe, and *T.* the hinder part of temporal lobe. *X.* is the position of the "sylvian point."

that the bone is as much concerned in forming the back as the side of the skull, so that the fissure of Rolando is only about 1½ inches behind the line of the coronal suture. The position of the fissure and lobes is approximately shown in Fig. 168.

Development.

From two centres appearing in membrane in the eighth week and rapidly fusing : sometimes apparently only a single centre, though this may be only an instance of very early union. The centres are generally upper and lower, and occur in the region of the parietal eminence ; from thence ossification radiates out toward all the margins. The four corners of the bone are not ossified at birth, and the spaces left in this way constitute, with the neighbouring bones, the fontanelles (see p. 266 for account of fontanelles).

OCCIPITAL.

A tabular bone situated in the lower and back part of the skull and resting on the atlas, so that its cerebral surface is markedly concave and its superficial surface convex. It is divisible, as we have seen, into parts grouped round the foramen magnum.

Skull and Hyoid

Seen from below, the foramen magnum is oval in shape, but its outline is encroached upon (Fig. 169) in its front half by the inner margins of the articular condyles; other-

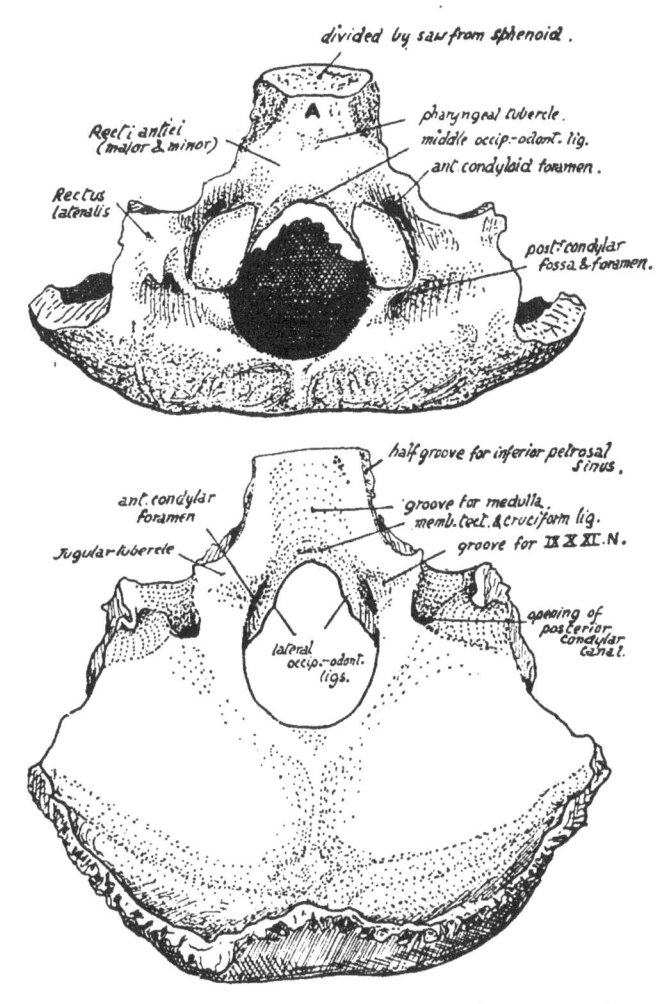

FIG. 169.—Occipital bone. *A.*, basi-occiput in relation with roof of naso-pharynx.

wise the margin is thin, giving attachment to anterior and posterior occipito-atloid ligaments and to the capsule on each side of the condylar projection.

There is often a small bony point in the middle line on the front margin, that marks the attachment of the middle occipito-odontoid ligament.

·The *condyles*, convex from before backwards but twisted on their long axis, look

Anatomy of Skeleton

on the whole downwards and backwards and outwards; each is raised from the surface on a block of bone traversed by the *anterior condylar canal*, the lower opening of which is above the antero-lateral margin of the condyle. A *posterior condylar fossa* and *foramen* (not always present) are placed behind the condyle.

The long axes of the condyles form an angle of 65 degrees with each other; they permit a nodding movement through an angle of about 45 degrees, and a certain amount of lateral rocking, but no rotation is possible (see p. 18). Very rarely there is an articulation in front for the atlas or odontoid; the only condyle present in birds and reptiles is an anterior median one.

Looking into the foramen from above, the prominence of each condyle is more striking, and above it is seen the upper opening of the anterior condylar canal: below this is a small bony tubercle marking the attachment of the lateral occipito-odontoid ligament.

Basi-occiput: in front of the foramen, a thick, short, rod-like piece of bone.
Lower aspect shows :—

(a) A *pharyngeal tubercle* for the attachment of the mid-point of the posterior wall of the pharynx.

(b) Markings behind and on each side of the tubercle for insertion of the anterior Recti.

(c) A small area in front of the tubercle in relation with mucous membrane and pharyngeal tonsil (*A*. in Fig. 169).

Upper aspect is grooved for reception of medulla and vertebral arteries, these being supported by a cushion of subarachnoid fluid. Deep to these the cruciform ligament and membrana tectoria are attached a little above the margin of the foramen, where their markings may be found by the finger. Along each lateral border of this aspect is a small half-groove for the *inferior petrosal sinus*, completed by the petrous bone in the articulated skull.

The *exoccipital* is that part of the bone that lies outside the foramen magnum and forms all its margins except the extreme front and back parts; it carries the hinder three-quarters or so of the condyle, and the line of its junction with the basi-occipital is marked by the anterior condylar canal. It forms the posterior wall of the jugular foramen and has on its outer side a roughened articular surface that is separated from the petrous bone by a layer of cartilage which ossifies in middle age. On the upper aspect of its junction with the basi-occiput is seen (Fig. 169) a jugular tubercle and, behind this, a groove for the ninth, tenth and eleventh nerves as they run to the jugular foramen.

The upper surface of the exoccipital has a prominent hooked process standing up at its outer end, and round this the deeply-cut groove for the sigmoid sinus turns sharply forward: the "posterior condylar canal" opens into the front and lower part of this groove. The lower surface (Fig. 169) shows rough markings for Rectus lateralis inserted here, and one of these ridges may be produced downwards as a *par-occipital process*, which may even articulate with the atlas.

Now articulate the bone with the temporal and look at the jugular foramen between them; it has a larger part externally for the vein and a small inner part for nerves, and these two parts are often separated in part or wholly by a small bony projection from the occipital above the anterior condylar foramen, called the "intra-jugular process."

The tabular or squamous part can be said to consist of a *post-* or *infra- occipital* and a *supraoccipital* portion: looking at the complete skull it can be seen that the part immediately behind the foramen magnum looks down towards the neck and is marked

206

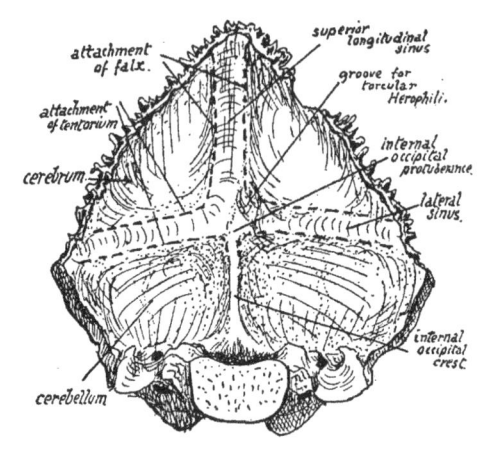

FIG. 170.—Cerebral aspect of occipital bone.

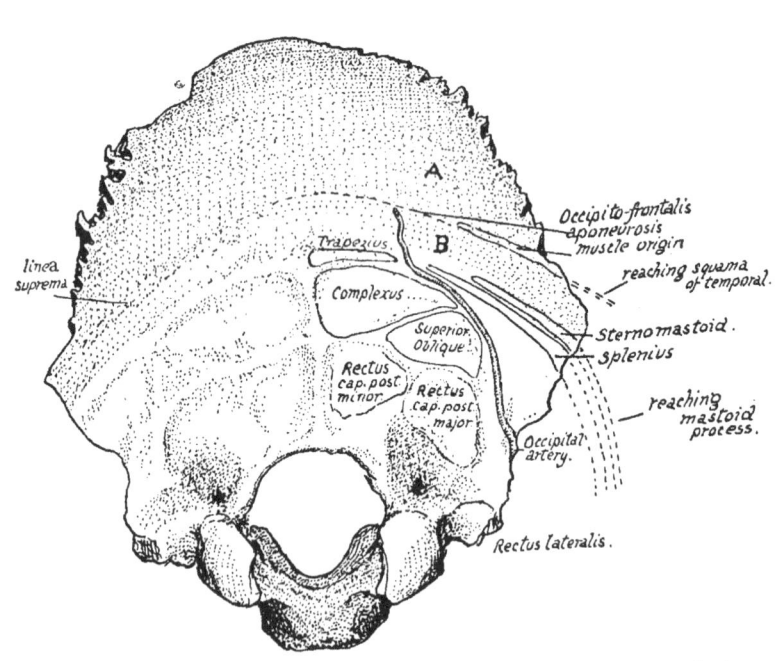

FIG. 171.—Attachments, etc., on occipital. *A*, covered by Occipito-frontalis; *B*, subcutaneous area crossed by artery and nerve.

by nuchal attachments, while the part above and behind this looks backwards and lies under the scalp; the latter is the (membranous) supraoccipital and is between the parietals,* and the former is the (cartilaginous) post-occipital and abuts on the petro-mastoid laterally. The borders of the squamous part show the distinction, for those on the supraoccipital are strongly serrated, but those on the post-occipital are not serrated at all.

The cerebral surface (Fig. 170) has (a) The *internal occipital protuberance*, with the *internal occipital crest* running down from it to the margin of the foramen magnum : the crest is sometimes replaced by a groove, containing the occipital sinus. (b) *Grooves* running out transversely for the *lateral sinuses*. (c) A groove ascending from the tuberosity for the *superior longitudinal sinus*, usually to the right of the middle line and continuous with the right lateral groove : the left lateral groove appears to start at the tuberosity, because the left sinus is generally continuous with the straight sinus, which is in the tentorium cerebelli. The junction between left and right systems is usually small and lies on the protuberance, but occasionally makes a groove on the bone, as in that figured ; it is called the *torcular Herophili*.

The lateral and vertical grooves have tentorium and falx respectively attached to their margins, hence the cerebral hemispheres are in relation with the supraoccipital and the cerebellum with the post-occipital. The falx cerebelli and occipital sinus are on the internal crest. There are marks for cerebral gyri in the upper quadrants, but the cerebellum and cisterna magna cause no marking on the bone, and ridges on this part are due to posterior diploic veins running to the sigmoid sinus, where their terminal openings can be seen.

The superficial convex surface of the tabular portion has (Fig. 171) its most promi nent part, the *external occipital protuberance*, a little below its centre. The *superior curved lines* run out from this to reach the mastoid process : below this, on the post occipital, are a vertical *external occipital crest* and a transverse *inferior curved line :* between them are rough markings made by the thick areolar fasciæ round the muscles of the suboccipital region.

The attachments (Fig. 171) are :—

To *upper curved line :* Trapezius, Splenius, Sternomastoid ; the last two reaching mastoid.

To *lower curved line :* Fascia separating posterior Recti from Complexus and Superior Oblique.

Between the lines : Complexus and Superior Oblique.

To *protuberance and median crest*—Ligam. nuchæ.

There may be a *groove for the occipital artery* outside the inferior curved line and the area of Superior Oblique.

The occipital artery reaches the base of the skull by crossing the carotid sheath in an upward, outward, and backward direction from its origin from the external carotid on the pharyngeal wall. In so doing, it lies on (see Fig. 22) the internal carotid, hypoglossal, internal jugular, and spinal accessory, and reaches the outer edge of Rectus lateralis, along which it passes to the skull : it is covered by the Digastric. The place where it first touches the skull, therefore, is between the lateral Rectus, on the exoccipital, and the digastric groove on the petro-mastoid—*i.e.*, it is usually to be found as a groove on or by the suture between these two parts of the skull. It now runs upwards, inwards, and backwards, in more or less close association with the occipital (Fig. 171) and round the outer side of the muscles forming the suboccipital triangle : it is covered at first by the mastoid process and its attached muscles, but as it runs back it gets beyond the process

* Hence frequently termed the " interparietal " part of the bone.

and the Digastric, then beyond the Trachelomastoid, and then, beyond the level of the posterior border of the Sternomastoid, is only covered by Splenius capitis, and makes for the interval between this muscle and the Trapezius. It lies on fibres of Complexus at this interval, then crosses the bone between the upper curved line and linea suprema, to pass on to the Occipito-frontalis. Before reaching the Complexus it gives off its *princeps cervicis* branch, which thus lies over the suboccipital muscles and can divide into deep and superficial branches in relation with the Complexus.

The supraoccipital or part of the bone above the superior curved line is covered by the occipito-frontalis, which is attached to the bone a little distance above the upper curved line on the so-called *linea suprema ;* the central part of the line is usually better marked because the aponeurosis is fastened here, the muscle fibres arising further out.

The bone between the linea suprema and linea superior is crossed by the occipital artery as it runs up to lie on the aponeurosis after emerging from between Trapezius and Splenius (Fig. 171): this part of the bone is sometimes prominently bulged, being then known as the *torus occipitalis transversus*.

The blunt prominence of the external protuberance is easily felt from the surface, and the bone can be easily palpated through the movable scalp above this level: the apex of the lambdoid suture is about $2\frac{1}{2}$ inches above the prominence. Below this level, however, the bone is deeply covered by the thick muscles of the neck and cannot be felt. As might be expected from the absence of connection between the factors responsible for them, the outer and inner prominences do not by any means necessarily correspond in position or level on the bone. This is a fact to be borne in mind when using the outer projection for purposes of locating the deep structures thus, for example, the lateral sinus may have the surface relations seen in Fig. 168, or may be higher or lower compared with the upper curved line and outer protuberance; as a rule the groove is well above the level of the superior line.

The front ends of the condyles are in a line joining the middle of the two outer meatuses, but on a lower level. The middle of the foramen magnum is very little behind a line joining the tips of the mastoid processes, but above its level. Observe, moreover, that in the skull in the ordinary position the lower surface of the condyles is about on a level with the hard palate.

Development.

The basi-, ex-, and post- occipital are laid down in cartilage, but the supraoccipital is a membrane bone. The basi-occiput begins to ossify first, quickly followed by the post-occipital, in the sixth to seventh week, while the centres for the exoccipital appear about the eighth week. Centres for the basi-occiput are paired, fusing rapidly to form one: in addition an anterior centre has been described, looked on by some as an epiphysis on the bone. Each exoccipital has a single centre. For the cartilaginous part of the post-occipital two or four centres are described, while for the upper membranous part four centres appear to be the rule, appearing first about the beginning of the third month. Of the four membranous centres the two lower are termed the *interparietal* and the smaller upper ones the *preinterparietals*. At birth the bone is in four pieces (see Fig. 214). The tabular and exoccipital parts unite in the third year, and the basal portion joins a year or two later. Looking at the figure it is seen that the exoccipitals form the greatest part of the margin of the foramen, the other elements of the bone only contributing a small piece each in front and behind: the hinder end of the margin may, however, have a separate small centre constituting the "*ossicle of Kerckring.*"

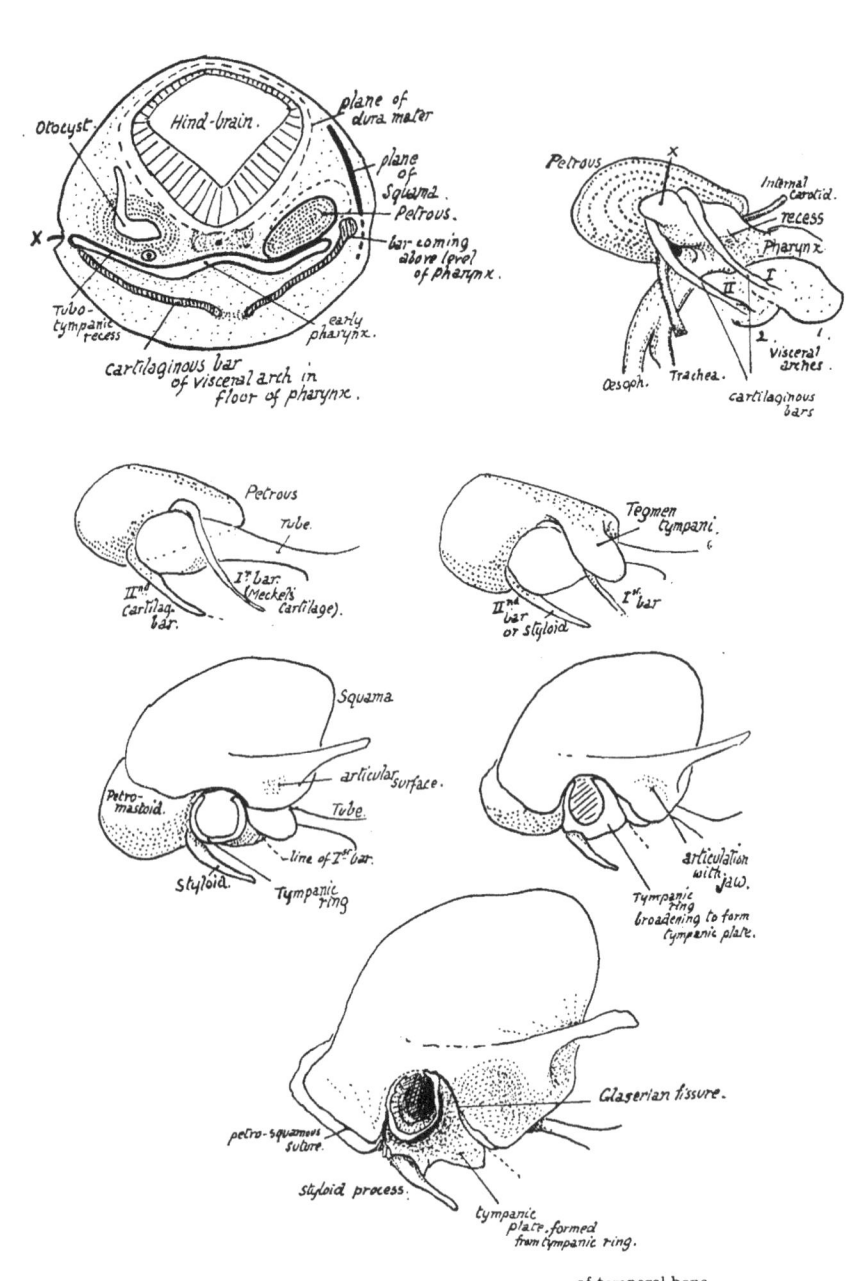

FIG. 172.—Schemes to illustrate development of temporal bone.

Skull and Hyoid

TEMPORAL

A bone consisting of a tabular or *squamous* part visible on the side of the skull, but with its more important *petrous* part, carrying the organ of hearing, embedded in the base of the skull.

We have seen (p. 193) that the bone consists of four distinct morphological elements, *squamous, petrous, styloid process,* and *tympanic plate,* and decidedly the best way to become acquainted with the plan of build of this complicated bone is to obtain first of all an idea of the reason why these constituents of the bone come into relation with each other, and what position in regard to one another they must bear.

Fig. 172 is intended to illustrate these points schematically. The first drawing is a scheme of a transverse section through the head of an embryo, showing on the left side how the otocyst, from which the inner ear is formed, lies on the roof of the recess which runs out from the early pharynx and from which the Eustachian tube and tympanum are formed: the cartilages of the first and second visceral arches lie under the floor of the recess. This tubo-tympanic recess comes into contact with the surface at one point, x; behind and in front of this point it lies some distance from the surface. The first arch cartilage, or *Meckel's cartilage*, comes up above the level of the pharynx in the interval in front of, and that of the second arch behind, the point of surface contact. The otocyst is surrounded by a mass of condensed mesenchyme: this chondrifies later and ossifies, making the petrous. The squama is a separate development in the tissues superficial to the rest: the right side of the first figure indicates the planes of the different structures, and it can be seen that when the upper end of a bar comes above the pharyngeal level, it must lie *between the petrous and squamous*, covered on the surface by the latter. No. 2 represents a view from the right side, in which the first and second cartilaginous bars are seen passing up in front of and behind the surface point x. The tubo-tympanic recess is directed upwards, outwards and backwards and slightly twisted, so that the petrous rudiment lies above, behind and internal to it: thus the upper end of the second bar is between the petrous and the recess.

The third figure represents these relations, with the recess drawn out and showing a dilated *tympanic* end with a narrower *tubal* part in front. The petrous projects back beyond the end of the recess. A wing is thrown out from the petrous over the tympanum: this is called the *tegmen tympani* and is shown in No. 4. But the squama is in relation with the upper part of the outer surface of these structures, as shown in No. 5. A *tympanic ring*, incomplete above, and developed in superficial tissues, is also shown surrounding the surface area of the tympanum. Thus part of the tegmen and the petro-mastoid are exposed below the squama, and the ring is partly in contact with them. The second bar joins the petro-mastoid at the back of the tympanum, and can now be termed the styloid. The first bar has its upper end under cover of the tegmen, and it disappears below this: the upper end forms the malleus. So it is not seen in No. 5, but its former position is indicated. In No. 6 growth has led to the squama covering more of the tegmen, and the tympanic ring is beginning to broaden into a plate, so that the margin of the tegmen lies below and in front of it, between it and the squama, while it partly covers the upper part of the styloid below and behind. The lower jaw articulates with the part of squama that is covering the tegmen.

No. 7 shows the further progress to the adult state. A margin of tegmen is still seen between the articular cavity of the squama and the tympanic plate; the plate has much increased in size, making the bony meatus and piling itself up against the styloid. Where the front border of the plate opposes the squama the "Glaserian fissure" is formed, so that the edge of the tegmen can be described as visible in the inner or lower part of the Glaserian fissure. The mode of formation of the mastoid process is also indicated, the back part of the squama being drawn down over the petro-mastoid: thus a " petro-squamous suture " is visible on the surface of the process.

As soon as the student comprehends the way in which the complete bone is put together, he can proceed to recognise the various parts and their necessary relations in the adult bone. Look at the bone from below; the Glaserian fissure (Fig. 173) is easily recognised at the bottom of the hollow where the jaw articulates. Behind it is the tympanic plate forming the bony meatus; in front of it is the articular surface of the squama, while the edge of the tegmen appears in the inner part of the fissure. The styloid process has the inner and back part of the plate heaped up against its outer and front aspect, constituting the *vaginal process*. On the inner side of the styloid and tympanic plate the mass of the petrous is directed forwards and inwards. It is clear, from study of Fig. 172, that the Eustachian tube must emerge from the bone *between*

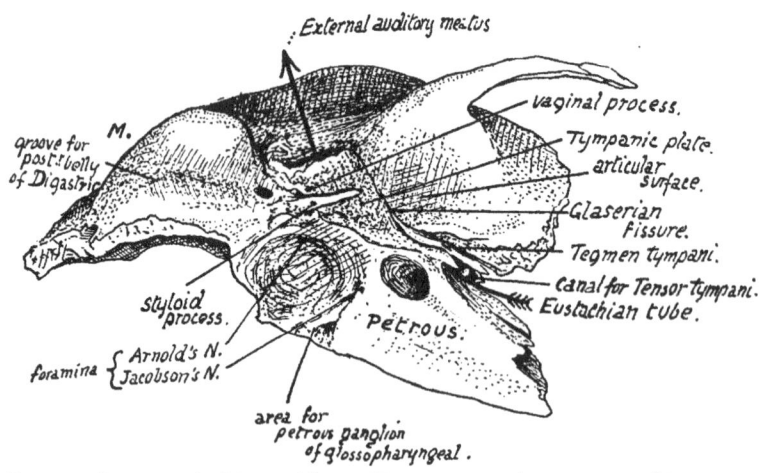

FIG. 173.—Lower aspect of temporal bone. The arrow passes in at the Eustachian opening and out at the external meatus; such a passage would only be possible when the tympanic membrane is absent. *M.*, mastoid process.

petrous and tegmen, separated from the squama by the tegmen, and above and internal to the inner part of the tympanic plate; examine the bone (Fig. 173), and the opening for the tube can be found in this position, in the angle, roughly speaking, between the petrous and squama, but really separated from the latter by the thin lamina of the tegmen. There is really a double opening here, both leading to the bony tympanum and separated by a very thin layer of bone, the *upper canal for Tensor tympani* and the *lower for the Eustachian tube*. Now follow the margin of the squama where it is applied to the superficial aspect of the petrous bone, comparing the complete structure with the schemes in Fig. 172. Starting above the external meatus, the margin can be traced down in the Glaserian fissure, being here a squamo-tympanic suture, but in the inner and lower part of the fissure it becomes petro-squamous. At the inner end of this part it turns up on the upper surface, still separating the squama from the tegmen (Fig. 162), and passes back along here as the (upper) petro-squamous suture to the petro-mastoid region: it is generally obliterated wholly or in part in this region in the adult bone, but can be picked up again on the outer side of the mastoid process, running downwards and forwards (Fig. 175) towards the apex, near which it turns up once more and runs round the meatus, becoming squamo-tympanic. If the squama and tympanic plate were removed, the tympanic cavity would be exposed, partly covered by the Tegmen tympani, as can be understood from the schemes in Fig. 172. Thus the cavity would have its inner wall made by the outer side of the petrous, its outer wall by the tympanic ring and the membrane which this holds in position, its roof by the tegmen, and the ossicles would be enclosed between these structures: we have already seen that the upper end of Meckel's cartilage forms the malleus, the stapes is developed from the upper end of the styloid bar, and the incus is made from a condensation continuous with that of the malleus. The backward extension of the cavity that forms the *mastoid antrum* goes back beyond the region of the ring and Membrana tympani, so that its outer wall is formed by squama.

The *outer* surface of the petrous portion, where it is not in relation with the tympanum and antrum, has the squama applied to it. The styloid process appears on the surface between the tympanic plate and the petrous: the plate in its growth spreads along the process and thus makes the "vaginal process," which therefore is outside and in front of the styloid process.

The student must become thoroughly familiar with these general relations of the parts of the temporal bone to each other, working them out from different points of view without troubling himself about the numerous foramina, etc., that he observes on the bone, before proceeding to study the specimen in detail, for only by this means will he be able to understand it and have the key to the many-sided intricacies of its structure.

The *petrous part* of the temporal is somewhat wedge-shaped, with its *apex* projecting forwards and inwards and its *base*, cut obliquely, represented by the thick petromastoid region of the bone. It has four sides: (*a*) *outer* or *anterior*, covered by squama save where it forms the inner wall of tympanum; (*b*) *lower*, seen on the basal surface of the skull; (*c*) *upper*; and (*d*) *inner* or *posterior*. Dura mater covers (*c*) and (*d*), which are respectively seen in the middle and posterior fossæ of the cranial cavity. The apex fits in between the great wing of the sphenoid and the region of the basi-sphenoid and basi-occiput. The *lower surface* is really infero-internal; compare the separate bone with the complete skull, and it is seen that the inner part of the surface is overlapped by the edge of the occipital, the remainder being visible from below. Thus we find an articular surface internally, coated by a thin layer of cartilage in the recent state, for the occipital, and behind this a *jugular fossa*, completed by the occipital, and converted into the jugular foramen. In front of the fossa, on the exposed part of the surface, is the rounded opening of the *carotid canal* that receives the internal carotid and the ramus caroticus of the superior cervical ganglion.

The artery is taken into the petrous bone secondarily; it is at first below the otic capsule, as shown in the first two drawings in Fig. 172, running forward over the tubo-tympanic recess. Later, as the ossification spreads, it encloses the artery, but the vessel retains its relation to the tubo-tympanic region (see later, p. 226)

The jugular fossa is frequently very deep, lodging the "bulb" of the vessel: the plate of bone that makes its floor is concerned, with the tympanic plate, in forming the floor of the tympanic cavity.

In front of the carotid opening and rather internal to it the surface is related to the lateral recess (fossa of Rosenmüller) of the pharynx, and some fibres of Levator palati arise here: immediately outside this is the outer surface of the bone, and here the cartilaginous part of the Eustachian tube is running downwards, forwards and inwards

14—2

212   Anatomy of Skeleton

from its bony opening. This portion of the tube is fastened by firm connective tissue to the bone in the neighbourhood.

The minute foramen that transmits the tympanic branch of the glosso-pharyngeal nerve is on the bony crest between the arterial and venous foramina : the ninth nerve, with the tenth and eleventh, lies in the front part of the jugular fossa, therefore just behind the artery immediately below the base of the skull. The little petrous ganglion of the ninth nerve lies in contact with the petrous on a small triangular surface of bone by the antero-internal angle of the jugular fossa (Fig. 173).

The petrous and tympanic plate meet along the outer margin of the jugular fossa, and thus a crest is formed here by their growth in apposition : this makes the vaginal process, so that the base of the styloid process and its vaginal sheath may be said to project immediately external to the jugular fossa or foramen.

A small *foramen for the posterior auricular branch of the vagus* is near this outer

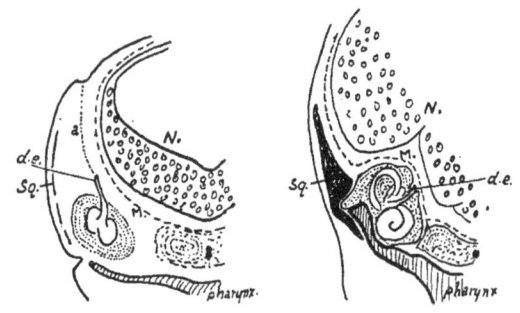

FIG. 174.—To show how the ductus endolymphaticus comes to lie on the inner side of the petrous. 1 is a section like Fig. 172, and in it the early brain N. is seen with its membranes M. Outside this the otocyst has come to rest on the roof of the tubo-tympanic recess of the pharynx, having descended from the dorso-lateral surface where the notch is seen ; its track is marked by a, the part which disappears, and $d.e.$, which is the drawn-out stalk of the otocyst and becomes the ductus. In 2 it is seen that the growth of the upper and back part of the otocyst to form semicircular canals has, with the accompanying growth of the capsule (petrous), left $d.e.$ on the inner side, while the growth of N. has pushed M. against and above the capsule so that the ductus now lies between bone and membrane. $Sq.$, plane of squama.

part   in the fœtus the nerve is not enclosed in the bone, but is taken in later as the tympanic plate and petrous increase in size.

The remainder of the lower surface extends backwards and outwards behind the jugular fossa to form the basis of the mastoid region. The thick inner margin of this region is also slightly overlapped by the occipital, so that the articular area immediately behind the fossa looks downwards as well as inwards : this is the part that in the first half of life is separated from the outer end of the jugular process of the occipital by a thin plate of cartilage.

The lower aspect of the mastoid portion shows a deep *groove for Digastric* just internal to the mastoid process, and part of a *groove for the occipital artery* internal to this along the sutural margin. The artery runs to the base of the skull along the deep surface of the posterior belly of the muscle, so its groove begins just internal to that for the Digastric. At the front end of the digastric groove is the *stylo-mastoid foramen.*

Skull and Hyoid

for the exit of the facial nerve and entrance of stylo-mastoid artery (from posterior auricular). This foramen is just behind the styloid process.

The seventh is the nerve of the second arch, of which the styloid is the bar   the

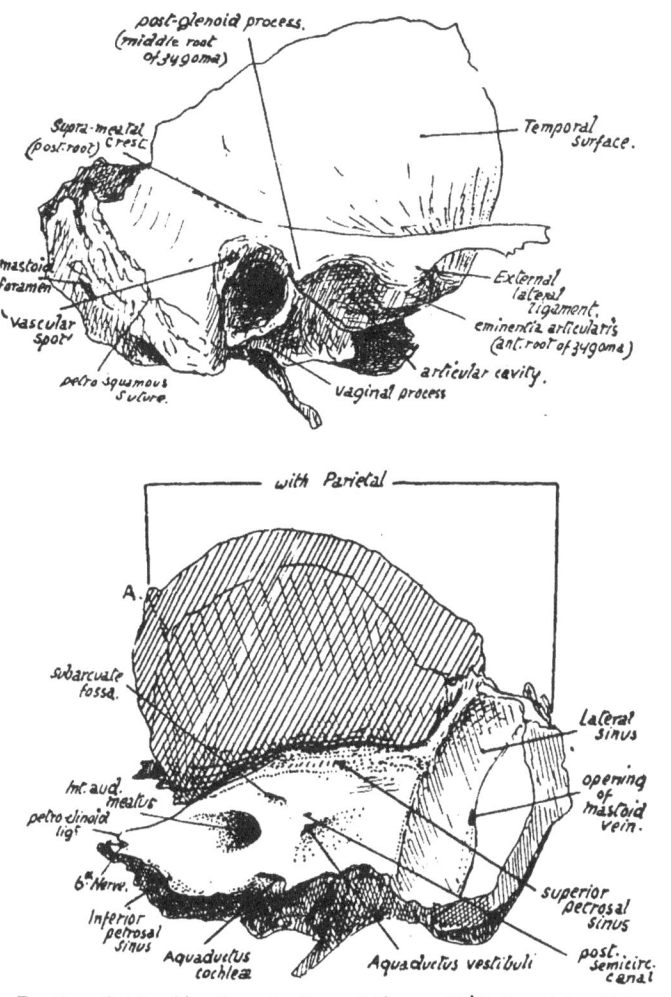

FIG. 175.—Outer and inner aspects of temporal bone. Right side. *A.* in lower figure marks junction of parietal and sphenoidal articular surfaces on margin.

typical arrangement is that the nerves run into their arches behind the bars and then turn forward round them to reach the front of the arches, and the facial nerve shows this relation, issuing behind the styloid and then turning forward over it.

The *inner* or *posterior surface* of the petrous looks backwards and inwards into the

Anatomy of Skeleton

posterior fossa and is nearly vertical in plane. It presents: a large and deep *internal auditory meatus* for the entrance of the seventh and eighth nerves, pars intermedia, and auditory branch of basilar artery: about half an inch behind this the *aquæductus vestibuli*, a narrow slit containing the blind terminal part of the ductus endolymphaticus.

This is the remains of the stalk of the otocyst, which (Fig. 174) is drawn out as a result of the descent of the otic vesicle: the growth of the vesicle to form the inner ear, and of the surrounding petrous to enclose it, leads to the undeveloped " duct " being embedded in the inner aspect of the bone, and between it and the dura mater.

Above and between the meatus and aqueduct is the *subarcuate fossa*, another slit-like depression, containing a fold of dura mater, the remains of a larger fossa present in the young skull and described later: below the meatus, close to the lower border of the surface, the opening of the *aquæductus cochleæ*, immediately above the fossette that lodges the petrous ganglion of the glosso-pharyngeal nerve (Fig. 175).

This foramen transmits a minute vein to the inferior petrosal sinus, also the " ductus cochleæ " from the scala tympani of the bony labyrinth to the subarachnoid space; the dura mater turns into the foramen to make a tunnel for the passage of the ductus.

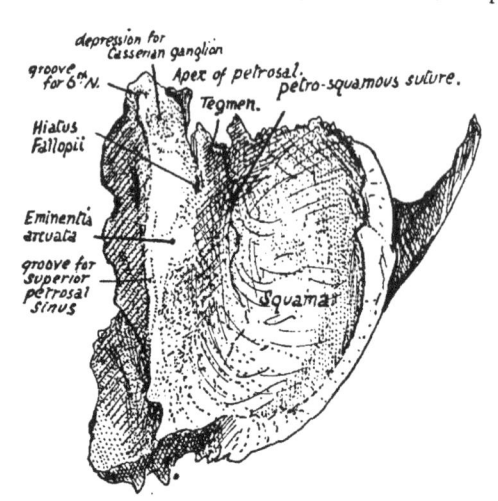

FIG. 176.—View of right temporal bone from above.

Along the lower border of the inner surface, in front of the aquæductus cochleæ, there runs a half-groove for the *inferior petrosal sinus* (see Fig. 175). The groove for the *lateral sinus* turns down behind the projecting part of the bone, on the mastoid portion, and this part of the bone is curved inwards as it helps to form the outer part of the back wall of the posterior fossa of the skull.

The whole of the lower margin of the inner surface articulates with the occipital, except where the jugular foramen occurs, but a thin layer of cartilage is between them from the front end to the lateral sinus.

The *upper surface* (Fig. 176) of the petrous looks also outwards and forwards. It presents: a *hollow* near the apex, *for the Gasserian or semilunar ganglion and trunk of the fifth nerve*: the *eminentia arcuata* marking the position of the superior semicircular canal: in front of this, and rather outside it, the *hiatus Fallopii*, a foramen for the exit of the great superficial petrosal nerve and entrance of the petrous branch of the middle meningeal artery, and a groove leading downwards, inwards, and forwards from the hiatus to the foramen lacerum for the reception of the nerve: outside the hiatus another smaller *opening* and groove for the *exit of the small superficial petrosal nerve*: another groove may usually be seen running from this opening to the neighbouring border, for the nerve. Outside these openings, between them and the petro-squamous suture, the surface is made by the *Tegmen tympani*, as a short examination of the bone will show.

Skull and Hyoid

In addition to the above, the upper surface has ill-defined markings, due to the pressure of the lower convolutions of the temporo-sphenoidal lobe of the brain.

The upper and inner surfaces are separated by the *upper margin* to which the *Tentorium cerebelli* is attached : the *superior petrosal sinus* runs in a groove along it between the layers of the tentorium, the groove extending from the lateral sinus behind to the depression for the ganglion in front. The sinus runs forward above the level of the fifth nerve, which is below the tentorium, so it does not touch the petrous bone at or in front of the depression for the nerve.

The *apex of the petrous* is rough with bony projections : it forms the posterior boundary of the foramen lacerum, and the size of the foramen depends on the extent to which the ossification of the petrous has extended forward. A ligamentous band, *petro-clinoid*, passes from the topmost point of the apex to the posterior clinoid process, and the sixth nerve passes under the outer part of this ligament ; ossification may extend into the ligament, which then forms a little point of bone on the apex (Fig. 176), overhanging a groove in which the nerve rests.

The *outer border* of the upper surface corresponds with the superior petro-squamous suture.

Dura mater is applied to the upper and inner surfaces of the petrous. Between it and the bone of the *upper* surface lie the ganglion (in its cavum of dura mater) and the superficial petrosal nerves. On the *inner* aspect the dura mater is turned into the internal meatus with the nerves, which pierce it in the meatus, and also into the aquæductus cochleæ, where the ductus cochleæ passes through it. It has the ductus endolymphaticus between it and the bone, also the petrosal and lateral sinuses and the small vein issuing from the cochlear foramen : it turns into the front part of the jugular foramen to form sheaths for the nerves passing through this opening.

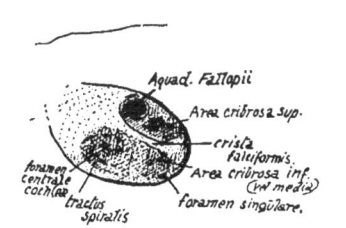

FIG. 177.—Plan of fundus of right internal auditory meatus. This is divided by a "crista falciformis" into upper and lower areas. The upper area has an opening for the seventh nerve in front, and others behind for nerves to the upper and outer semicircular canals and utricle ; in the lower part are foramina for cochlear nerves in front, and for nerves to saccule and posterior canal (foramen singulare) behind. The fundus may be barely visible, or, as in the specimen from which this was drawn, may be seen without much trouble.

The *styloid process* projects downwards, inwards and forwards from the lower aspect of the bone, emerging between the tympanic plate and the petrous, so that the jugular foramen is just internal to it and the stylo-mastoid foramen immediately behind it. It varies in length and has the stylo-hyoid ligament continuous with its extremity. Its base is fused with the petrous bone in the back wall of the tympanum.

This embedded portion is termed the *tympano-hyal*, the bony process is the *stylo-hyal*, the ligament the *epi-hyal*, and the lesser cornu and upper part of body of hyoid are the *cerato-hyal* and *basi-hyal* respectively ; these are all modifications of the bar of the second visceral arch in continuity.

The Stylo-pharyngeus arises from the inner side of the base of the process—therefore in relation with the internal jugular vein ; the Stylo-hyoid arises from its posterior aspect about half-way down, and the Stylo-glossus from its tip and from the Stylo-hyoid ligament. The process with its muscles and ligament constitute a mass of tissues passing with a general direction downwards, forwards and inwards and consequently (Fig. 205) lying in antero-lateral relation to the carotid sheath, a fact that can be at once appreciated on looking at the skull.

The *glenoid fossa* is the deep hollow that lies in front of the bony outer meatus. It is crossed by the *Glaserian fissure*, and can be divided into a *mandibular* or *articular* part in front of this fissure, and made by the squama, and a *posterior or non-articular* part composed of the tympanic plate. The articular part is the deepest portion of the whole fossa, concave in all directions and longer from side to side : its long axis, however, is not quite transverse, but slightly oblique. It is bounded in front by the *eminentia articularis*, which forms the anterior root of the zygoma and is convex from before backwards and very slightly concave from side to side : the articular surface is carried forward with its cartilaginous lining on to the eminence, for the play of the meniscus in the joint, and thin capsular fibres are attached round the articular surface. The nerve to the Masseter runs out under the anterior margin of the eminence in contact with the anterior capsular fibres * (see Fig. 182).

The posterior part of the glenoid fossa, formed by the tympanic plate, extends inwards and backwards to reach the styloid process and the situation of the carotid sheath : it is concave in all directions, the bony plate looking forwards, downwards and outwards, and contains some fibro-fatty tissue, or some upper lobules of the parotid if that gland is large.

The *Glaserian fissure* is double in its inner part owing to the appearance in it of the edge of the tegmen tympani : in this part between the tegmen and the tympanic plate, the fissure contains the *process gracilis of the malleus* (with its *inferior Meckelian ligament*), and affords passage to the *chorda tympani* and the *tympanic branch of the internal maxillary artery*. The canal of exit of the nerve, termed sometimes the *canal of Huguier*, is near the inner end of the fissure and practically marks the spot where Meckel's cartilage lay before its disappearance (see Fig. 172).

The *external auditory meatus* is a bony canal running inwards and very slightly forwards, about three-fifths of an inch long, and continuous with the tympanic cavity in the dry bone, but separated from it really by the membrana tympani. The opening and canal are oval on section with the long axis nearly vertical. Its front and lower walls are formed by the tympanic plate, its back wall by this plate fused with the mastoid downgrowth of the squama, and its roof by the lower surface of the squama. The cartilaginous pinna is fastened round the margins of the orifice.

The *squamous part of the bone* makes with its lower surface the roof of the meatus, the articular cavity, and the articular eminence : these have been already examined. Above these it forms a thin bony plate that is seen on the side of the skull. The concave *cerebral surface* of this plate is marked by cerebral gyri, crossed by *grooves for middle meningeal vessels*, and ends below by turning in to meet the petrous at the petrosquamous suture.

The *outer surface*, slightly convex where it helps to form the temporal fossa, is smooth for origin of Temporal muscle fibres : this part is separated below and behind by the *posterior or ascending root of the zygoma* from a slightly concave part of the squama drawn down on the mastoid process (Fig. 175).

The *zygomatic process* springs from the lower part of the outer surface of the squama : it has *upper* and *lower borders* and articulates at its extremity, which is serrated, with the malar. The inner surface is smooth and lies on the Temporal muscle, and is continuous behind with the outer surface of the squama by a widened and obliquely-sloped area over which the posterior margin of the Temporal tendon plays. The Masseter arises from its lower border and by muscular fibres from its inner surface. The process has a broad basis of attachment to the squama, which exhibits three

---

* For description of joint, see under " Lower Jaw."

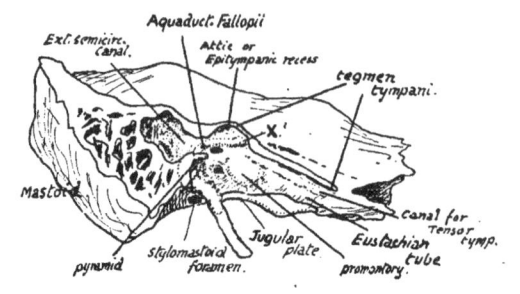

FIG. 178.—A diagrammatic sketch showing the outer side of the petrous bone after removal of the squama and the tympanic plate: the area over which the plate lies against the petrous is coloured pink, whereas the area of petro-squamous junction is shown uncoloured. The meeting of the tympanic plate with the jugular plate of the petrous bone makes the floor of the *tympanic cavity* which the removal of the superficial bones has exposed, and the tympanic plate also forms the floor of the bony outer part of the Eustachian tube. The inner wall of the tympanum forms the outer covering of the inner ear; the first turn of the cochlea makes an elevation, the *promontory*, on the front part of the wall, the vestibule is situated behind this and has the *fenestræ* associated with it, and the semicircular canals, which constitute the hinder part of the inner ear, cause a slight prominence (external canal) in the *aditus* which leads from the tympanum to the mastoid antrum. The seventh nerve runs out behind and above the cochlea, and thus comes to the inner wall above and behind (X.) the promontory: here the "Fallopian aqueduct," the bony canal which contains the nerve, turns back along the inner wall, above the fenestra ovalis, to reach the posterior wall, down which it turns to end at the stylomastoid foramen. The relative position of the facial nerve and the various parts of the inner ear. as seen in the tympanum, is also illustrated in the arrangement of foramina in the floor of the internal auditory meatus (Fig. 177).

so-called " roots " : anterior (articular eminence), middle (post-glenoid tubercle) and posterior or ascending (supramastoid crest) (Fig. 175).

The outer end of the articular eminence is roughened by the attachment of the external lateral ligament of the joint, and the temporal fascia is fastened to the upper border of the free process and along the supramastoid crest.

The prominent *mastoid process* projects downwards and forwards immediately behind the meatus : between it and the tympanic plate is the point of emergence of the posterior auricular branch of the vagus. The process is made by a mass continuous with the petrous, with a part of the squama covering its upper and front area externally, so that a petro-squamous suture may be apparent on this surface of the bone. Beside this it presents : (*a*) attachment of Retrahens aurem ; (*b*) some fibres of origin of Occipito-frontalis ; (*c*) attachments of Sterno-mastoid, Splenius, and Trachelo-mastoid ; (*d*) a *mastoid foramen* near its posterior border, opening internally into the groove for the lateral sinus ; (*e*) digastric groove, etc., already considered. The process is not present at birth, may owe its existence to the pull of the muscles inserted into it, and is usually better developed in men than in women.

The upper part of the region between the process and the external meatus exhibits several vascular foramina and may present a small *pos'-meatal spine* projecting from the squama : this depressed area (*vascular spot*) is of surgical importance, as it marks on the surface the position of the mastoid antrum.

On the front border the squama articulates with the great wing of the sphenoid and with the parietal in the upper and back part. The meeting of these two surfaces is marked on the articular bevelled area of the squama by a definite ridge (A in Fig. 175). In front of this ridge the *sphenoidal* surface extends to the apex of the petrous, while behind it the *parietal* extends to the prominent posterior angle on the mastoid margin. The *occipital* articulates along the lower and inner margin from this angle to the apex of the petrous. The *malar* joins the zygomatic process, and the *lower jaw* articulates with the squama. The occipital articulation is largely through the medium of a thin layer of fibro-cartilage : in this are sometimes found small ossifications, the *ossicles of Riolan*.

The outer side of the petrous is covered by the other elements of the bone and cannot be seen without removing these : this is impossible without the saw in the adult bone, though easily effected in the young specimen. A semi-diagrammatic view of the outer side is given in Fig. 178. The tympanic cavity is exposed by this removal of the squama and tympanic plate : the styloid process is left and its upper end is seen embedded in the posterior wall of the tympanum (tympanohyal). The tympanic plate covered this in ; its area of articulation is shown. The rest of the cut surface represents the extent to which the squama rested on the petrous.

The cavity of the bony *Eustachian* tube is opened, leading to the tympanum, and behind this again a narrower *aditus* opens into the *mastoid antrum*. Evidently these cavities are covered in externally by the squama and tympanic plate ; the latter forms the outer wall of the tympanum. The plate meets the jugular plate below and thus completes the tympanic floor ; it also forms the floor of the bony tube. The canal for the Tensor tympani is above the tube and also leads to the tympanum. *Mastoid air-cells* are seen behind the antrum : these are very variable in extent of development and may be found throughout the whole mastoid region, communicating with the antrum.

The inner ear is close to the inner wall of the tympanum, and the first turn of the cochlea causes a bulging, the *promontory*, on the inner wall of the tympanum : immediately behind this are the *fenestra ovalis*, for the footpiece of the stapes, and the *fenestra rotunda*, which is closed by membrane : behind and below these is a depression, the *sinus tympani*. The facial nerve, having run outwards from the inner meatus to the point X in the figure, turns back sharply here along the upper part of the inner wall, in the bony canal known as the " aquæductus Fallopii," to the lower and inner side of the aditus ad antrum, where it turns down along the posterior wall to emerge at the stylo-mastoid foramen. Outside the descending part of the canal is the " pyramid," a small bony eminence containing the Stapedius muscle, whose tendon emerges from its apex : to the outside of this and a little lower the *iter chordæ posterius* allows the chorda tympani to leave the facial canal. The tympanic branch of the ninth nerve comes through the floor just below the promontory, on to which the nerve runs and ramifies.

The promontory marking the position of the cochlea, it follows that the fenestræ must be in the region of the bony vestibule, into which they open, and the semi-circular canals must lie behind this. These canals cause prominences on the surfaces of the bone, the ampulla of the *external canal showing in the aditus*, above and behind the facial canal, the upper curve of the *superior semi-circular canal making the arcuate eminence*, and the upper arc of the *posterior one causing a rounded elevation between the subarcuate fossa and the aquæductus vestibuli* on the inner aspect of the bone.

An incomplete bony septum separates the bony tube from the canal of the Tensor tympani, ending behind in a hook-like process round which the tendon of the muscle turns.

Development.

At birth the bone (Fig. 214) is practically in the condition shown schematically in the fifth and sixth drawings in Fig. 172. The four elements of the bone are easily separated, the styloid process being as a rule cartilaginous, although its embedded part is bony. The tympanic plate is represented by the incomplete ring, so that there is no bony meatus, but the membrane is practically on the surface of the bone. It is unnecessary to give a further account of the origin of the four parts of the bone already illustrated in Fig. 172, and it only remains to deal with the ossification : this occurs in membrane in the squama and tympanic ring, but in cartilage in the petrous and styloid.

Squama.

One centre, above the root of the zygoma, about the seventh or eighth week : some observers maintain that there are two additional centres, one above and one behind this.

Tympanic Ring.

One centre, appearing in the middle of the third month in the thick tissue by the wall of the cavity and extending from this backward and upward to make the ring.

The Petrous.

This ossifies from four or more centres that appear in the fifth month and are indistinguishable as separate centres in the sixth month. These are :—

(1) *Opisthotic* : the first to appear, forms that part of the petrous that lies below the level of the internal meatus : this includes the lower part of the inner wall of the tympanum and, in front of this, the carotid canal.

(2) *Pro-otic*, the main centre, forming the part of the petrous that lies above the internal meatus, as far forward as the apex and back to the mastoid region : the centre appears near the eminentia arcuata.

(3) *Pterotic* : a small centre responsible for the roof of the tympanum and tegmen tympani ; there may, however, be a separate centre for the latter.

(4) *Epiotic* : one or more from which the mass of the mastoid region is ossified.

Styloid.

The tympano-hyal centre makes its appearance late in fœtal life : the stylo hyal centre appears shortly after birth. The process remains attached by cartilage until nearly middle life, when it unites with the tympano-hyal.

Skull and Hyoid

At birth only the central part of the carotid canal is covered in : subsequent growth of the opisthotic and of the tympanic plate completes the process. The growth of the plate outwards is brought about largely by the spread of ossification from two main points of the ring, which by their ultimate fusion form the floor of the external meatus : when they first join a small hiatus may exist between them and is normally present (*foramen of Huschke*) up to the age of five or six, and may persist beyond this. Other points of interest in the young temporal bone are mentioned in the section on the " Fœtal Skull."

SPHENOID.

A single bone situated in the base of the skull, but reaching the side walls : consisting of a centrally placed *body* from which two *wings*, *great* and *small*, project on each

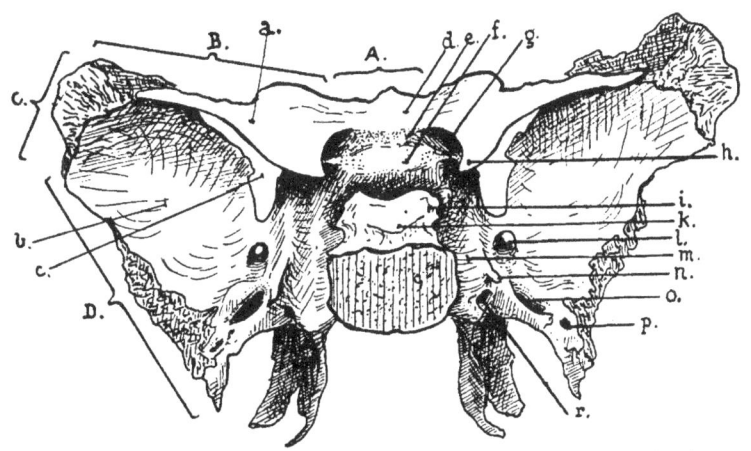

FIG. 179.—Sphenoid seen from above and behind. *A*., marginal articulation with ethmoid ; *B*., with frontal ; *C*., with parietal ; *D*., with squama of temporal. *a*., Small wing ; *b*., great wing ; *c*., sphenoidal fissure ; *d*., jugum sphenoidale ; *e*., optic groove ; *f*., olivary eminence ; *g*., optic foramen ; *h*., anterior clinoid process ; *i*., posterior clinoid process ; *k*., dorsum sellæ ; *l*., foramen rotundum ; *m*., carotid groove ; *n*., lingula ; *o*., foramen ovale ; *p*., foramen spinosum ; *r*., Vidian foramen.

side, and two *pterygoid plates* project downwards on each side, the *inner* from the body and the *outer* from the under surface of the great wing.

The body, on the whole cuboidal, articulates behind with the occipital, with which it is synostosed about twenty-five, and in front carries the ethmoid : it is hollowed out by the *sphenoidal air sinuses*,* which extend into it from the nasal fossæ within a few years after birth, and when fully developed occupy practically the whole of the body and may extend even into the attached parts of the wings and pterygoid processes. The two sinuses are separated by a septum, which may be incomplete and is usually asymmetrical : there are as a rule two partial septa also present, one in each sinus near its upper and outer part. The sinuses are partly closed by two *sphenoidal turbinals*,* thin paper-like bones, that have openings above their centres through which the cavities are connected with the ethmo-sphenoidal recesses in the nasal fossæ.

* For further details, see p. 249 (sphenoidal sinus), and p. 227 (sphenoidal turbinals).

Anatomy of Skeleton

The *upper surface of the body* presents a deep hollow open at the side, the *pituitary fossa*, for lodging the pituitary gland and its surrounding vascular plexus; in front of this the *olivary* eminence separates it from the *optic groove*, which runs into the optic *foramen* at each side. In front of the groove the flatter *jugum sphenoidale* stands at a higher level than the rest of the body and is continued laterally into the lesser wings.

The optic chiasma is attached to the front of the infundibulum of the pituitary, and is therefore raised above the level of the bone and does not lie in the optic groove, but the optic nerves run from it into the optic foramina accompanied by the ophthalmic arteries. The pituitary fossa is roofed in by the *diaphragma sellæ*, a curtain of dura mater that is pierced centrally by the stalk of the gland. The diaphragm (Fig. 181) is attached to the front and back margins of the fossa, and at the sides to the *clinoid processes* and interclinoid ligaments: the *an'erior clinoid process* is a thick angular projection from the base of the small wing towards the fossa, and the *posterior process* is the prominent upper and lateral angle of the *dorsum sellæ* that overhangs the back of the fossa. The *middle clinoid process*, if present, is a low point of bone situated between and below the other two at the side of the fossa. *Interclinoid ligaments* ccnnect these processes.

FIG. 180.—A specimen in which the ligamentous bands joining the three clinoid processes have ossified, with the result that foramina exist which are not present with bony boundaries in the sphenoid as usually seen. The carotid passed through the arterial foramen and the circular sinus through the venous foramen. A carotid foramen is not uncommon; the double structure is more rare.

At the sides, where it slopes down to be continuous with the great wings, the body supports the cavernous sinus on each side (Fig. 182), and the system of venous spaces (known as the circular sinus) that surrounds the pituitary body joins the two cavernous sinuses. The internal carotid artery reaches the bone here, deep to the sinus, from the foramen lacerum (see Complete Base), and runs along it in the *carotid groove*, to turn in under the overhanging anterior clinoid process and then up to the inner side of the process behind the optic nerve; here it gives its ophthalmic branch. Thus the artery passes in front of the ligamentous connection between the anterior and middle clinoid processes, which may be ossified, so that there may be foramina at each side of the upper surface of the body, as shown in Fig. 180, for the optic nerve, the carotid artery, and the circular sinus.

The *lower surface of the body* has a thick, wedge-shaped bar running along its front part, supporting the septum between the air sinuses above and ending in front in the prominent *rostrum*: the hinder end of this bar of bone is continuous with the rest of the body and is partly overlapped by the vaginal processes that turn in from the internal pterygoid plates. This aspect of the body, as can be seen in the skull, carries the vomer, and the two alæ of that bone open out below the sphenoidal body and fit in under cover of the vaginal processes. Behind this region the under surface is lined by mucous membrane of the roof of the naso-pharynx.

On each side the body is continuous with the great wing: the carotid groove commences on the back of this junction, and here there is a projection on the outer

220

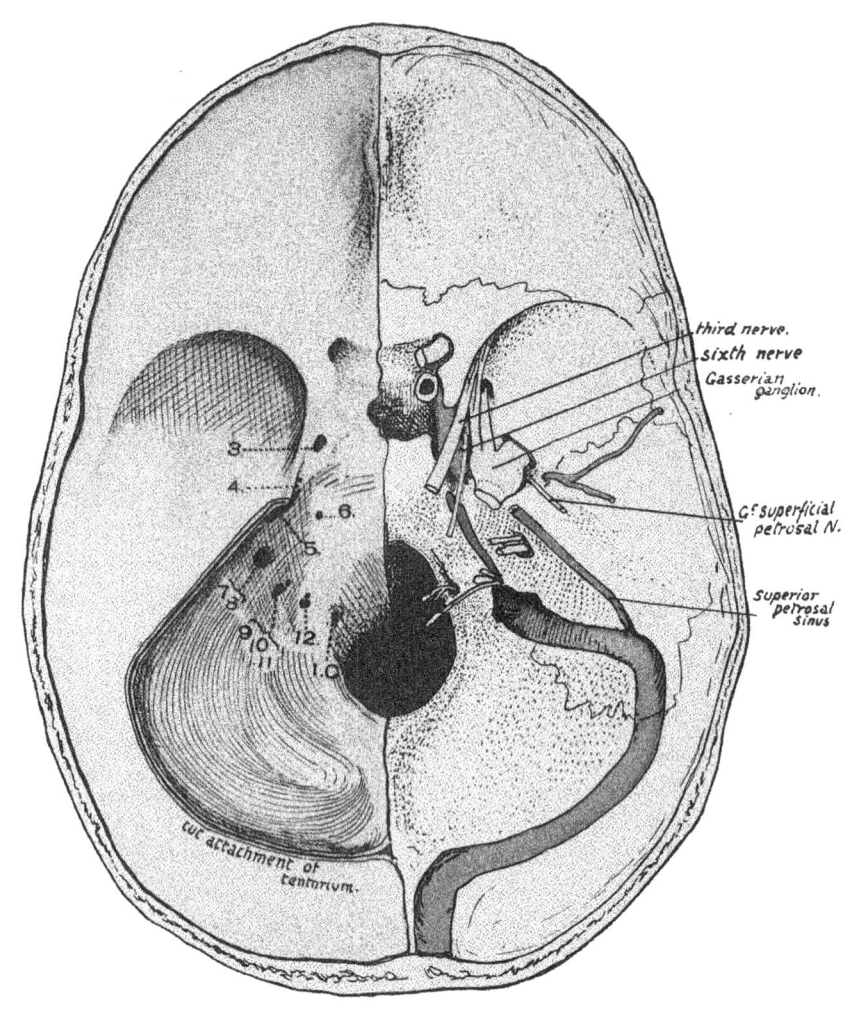

FIG. 181.—Cranial surface of base of skull, with dura mater in position on the left side, the tentorium cerebelli being cut away. 3—12, foramina made by the corresponding cranial nerves piercing dura mater 1. C. first cervical nerve foramen. Observe that when the dura is in position the margins of the foramen magnum are not defined, owing to the continuation of the dural slope on to the odontoid, consequently the foramina have the appearance of being higher than they really are in the skull. The membrane covers in the pituitary fossa, forming the *diaphragma sellæ*, attached to the inter-clinoid ligaments. The structures represented on the right side are sufficiently described in the text. Compare with Fig. 182 and p. 225.

side of the groove known as the *lingula*. This appears to be only present as a guard for the artery, but its morphological value is doubtful ; it has a separate centre of ossification, and is laid down as a separate cartilaginous structure in the embryonic skull.

The *great wing* or *alisphenoid* supports the anterior pole of the temporo-sphenoidal lobe of the cerebrum, is therefore concave on its upper surface in all directions, forming the front part of the middle fossa on each side, and exhibits impressions for the cerebral gyri. The wing is thickened in front and externally, so that, in addition to its *upper* and *lower* surfaces, it presents an anterior or *orbital* and an outer or *temporal* aspect.

The *upper* concave cerebral aspect is in a very general way triangular, attached to the body by the internal angle : the front border, raised, is separated from the small wing by the sphenoidal fissure and, outside this, articulates with the frontal : the outer and upper angle, raised, is blunt and articulates with the parietal : the outer or posteroexternal border descends rapidly from this angle in a concave line that articulates with the temporal squama and runs downwards, backwards, and inwards : the posterior angle projects well back beyond the level of the rest of the bone and fits in between the squamous and petrous temporal : the inner or postero-internal border, articulating with the petrous, runs from this angle towards the body, where it is joined by the lingula and marked by the carotid groove. This surface presents (Fig. 179) :—

(*a*) *Foramen spinosum*, in posterior angle, for meningeal vessels and recurrent branch of mandibular nerve.; from this a groove for the anterior division of the artery is usually found running along the postero-external border, but this is occasionally on the temporal side of the suture. A small *canalis innominatus* is occasionally seen internal to the foramen spinosum, for the passage of the small superficial petrosal nerve.

(*b*) *Foramen ovale*, for the mandibular division of the fifth nerve, its motor root, and the small meningeal artery.

(*c*) *Foramen rotundum*, for the superior maxillary division of the nerve. Notice that this opening is just below the inner end of the sphenoidal fissure, and it may be looked on as part of this fissure cut off by extending ossification.

Articulate the bone with the temporal and observe the position of the Gasserian ganglion (Fig. 181). It lies on the petrous and over the outer part of the foramen lacerum, just behind the lingula and internal to the foramen ovale, into which its largest offset goes.

The other two divisions of the nerve, running to the foramen rotundum and sphenoidal fissure respectively, are therefore internal to and above the mandibular division and are in relation with the outer wall of the cavernous sinus.

The bone is covered on this aspect by dura mater, and the nerves and artery lie between it and the membrane, which extends from the front border of the wing on to the lesser wing and thus covers in the sphenoidal fissure.

The *lower* or *basal aspect of the wing* is small, and corresponds with the inner part of the wing and with that portion that extends back to form the posterior angle : the outer part of the wing is turned up, forming a continuous curve with the remainder on the cerebral aspect, but being marked off from the lower surface externally by a definite border or ridge, the *inferior temporal crest*. The basal surface can thus be said to lie between this crest and the external pterygoid plate, while the temporal surface lies above the crest.

The basal surface gives origin to External Pterygoid from the whole of the surface outside the pterygoid plate, the area of origin being bounded in front by the *pterygoid ridge*, which turns outwards from the front of the plate and usually shows a *pterygoid spine* to which some aponeurotic fibres on the superficial surface of the muscle are attached. But behind the level of the external plate, where the bony surface does not give origin to the muscle but is in contact with it, the wing is seen to narrow sharply to the posterior angle, where it bears the *spine*, to which the long internal lateral ligament of the mandible is fastened. Along the postero-internal edge the Tensor palati arises, and immediately external to this are the *lower openings* of the *oval* and *spinous* foramina.

Examine this region on the skull. The basal aspect of the bone is continuous with that of the eminentia articularis on the temporal squamous, and the inferior temporal crest is continued back on the temporal to run into the front border of the anterior zygomatic root : evidently all this basal area is in relation with the External Pterygoid as this passes to its insertion into the neck and capsule of the lower jaw, and the muscle covers the bone as far out as the lower temporal ridge : hence the Temporal muscle, coming down over the ridge, makes the immediate outer relation of the External Pterygoid (Fig. 182).

FIG. 182.—The upper left figure shows on the base of the skull the course of some of the branches of the mandibular nerve. *A* is the basal surface of the great Wing, in relation with the upper surface of external Pterygoid, and the middle deep temporal and masseteric nerves are seen crossing this surface, therefore between it and the muscle. The temporal nerve reaches the inferior temporal crest and thus the deep aspect of the Temporal muscle, but the nerve to the Masseter, lying a little further back, passes behind the Temporal tendon (giving a posterior deep temporal nerve to the muscle) and lies along the eminentia articularis, in contact with the articular capsule ; it thus reaches the back part of its muscle. The auriculo-temporal nerve runs backwards and outwards, between the sphenoidal spine (long internal ligament) and the capsule (short internal lig.), and then turns out in contact with the back of the capsule, below the Glaserian fissure. The Tensor palati separates the issuing mandibular nerve from the Eustachian tube which issues just internal to the spine. The figure on the right is a scheme of the planes, on coronal section, in the pterygoid region. *A* is the plane of the deep temporal vessels, *B* of the (deep) temporal nerve, and *C* of the inferior dental and lingual nerves. Observe that, as can be appreciated in the previous figure, the external Pterygoid is in relation externally with the Temporal tendon and coronoid process of the mandible : thus the long buccal nerve, piercing the muscle, comes into relation externally with the Temporal and can give an anterior deep temporal nerve to its front part ; when it comes forward from under the tendon it appears superficially from under the Masseter, as can be seen from the figure, and lies on Buccinator (see Fig. 206). The lower figures show the relations of nerves, etc., in the region of the foramen lacerum and cavernous sinus. The sinus is removed in one drawing to show the structures deep to it, and its formation is shown in the other. Compare with Fig. 181. Description in text, p. 225.

Now follow the course of the nerves that run *directly out* from the mandibular nerve : *they must lie between the bone and the Pterygoid muscle*. The arrangement of the deep temporal filaments is very variable, but a typical one would have a *middle* deep temporal running by itself outward and a little forwards from the foramen ovale ; a line showing its course would cross the basal aspect of the great wing and would just lie on the temporal bone as it reached the temporal ridge, turning round this to run up on the deep surface of the Temporal muscle. The bone is sometimes grooved by the nerve near the ridge. The masseteric nerve runs behind the other and more directly outward, thus passing on to the front part of the eminentia articularis after a shorter course on the sphenoid : on the eminence it runs out to the back of the Masseter, passing behind the Temporal tendon and giving a *posterior* deep temporal filament up along the tendon (Fig. 182).

The structures passing through the foramina are evidently placed between the Tensor palati and the External Pterygoid.

Now examine the position of the spine of the sphenoid. Being on the extreme angle of the wing, it lies immediately in front of the inner end of the Glaserian fissure, and the chorda tympani coming out of the fissure runs on to the surface of the Tensor palati by passing to the *inner* side of the spine, where it may make a groove (*groove of Lucas*). Thus the spine has the long ligament fastened to its tip, the chorda tympani on its inner side and the auriculo-temporal nerve passing back between it and the proper capsule of the joint (Fig. 182).

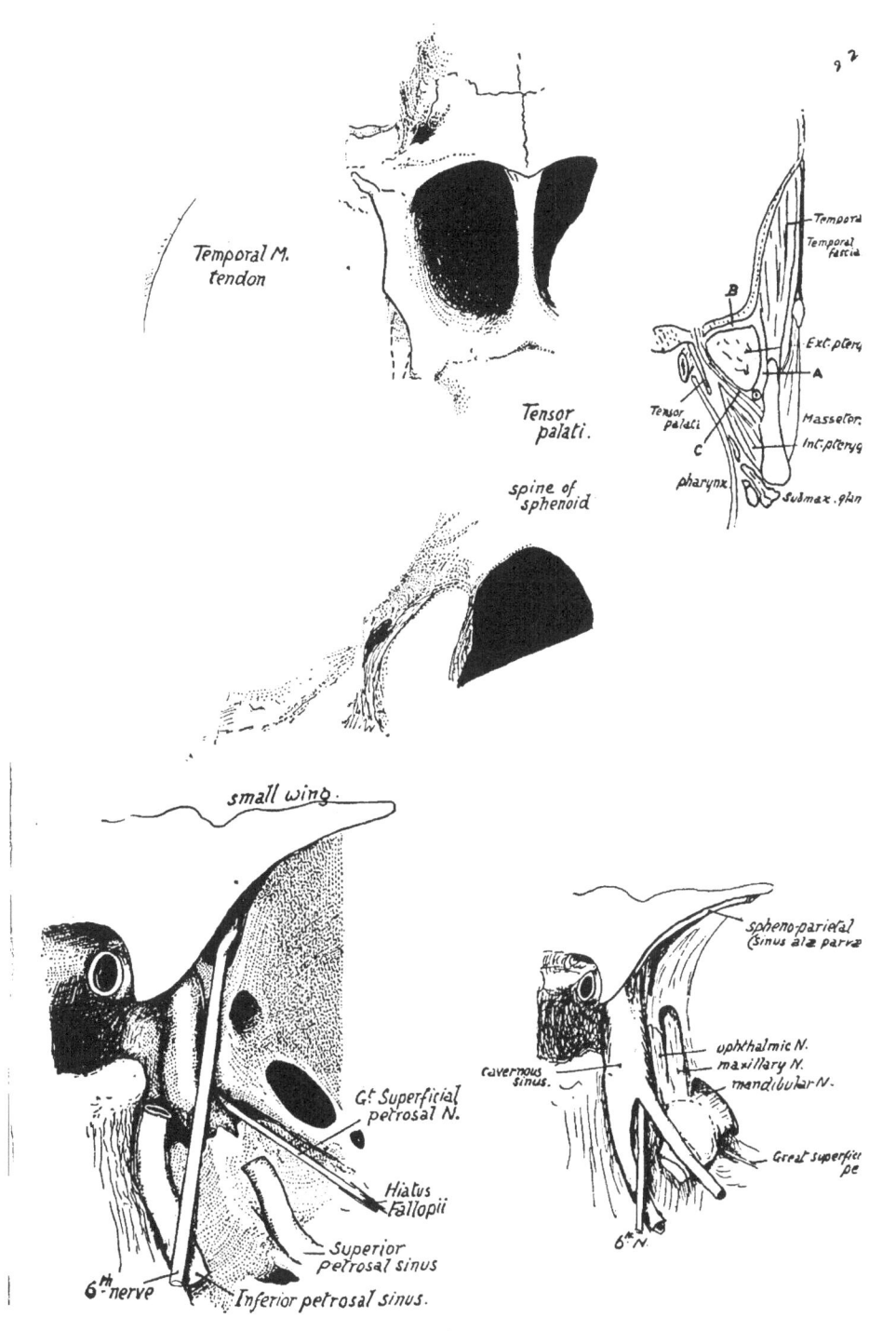

FIG. 182.

Skull and Hyoid

Again, look at the Eustachian opening : it is situated just internal to the position of the spine, and the tube passes downwards, forwards and inwards from it, and is thus only separated by the Tensor palati from the structures passing through the foramina in the great wing.

The *temporal surface* articulates by its front edge (Fig. 183) with the malar and is therefore concave from behind forwards as it enters into the formation of the front wall of the temporal fossa as well as its floor. It is covered by and gives origin to the Temporal muscle. Its margin articulates with the malar in front, above this the frontal, the parietal along the top and the squama behind.

The *orbital* or *front surface* of the wing, nearly flat, and deepening as it is followed out, forms a great part of the outer or postero-external wall of the orbit, and is in relation with the external Rectus (which gains a small additional origin from its inner part) and the lachrymal vessels and nerve. Its upper and inner border is the lower

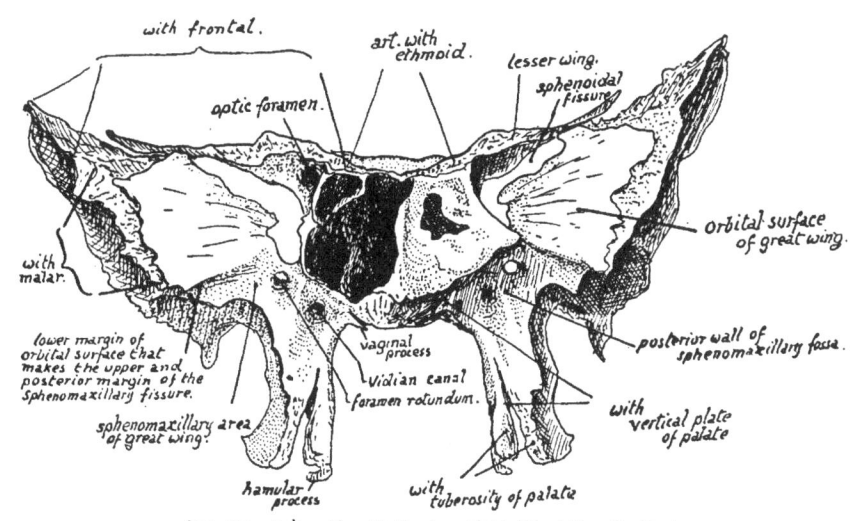

FIG. 183.—Sphenoid and left sphenoidal turbinal from the front.

margin of the sphenoidal fissure and articulates with the frontal outside this ; its outer border is practically the front border of the temporal surface (Fig. 183) and carries the malar, while its lower border is separated from the maxilla by an interval, the *spheno-maxillary fissure*, of which it therefore forms the upper and posterior margin. On the bone this lower border separates the orbital surface from a district of variable size that really belongs to the basal area but is cut off from the pterygoid or true basal region by the pterygoid ridge : this district is *spheno-maxilla*ry and will be considered with the pterygoid processes.

The *small wing* or *orbito-sphenoid* projects outwards and a little forwards from the body. It has a smooth *upp*er surface that is covered by dura mater and forms part of the anterior fossa ; a smooth rounded and concave *posterior edge*, to which dura mater is attached after it has covered the back of the sphenoidal fissure and which marks the position of the Sylvian fissure, and along which the spheno-parietal venous sinus runs, deep to the dura, to reach the cavernous sinus ; a rough *front border* that articulates with the frontal ; a smooth *lower* surface which forms the upper boundary of the sphenoidal fissure and, in front of this, the extreme back and inner part of the roof of the orbital cavity.

The posterior margin is continued into the anterior clinoid process, and the widened wing is attached here by two " roots," which enclose the optic foramen between them. The orbital muscles arise from the front and lower surface of the wing round the anterior opening of the optic foramen.

The *Pterygoid Processes*—each *inner* process is fused along its front and upper part with the corresponding *outer* process, but they are separate below. They diverge from each other behind, the plane of the inner plate being practically parallel with that of its fellow, whereas that of the outer plate is nearly at a right angle with the plane of the outer plate of the opposite side. The deep interval thus left between the inner and outer plates behind is termed the inter-pterygoid or *pterygoid fossa :* its floor is *completed below by the tuberosity of the palate bone*, which articulates with the front of the two plates in their lower halves and thus closes the gap that exists here between them in the separated bone. A small area is cut off from the upper or proximal part of the pterygoid fossa by an oblique ridge, and constitutes the *scaphoid fossa :* the front part of the Tensor palati arises here, and its area of origin extends from this backwards and outwards along the margin of the great wing internal to the foramina, nearly as far as the spine of the sphenoid (Fig. 182). Otherwise the pterygoid fossa is occupied by the Internal Pterygoid muscle, which arises from the *inner* side of the *outer* plate and from the tuberosity of the palate between the plates. The External Pterygoid takes origin from the outer side of the outer plate, an area continuous with that on the basal surface of the great wing. Thus the external plate may be looked on as a muscular process, probably developed only in association with such function ; but the inner plate is a bone developed in the pharyngeal wall and only gives origin to one muscle, the upper Constrictor of the pharynx, which arises from the lower half of its posterior border. It is covered by mucous membrane on its inner surface, where it forms the outer wall of the posterior opening of the nasal fossa, and its outer surface is in relation with Internal Pterygoid and Tensor palati, but does not give origin to them. Its lower end is prolonged into a hooked *hamular process* directed backwards and outwards and showing a deep notch on its outer and front aspect, round which the tendon of Tensor palati turns in to enter the soft palate.

The posterior margin of the inner plate has the Constrictor arising from its lower half ; above this the pharyngeal end of the cartilaginous Eustachian tube rests against the border just external to the pharyngeal opening, and is fastened here by a strong fascia ; the small *Eustachian spine* on the border marks the junction of the tubal and constrictor regions.

A projection, more or less prominent, on the posterior edge of the outer plate a little way down, may mark the anterior attachment of a " pterygo-spinous " ligament extending to the base of the sphenoidal spine. The ligament may be short, fastened higher up on the plate, or two ligaments, long and short, may be present : ossification may extend some way into these, so that a bony bar may be present here. The nerves issuing from the foramen ovale have varying relations with these bands, which are probably modified fibres of the external pterygoid muscle.

The *front* border of the inner plate, thick below, is in articulation with the vertical plate of the palate bone ; it turns in under the body of the sphenoid, and here it is interrupted by a groove that marks the *pterygo-palatine canal*. The inner side of the plate here has the projection of the *vaginal process*, which receives, between it and the body, the ala of the vomer (Fig. 183).

Skull and Hyoid

The front of the external plate widens in its upper part; this is the part that lies behind the upper jaw and forms the back wall of the spheno-maxillary fossa—of which the vertical plate of the palate, articulating with the inner pterygoid plate, forms the inner wall, and the tuberosity of the palate, fitting in between the pterygoid plates behind and the jaw in front, makes the lower boundary. This spheno-maxillary surface of the external plate is continuous with the area already noticed immediately below the lower edge of the orbital surface of the great wing, on to which the foramen rotundum opens.

The two pterygoid processes are completely separate in the embryonic skull, and vessels and nerves pass between them below the base of the skull; later, as the processes are formed and fuse together, this vasculo-nervous bundle is enclosed in a canal, the *Vidian canal* or foramen, which passes between the two plates high up, just below the junction of body and great wing. The anterior opening of this canal is thus into the spheno-maxillary fossa and its posterior opening is in the thickness of the fused mass of great wing, body, and pterygoid plates, that is, it is in the lower part of the front wall of the foramen lacerum, and just above and internal to the scaphoid fossa and below the commencement of the carotid groove.

We can now examine with greater care the region of the foramen lacerum and cavernous sinus, choosing a skull, if possible, in which the foramen is of fair size, not closed in too much by petrous ossification (Fig. 182). The carotid artery is directed forwards, inwards and upwards from the apex of the petrous, but on reaching the sphenoid it loses the inward inclination, runs upwards and forwards in the groove on the bone, and then turns sharply inwards and upwards under the overhanging anterior clinoid process, to pierce the dura mater internal to this by turning upwards to the brain. It has the carotid plexus on its outer side as it leaves the petrous and lies in the foramen lacerum. The great superficial petrosal nerve, leaving the hiatus Fallopii, lies in a groove that is directed downwards and forwards and inwards to the foramen lacerum, so that the nerve enters the foramen *above and outside* the issuing artery; but, because it is directed downwards while the artery is running upwards, the nerve is very quickly below the level of the artery, and when the artery turns forwards on to the sphenoid it lies *above* the nerve. The nerve has run downwards and inwards to the lower part of the front wall of the foramen, in a straight line for the Vidian canal, which opens here, and in doing so it runs obliquely along the outer side of the artery, in contact with it and the carotid plexus, and receives a short branch from this plexus, the *great deep petrosal*, so that the combined fibres run on as the Vidian nerve to the spheno-maxillary fossa and ganglion. In addition to these fibres are some backward-running ones from the ganglion that complete the nerve in the canal.

The sixth nerve runs nearly horizontally forward from the apex of the petrous to the lower part of the sphenoidal fissure. It pierces the dura a little distance behind and below the apex, and reaches this by passing *outside* the inferior petrosal sinus and *below* the superior petrosal sinus and petro-clinoid ligament. Running forward from this it must cross obliquely above the carotid artery in the foramen and lie above and outside the artery when this vessel turns forward on the sphenoid; after this the artery is ascending continuously while the nerve goes forward, so that the nerve is lying below and outside the vessel when this turns under the clinoid projection.

If now we look on the cavernous sinus as formed posteriorly by the junction of the two petrosal sinuses, it is apparent that the artery and the sixth nerve must lie altogether deep to the back part of the sinus, for the superior petrosal sinus comes above the nerve and the inferior one runs up along its inner side, between it and the posterior clinoid process, to join the upper sinus; further forward, however, the artery comes into relation with the inner wall and roof of the sinus while the nerve comes against the outer wall, for this is directed inwards as well as forwards.

The cavernous sinus lies outside and below the interclinoid band that joins the anterior and posterior processes, and is thus roofed in by the triangular area of dura mater (Fig. 181) that lies here and is made by the edges of the tentorium cerebelli. The third nerve pierces the dura in this situation and the fourth nerve a little further back, so these nerves lie in the roof of the sinus. A glance at the bone will show that the ophthalmic and maxillary divisions of the fifth nerve must lie against its outer wall, and the ganglion itself is a lateral relation of the posterior part of the sinus.

F.A. 15

From the account just given it can be seen that the foramen lacerum is filled in large part by the carotid artery and plexus and petrosal nerves, but these do not pass altogether through it ; in fact, there is only one structure, a meningeal branch of the ascending pharyngeal, that passes quite through it, from below upwards. The remaining parts of the foramen are filled by fibro-cartilaginous tissue, and its outer and posterior part, that receives the great superficial petrosal nerve, is under cover of the Gasserian ganglion.

Looking at the foramen from below (Fig. 182) we can see that the *Eustachian tube lies under it*. The cartilaginous portion of the tube is attached to the petrous bone internal to the spine of the sphenoid and passes downwards, inwards and forwards from this to the upper part of the posterior margin of the internal pterygoid plate. Lying in such a situation, it must pass under the foramen lacerum, and its relations are in fact apparent on the skull : above it is the petro-sphenoidal articulation and the foramen and basisphenoid, behind and internal the origin of Levator palati and lateral recess of the pharynx, which reaches the lower aspect of the petrous : outside and in front the Tensor palati and internal pterygoid plate ; below it the fibres of the Constrictor, passing from the internal pterygoid plate towards the pharyngeal tubercle on the basiocciput. As it lies below the foramen it necessarily has the carotid artery and Vidian nerve above it, and thus the backward-running branches of the nerve, from the spheno-maxillary ganglion, can reach the tube and adjoining pharynx which they supply.

A small triangular surface of bone outside the posterior Vidian opening forms part of the wall of the foramen and is grooved by the nerve entering the foramen : its lower border gives attachment to the fibrous tissue of the tube.

Articulations.—*Great Wing.*—Front margin, malar ; upper and front margin frontal ; upper margin, parietal ; postero-external, squamous temporal ; postero-internal, petrous temporal.

Pterygoid Processes.—Front margin of inner plate, vertical plate of palate ; lower front of both plates, tuberosity of palate.

Small Wing.—Front border, frontal.

Body.—Front surface, ethmoid ; back surface, occipital ; lower surface, vomer and palate (vertical plate).

Development.

The sphenoid is partly preformed in cartilage : the *cartilaginous* portion includes the *body*, the *small wings*, and the *inner parts of the great wings*. The rest of the alisphenoids and the pterygoid processes, with the exception of the hamular processes, are ossified in *membrane*.

The bone is formed by the coalescence of fourteen centres or more. The *basisphenoid* has four centres—on each side one for the floor of the sella turcica (two according to some authors) and another lateral one, which is really an extension inwards from the centre for the *lingula*. The *presphenoid* has two centres, one on each side. The *small wings* one centre each, which meet above the presphenoid at a later stage to form the jugum.

Great wings : one centre for each cartilaginous part, which also forms the *external pterygoid plate* The outer membranous part may be ossified by extension from this, or may have a separate centre for its margin (*os intertemporale*), in which case this portion may join the frontal or the temporal instead of the sphenoid, or may remain separate (*os epiptericum*).

Skull and Hyoid

Pterygoid processes: the *outer* is developed in continuity with the great wing. Each *inner* plate has a separate centre of ossification and fuses secondarily with the mass of the bone.

The centres for the great wing and internal plates appear towards the end of the second month, those for the presphenoid and small wings early in the third month, while the basisphenoid begins to ossify in the latter part of this month. The centre for each lingula is the last to appear, during the fifth month. The bony parts of the internal pterygoid plates and the two halves of the basisphenoid *unite* in the fourth month, and are joined some time later by the lingulæ. The orbito-sphenoids and their corresponding presphenoid centres also *unite* on each side in the fourth month, but it is not until the eighth month that the two halves of the presphenoid unite with each other and the basisphenoid, and even at birth the fusion is not complete, but is represented in part by a cartilaginous junction. In the dried specimen the situation of this cartilage is shown by one or two centrally or laterally placed fossæ or foramina, sometimes termed " cranio-pharyngeal canals," but they should not be confused with the canal properly so called that may result from partial persistence of the track of Rathke's pouch ; the latter is normally hardly distinguishable in the cartilaginous skull before ossification, and should be completely obliterated by extension forward of the basisphenoidal centres.

Sphenoidal Turbinate Bones.

These thin anterior coverings of the sphenoidal sinuses, often termed *conchæ sphenoidales* or *bones of Bertin*, are found as a rule attached to the sphenoid, but as they are developed as ossifications associated with the posterior cupola of the cartilaginous nasal capsule, quite distinctly separate from the sphenoid, they are considered separately here. Each bone fits as a cap on the front of the corresponding sinus and presents a rounded opening above the centre of its anterior surface : the outer and lower corner of this surface is prolonged outwards and downwards with a sharp angle, and the whole surface is applied to the back of the ethmoid.

The *lower aspect* is separated from its fellow by the projection of the rostrum, and forms a point which fits in between the rostrum and the internal pterygoid plate. The *inner, upper* and *outer* parts of the bone are applied to and fused with the sphenoidal septum and upper and outer parts of the body of the sphenoid respectively. The lower surface appears in the roof of the nasal fossa, and, outside this, articulates with the palate bone and forms the upper boundary of the spheno-palatine foramen : the outer surface is in the inner wall of the orbit between the ethmoid and the presphenoid.

Each bone is usually destroyed in part during disarticulation owing to its adhesion to the neighbouring bones : a fairly complete specimen is shown on one side in Fig. 183.

Each ossicle commences to *ossify* in the fifth month by a centre on the innner part of the cartilaginous cupola of the nasal capsule, followed about a couple of months later by a second centre on the outer part. These centres form the bone by fusion with each other and with additional lower centres in the first year or so of life. During the fourth and fifth year the cartilaginous structure is absorbed, and this absorption extends to the outer and back part of the bony structure, so that the commencing sphenoidal sinus comes into relation with the sphenoid, into which it commences to extend about the seventh year. The bones are not fused with the sphenoid until the cartilage is absorbed.

228 Anatomy of Skeleton

MAXILLA.

The bones that make the upper jaw form the main skeleton of the face : each bone is situated below the orbit and beside the nose, forming part of the bony walls of these cavities, and also makes the greater part of the roof of the mouth. It is placed in front of the pterygoid region of the sphenoid, but is of course much wider than this region transversely.

Each maxilla has a *body*, below which its *alveolar process* carries the teeth ; a *palatine* or *horizontal process* articulating with its fellow in the middle line below the nasal region ; a *nasal process* standing up from the front and upper part of the body to articulate with the frontal ; and a rough *malar process* externally that carries the malar, through which the maxilla is once more connected, indirectly, with the frontal as well as with the temporal.

The *body* is roughly a three-sided pyramid, with its base forming part of the outer wall of the nasal fossa and its apex supporting the malar, an upper *orbital surface*,

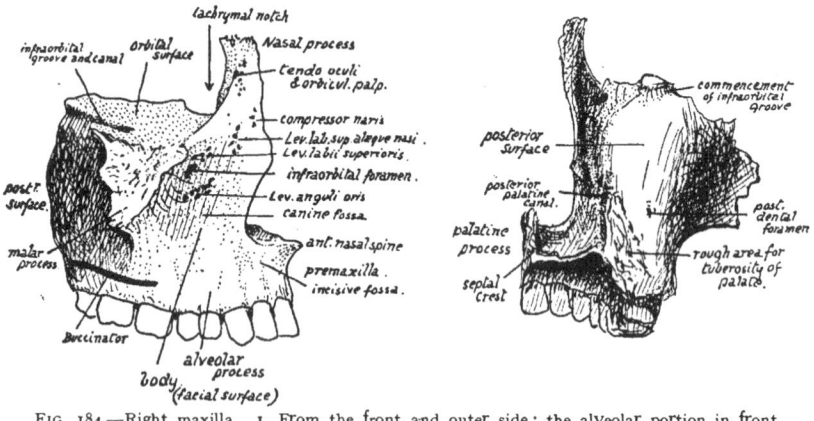

FIG. 184.—Right maxilla. 1. From the front and outer side: the alveolar portion in front, and below the level of origin of Buccinator behind, is covered by mucous membrane and can be examined from the mouth. 2. Posterior surface, showing the part which makes the front wall of the sphenomaxillary fossa and, outside this, the front wall of the zygomatic fossa.

an anterior *facial surface*, and a rounded *posterior surface*. The body is hollow, containing the large air sinus known as the *antrum of Highmore* or *maxillary sinus*, which opens into the nasal fossa.

The *orbital surface*, triangular in accordance with the pyramidal conception of the body, is nearly flat, but forms a sloping plane looking upwards and outwards and slightly forwards. It is *grooved* from behind, forwards and outwards, by the infraorbital nerve and vessels, the groove ending in front in the *infraorbital canal*. The surface is in contact with the inferior Rectus and inferior Oblique muscles, the latter arising from it, and is continuous externally with the rough area for the malar. Its anterior border forms a part of the lower orbital margin, the posterior border is the lower margin of the spheno-maxillary fissure, and the inner margin supports the ethmoid except where the orbital process of the palate, at the posterior end of the margin, reaches the floor of the orbit between the two bones : at the front end of the inner margin, just behind the nasal process, is the *lachrymal notch*, which is made into a complete canal or foramen by the lachrymal bone, and this bone articulates with the maxilla along the posterior border of the notch as well as along the nasal process.

The *facial surface* (Fig 184) of the body, antero-external, is rather concave from the presence of the malar process and the *canine fossa*, above which, and about one-third of an inch below the orbital margin, is the anterior opening (*infraorbital foramen*) of the infraorbital canal. The surface is directly continuous below with that of the alveolar process, which presents ridges and depressions corresponding with the position of the fangs of the teeth : the front surface of the body is bounded internally by the sharp margin of the nasal opening, but the alveolus is continued below this to the middle line to meet its fellow and to carry the incisors, and it presents here an anterior concavity, the *incisive* or *myrtiform fossa*. This part of the alveolus, frequently termed the *premaxilla*, is continuous behind with the palatine process, and is limited above by the lower margin of the nasal orifice, which is produced centrally into a *nasal spine*. The incisive slip of the Orbicularis oris takes origin from the incisive fossa.

The facial surface is covered in its upper part by fibres of Orbicularis palpebrarum, over the alveolar portion by mucous membrane and Orbicularis oris and Buccinator, and between these two areas by small facial muscles passing to the upper lip : the Levator anguli oris arises below the foramen, and the Levator labii superioris above the foramen, both under cover of the orbicular fibres of the lower lid ; while the Levator anguli oris alæque nasi and Compressor naris take origin further in near the root of the nasal process.

The *posterior or postero-external surface* (Fig. 184) is separated from the one last described by the prominent lower border of the projection that supports the malar, but the alveolar process is carried continuously to the back of the bone below this projection and affords origin to Buccinator. The posterior surface is rounded and forms the front wall of the fossa that lies deep to the zygoma and below the inferior temporal crest and outside the outer pterygoid plate : the *fossa* is termed the *zygomatic* or *infratemporal*, and is occupied by the External Pterygoid and, superficial to this, by the tendon of the Temporal muscle and the coronoid process of the mandible which receives this tendon. The inner part of the posterior surface, however, is in front of the pterygoid region of the sphenoid, and therefore constitutes the front wall of the spheno-maxillary fossa. This fossa is closed below by the tuberosity of the palate bone fitting in between the maxilla and the pterygoid plates, so that a rough *area for this tuberosity* is visible on the lower and inner part of the posterior surface of the maxilla. The tuberosity of the palate is always completely interposed between the two bones, and the maxilla never articulates here with the sphenoid, although the suture lines between the various elements are not as a rule visible on the complete skull. Impressions made on the bone by certain of the structures leaving the fossa can be easily seen : thus, the *posterior dental and alveola*r vessels and nerves reach *foramina* on the back aspect of the bone, the commencement of the *infraorbital* groove is seen above, and the *posterior palatine* groove, completed by the palate bone, is apparent low down on the back and inner aspect of the body in front of the area for the tuberosity of the palate.

The upper margin of the posterior surface corresponds with the posterior margin of the orbital surface and thus forms the lower boundary of the spheno-maxillary fissure, which therefore opens from the orbit into the zygomatic fossa in its outer part but into the spheno-maxillary fossa in its inner part : the spheno-maxillary fossa can be looked on as an inward extension, behind the jaw, of the zygomatic fossa.

The rounded part of the posterior surface is sometimes referred to as the " tuberosity " of the maxilla.

230   Anatomy of Skeleton

The *inner* or *nasal surface* (Fig. 185) of the body presents the large, irregular opening of the antrum. The surface may be taken on the whole as vertical, but its upper part is sloped outward and here the ethmoid rests on it and covers in the upper part of the opening of the antrum. The posterior third or more has the vertical plate of the palate applied to it, also closing in the antral aperture, and the upper end of the vertical plate passes up between the maxilla and the ethmoid, separating them and appearing between them in the extreme back part of the orbital floor. In front of the antral opening is the groove for the *nasal duct*, which is covered in from this aspect, in its upper part, by the lachrymal bone; this bone also encroaches to a slight extent on the upper and front part of the antral orifice. The lower part of the opening is filled by the down-turned maxillary process of the inferior turbinate bone, so that the opening, still further reduced to a very small size by mucous membrane, is situated in the outer wall of the nose above the inferior turbinate, *i.e.*, in the middle meatus.

Below the level of the inferior turbinal articulation the nasal surface is smooth and concave, forming the wall of the inferior meatus and becoming continuous with the smooth upper surface of the palatine process.

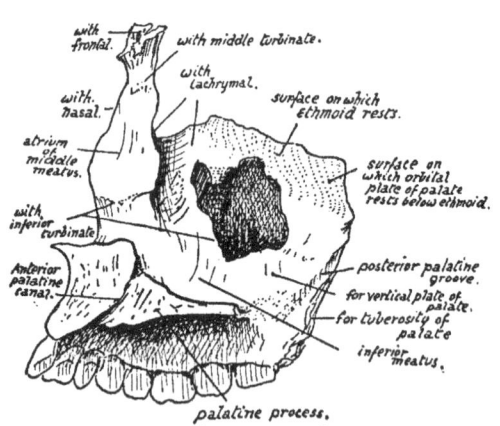

FIG. 185.—View of inner aspect of right maxilla.

In front of the lower end of the lachrymal groove the inner surface of the body is continuous with the inner surface of the nasal process. Here there is a ridge that carries the anterior part of the *inferior turbinate*, which, to reach it crosses the lower end of the lachrymal groove and forms its bony wall below the lachrymal bone.

The *nasal process*, by its posterior aspect, forms the front wall of the nasal duct and articulates here with the lachrymal. In addition it has :—

(*a*) An *inner surface*, forming the wall of the nose above the inferior turbinal ridge, smooth and concave and covered by mucous membrane of the atrium of the middle meatus, and *ridged* above for articulation with the *middle turbinate* (ethmoid).

(*b*) A *front border* articulating with the nasal bone.

(*c*) A rounded *outer border*, that forms part of the orbital margin and has the tendo oculi attached to it.

(*d*) A *facial surface*, covered by and giving origin to Orbicularis palpebrarum and continuous with the facial surface of the body.

(*e*) A rough *upper end* articulating with the frontal.

The thick *alveolar process* contains eight alveoli for permanent teeth, separated by interalveolar septa. The incisors are carried in the (premaxillary) portion below the nasal opening, the canine, bicuspids, and first molars below the facial surface of the body, and the other two molars below the region of the malar prominence and the postero-external surface. The inner alveolar wall, behind the premaxillary region, is in the plane of the inner surface of the body, while its outer wall is confluent with the superficial surface of the body, so that the alveolar process can be said to form the floor of the maxillary sinus or antrum : the fangs of the teeth, particularly of the first molar, may occasionally project into the cavity, covered by its lining mucous membrane.

The complete alveolar arcade made by the two maxillæ forms a semi-ellipse, contrasting with the parabolic curve of the mandibular alveolus ; as a result of this difference the back upper molars are somewhat out-turned to oppose the lower teeth.

The *palatine* or *horizontal process* is on the level of the lower part of the body and thus projects inwards above the alveolar level. It has a rough *median border* articulating with its fellow and forming with it a raised *septal crest* that supports the vomer; also a rough but thin *posterior border* articulating with the horizontal plate of the palate bone. It is thicker in front where it joins the premaxillary part of the alveolus, and at the junction on the inner articular surface, the *anterior palatine foramen or canal* is visible on the bone. The anterior palatine canal (naso-palatine) lies centrally between the premaxilla and the palate proper, and each maxilla has the half-impression of the canal ; the common lower opening of the canal leads to four smaller canals which form the upper part of the foramen, and these are : two antero-posterior for the naso-palatine nerves and two side by side for the terminal parts of the posterior palatine vessels.

The upper surface of the palatine process is concave between the septal crest and the outer wall of the nose and rises slightly in front on to the premaxilla. The lower surface is roughened by the mucous membrane and grooved by the posterior palatine nerve and vessels running forward.

The malar process has a smooth concave *facial surface*, a smooth *posterior aspect* looking toward the zygomatic fossa, a lower bracket-like ridge that extends to the alveolus between the first two molars, and a rough *upper and outer surface*, continuous with the orbital surface, that carries the malar and enables this bone to complete the outer part of the floor of the orbit.

The *antrum* has the alveolus as its floor, and is often irregularly ridged in this situation. The cavity may extend into the malar process and even a short way into the premaxilla. It may be partly or completely subdivided by bony septa. The infraorbital canal causes a ridge in its front and upper part, and other ridges may be seen running from this and containing the middle and anterior dental nerves ; the last nerve, going to the incisors, runs close to the anterior nasal margin.

This bone is further considered in association with the palate (p. 235), spheno-maxillary fossa (p. 236), nose (p. 244), and orbit (p. 253).

Development.

The bone is developed in membrane, laid down in the maxillary process of the mandibular arch and in the fronto-nasal process with which this fuses. Ossification commences in the sixth week on the outer side of the dental groove, in the future canine region, and extends from this forward into the incisor region and backwards towards the molars. The body of the bone and its nasal and malar processes are extensions upwards from this primary centre.

This "single" centre is apparently made by the rapid fusion of two distinct ossifications, which appear, in the body and premaxillary regions respectively, at the same time.

In the tenth week the ossification extends inwards and backwards above the teeth germs into the palate folds, which have met in the ninth week (Fig. 186). From this extension the inner walls of the alveoli are formed as secondary down-growths : before this takes place there may be a cartilaginous change here and there in the alveolar condensation.

The down-growth that comes from the premaxillary region to form the inner walls of the incisive alveoli are separated from the rest of the palatine in-growth by the position of the anterior palatine foramen, which marks the situation of junction of the paired palate-folds with the fronto-nasal process (Fig. 186), and these alveolar walls may not completely fuse with each other or with the canine alveolus, giving rise to the

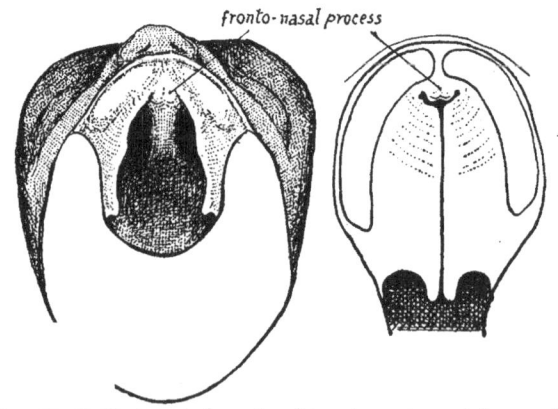

FIG. 186.—To illustrate the formation of the palate. On the left the roof of the mouth in an embryo of seven weeks is seen from below, the lower jaw being removed. The palate is represented by a pair of folds which project from the inner side of each maxillary process and, with that process, reach the fronto-nasal process in front. In the next figure the palate folds are shown closed, having met behind the fronto-nasal process or premaxilla ; thus the whole palate, though mainly formed from the folds, has a small piece of fronto-nasal process in its extreme front part. When the folds do not meet a *cleft palate* results, and the cleft may extend forward between the maxillary and fronto-nasal processes.

incisive sutures frequently seen in this region of the hard palate, and formerly supposed to mark off the separate premaxilla from the true maxilla.

About the time that the palatine process grows inwards a secondary centre appears in the upper part of the premaxillary region, from which the " vomerine crest " and front wall of the canals here are formed.

The infraorbital nerve and vessels are at first in a groove on the upper aspect of the body, but are covered in in part after the fourth month by a growth inwards of the outer part of the body : a suture usually persists in the adult above and internal to the canal, marking the extent of this inward growth.

The antrum is represented by a depression on the outer wall of the nose in the fourth month, and at birth it forms a fair-sized cavity in the inner part of the maxilla (Fig. 214) : it reaches the situation of the infraorbital canal during the first year, and after this grows fairly rapidly until it attains its full size ; its growth is of course conditioned by the size of the bone, and cannot be completed until the last molar comes into position. (For further account see later, p. 250.)

Skull and Hyoid

PALATE BONE.

Each palate bone consists of : (1) a *vertical plate*, which is applied to the posterior part of the inner surface of the body of the maxilla ; (2) a *horizontal plate*, which projects inwards from the lower end of the vertical plate, lies in the plane of the palatine process of the maxilla, and articulates mesially with its fellow of the opposite side ; (3) a *tuberosity* that projects downwards, outwards and backwards from the vertical plate, and so lies behind the lower part of the back of the maxilla and separates it from the pterygoid plates of the sphenoid, against which it is placed.

The *vertical plate* is not only applied to the inner side of the maxilla, but extends back behind the level of this bone to articulate with the front border of the internal pterygoid plate, and in this way its hinder part forms the inner wall of the spheno-maxillary fossa. These two articulations cause a difference in the direction of the back

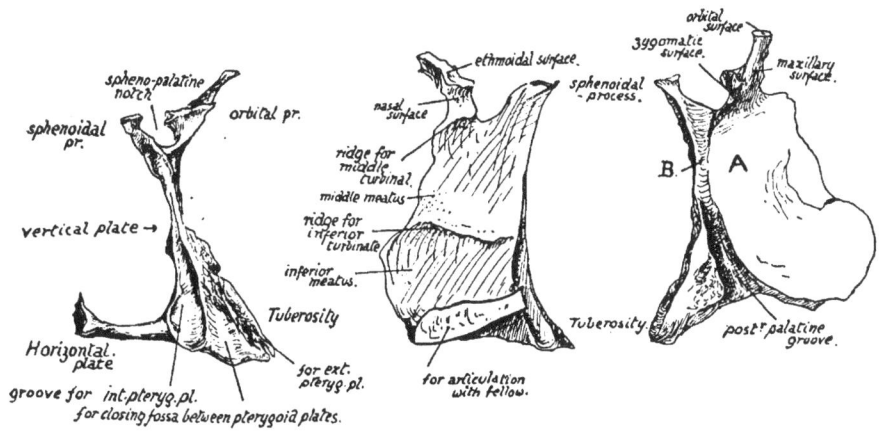

FIG. 187.—Right palate bone. 1. Posterior aspect. Observe the deep groove for internal pterygoid plate against which the bone is placed. 2. From the nasal or inner side. 3. From the outer side. The greater part, *A*., of this surface rests against the maxilla, but a small part, *B*., projects back beyond that bone and forms the inner wall of the spheno-maxillary fossa ; the post. pal. canal runs down from this surface, between *A*. and the tuberosity.

and front portions of the vertical plate : the front part (Fig. 185), being applied to the maxilla, turns outward at its upper end with the maxillary surface, and thus lies between the maxilla and the ethmoid and reaches the orbit, where it forms part of the floor ; whereas the posterior part, following the margin of the internal pterygoid plate, turns in at the top (Fig. 183) and comes to lie below the body of the sphenoid. The vertical plate can be described, therefore, as terminating above in an anterior *orbital process*, directed upwards and outwards, and a posterior *sphenoidal process*, directed upwards and inwards. Between these two is a deep *spheno-palatine notch*, converted into a foramen by the sphenoidal turbinate that lies above it. Sometimes a large orbital process may convert the notch into a complete foramen : this has also been seen double.

The *orbital process*, very variable in size, has :—

    (a) A *maxillary surface*, applied to the maxilla ;

(b) an *ethmoidal* (upper) *surface*, which supports the ethmoid and shows a crest *for the middle turbinal*, a part of the ethmoid ;

(c) a *spheno-maxillary* (or *zygomatic*) *surface*, looking outwards and backwards behind the maxilla and therefore into the spheno-maxillary fossa ;

(d) an *orbital* (terminal) *surface*, visible between ethmoid and maxilla in the floor of the orbit.

The *sphenoidal process* has :—

(a) An outer or *spheno-maxillary surface ;*
(b) an inner or *nasal surface* ,
(c) an upper *terminal part* applied to the sphenoidal body.

It is grooved behind by the margin of the internal pterygoid plate.

The vertical plate below the processes and notch has an *inner surface* that is purely nasal, crossed by a rough *ridge* that carries the *lower turbinate bone*, and an *outer surface* that is mostly maxillary, but has a small triangular area behind this and immediately below the notch (Fig. 187) which forms the inner wall of the spheno-maxillary fossa. Below this the *tuberosity* begins to grow out from the bone. On this surface of the plate, in front of the tuberosity, the *posterior palatine groove* is seen running downwards and forwards : this part of the bone is applied to the maxilla, which therefore completes the canal. The lower part of the groove may be partly or completely closed in by the tuberosity.

The *horizontal plate* has rough articular *front* and *inner margins* for the maxillary palate and its fellow respectively, a concave *posterior edge* to which the palatine aponeurosis (soft palate) is attached, with a prominent *posterior nasal spine* centrally, where the Azygos uvulæ muscle arises, a rough *lower surface*, and a smooth *upper or nasal surface* which forms a prominent median *crest* with its opposite fellow, for the support of the vomer.

The *tuberosity* extends in a direction downwards and outwards and slightly backwards. It has a rough antero-external *maxillary surface*, separated by the posterior palatine groove from the maxillary surface of the vertical plate, and a *posterior surface* which is applied to the front of the pterygoid processes. This surface (Fig. 187) shows a deep groove internally for the internal pterygoid plate, a rough articular area obliquely directed by its outer margin for the outer plate, and a shallow concave fossa between these that completes the floor of the pterygoid fossa (see Fig. 161) and gives origin to fibres of the Internal Pterygoid muscle. This origin extends on to the *lower surface* (Fig. 161) of the tuberosity and from this runs on to the *outer* part, which is visible between the external plate and maxilla.

This origin of part of Internal Pterygoid gives its front and lower fibres the appearance of being superficial to the lower fibres of the External Pterygoid.

Articulations.—With the maxilla (vertical plate, horizontal plate and tuberosity), ethmoid and inferior turbinal (vertical plate), sphenoid (vertical plate and tuberosity), vomer and its fellow (horizontal plate).

Development.

The bone is developed in membrane on the inner aspect of the cartilaginous nasal capsule, from a centre that appears during the eighth week in the region of the future tuberosity ; from this ossification extends into the palate fold as this meets its fellow, to form the horizontal plate, and more slowly upward to make the vertical plate. Owing to the relative small vertical measurement of the nasal cavities at birth, associated with the incomplete growth of the maxillæ, the palate bone at this time shows a vertical plate of only about the same length as the horizontal plate.

THE HARD PALATE.

The *hard palate*, as seen on the basal aspect of the skull, has its anterior two-thirds to three-quarters made by the palatine plates of the maxillæ, with the horizontal plates of the palate bones forming the posterior third or fourth. It is bounded by the alveolar process, which is highest at the sides. The region is concave as a whole in all directions. The anterior palatine foramen is just behind the incisor region ; this is covered by the incisive papilla in the recent state. Behind this the palate is rough and has the mucous membrane firmly adherent to it—the membrane becomes thicker further back and leaves the bone smoother, although there are more abundant glands in this situation.

The (large) *posterior palatine foramen* is seen immediately internal to the alveolus of the last molar tooth, and the suture line between maxilla and palate turns sharply back along the alveolus to run into the foramen, indicating that the canal is made by the articulation of the two bones. The main posterior palatine vessels and nerve emerge here and turn forward toward the anterior palatine foramen, through which the artery turns up on to the septum. The groove containing the structures is frequently double, in which case the artery occupies the groove nearer the alveolus. Behind the main opening another or perhaps two are visible ; these are for smaller nerves and vessels belonging to the same group but turning back into the soft palate. The aponeurosis of the Tensor palati reaches a ridge which runs transversely between the foramina. In this region the lower surface of the horizontal plate of the palate can be seen to pass directly into the lower surface of the tuberosity, behind the foramina. The posterior edge is thin and concave, being prolonged centrally into a *posterior nasal spine*.

The region of the palatine suture is occasionally raised in the whole or part of its length into a prominent ridge, known as the *torus palatinus*. A variety of palate is rarely seen in which the maxillæ are produced backwards between the palate bones centrally. The commonest abnormality of the palate is " cleft palate," in which the two palate-folds have failed to unite with each other, and if the cleft is complete, and extends to the face, the fold at fault has also failed in union with the fronto-nasal process : such failure may be on one or both sides. Fig. 186 shows how the deformity can occur, and it is evident that if the folds fail to meet the bones cannot join subsequently.

As already mentioned, the secondary down-growths of the hinder walls of the alveoli of the incisor region, toward the anterior palatine foramen, make the surface of this part of the bony palate, and one or other of these processes may fail to unite with its neighbour : this leads to the presence of a fissure extending out from the foramen. Such a fissure may appear between the incisors, or to the outer side of the lateral incisor, or in rare cases even to the outer side of the canine. The fissure was for many years considered to mark the limit of the " premaxilla," but its irregularity of position, and the fact that it has even been seen doubled on one or both sides, was sufficient argument against this view, and now it is known to be of quite secondary origin, as described. It is of historical interest to note, also, that the presence of these fissures at one time led to the adoption by many of the suggestion that the alveolar portion of the upper jaw-resulted from the fusion of three distinct parts, the *endognathion*, placed on the inner nasal process and carrying the central incisor, the *mesognathion* on the lateral process bearing the lateral incisor, and the *exognathion* further out. There is, of course, no justification in development for this view, which was introduced really to explain not only the occurrence of double fissures, but also the presence of teeth bearing unusual relation to clefts in the palate and alveolus. These extra teeth, however, are more rationally explained on the assumption of breaking up of the dental epithelial area over the situation of the cleft.

The palate is considerably narrower in female skulls than in those of the other sex.

Spheno-maxillary Fossa.

The student should now examine the spheno-maxillary fossa with great care, so that he may gain a clear comprehension of the way in which it is built up and of its relations and connections, for the key to an understanding of a great deal of the deep anatomy of the face is to be found in this region.

We have seen that the palate bone is applied to the inner side of the maxilla, but that it projects back further than this bone to articulate with the internal pterygoid plate, so holding the jaw away from the pterygoid region by the amount of this projection, and thus leading to the existence of an interval between them that is the spheno-maxillary fossa. Evidently the vertical plate of the palate must form the inner wall of this fossa and must separate it from the nose, for this plate is part of the outer wall of the nasal fossa. Because the vertical plate articulates behind with the internal pterygoid plate along the whole length of its front border it is equally plain that the posterior wall of the fossa must be formed by the *outer* plate (Fig. 183) except in its extreme upper and inner part, where the inner plate is slightly wider than its front edge that carries the palate bone : in other words, the front edge of the internal plate turns in under the body of the sphenoid and thus leaves a small piece of the plate exposed to form part of the back wall of the fossa just below and outside the body of the bone. This is the level of the Vidian opening, which is seen immediately outside this little piece of inner plate. Outside this the external pterygoid plate makes the chief part of the posterior wall : the plate broadens rapidly as it passes up and becomes continuous with the area of bone immediately below the orbital surface of the great wing.

In Fig. 183 the back wall of the fossa is shown, and the outer plate is seen to widen very much in its upper part and to open out into the spheno-maxillary surface of the great wing. The position of the vertical plate of the palate is also shown, and the small part of the inner plate that enters into the formation of the back wall. At the outer edge of the broad front surface of the outer plate the fossa becomes continuous with the zygomatic fossa, so it is evident that the breadth of the fossa depends on the width of the surface of the plate ; this varies in different bones, leading to a shallow or deep fossa, as the case may be. The opening, bounded by a margin of the plate, through which the fossa becomes continuous with the zygomatic fossa may be termed the *pterygo-maxillary fissure*, to distinguish it from the *spheno-maxillary fissure*, which opens into the orbit from the fossa. Observe that the anterior opening of the foramen rotundum is on the spheno-maxillary surface of the great wing, so that the nerve enters the fossa on leaving the foramen ; also notice that the Vidian opening is further in, and on a lower level, because it runs between the two plates and not through the great wing. A ridge of bone separates the two foramina and can be traced up to the side of the body of the bone. The *inner wall* of the fossa is made by the vertical plate of the palate. Articulate this bone with the maxilla and observe how the front portion of the vertical plate lies against the nasal surface of the maxilla, so that its orbital process can be seen from behind the jaw, at the top back corner of the bone, as it turns out to reach the orbit ; in this way the *orbital process* comes to possess a *zygomatic surface*. The spheno-palatine foramen separates this process from the sphenoidal process, which is directed in as it follows the internal plate under the body of the sphenoid.

Skull and Hyoid

It is clear, therefore, that the inner wall of the fossa is made by that part of the vertical plate that lies below the foramen and which forms the sphenoidal process behind this. The foramen is closed above by the body and turbinate of the sphenoid and thus opens into the nasal fossa *immediately below the roof*, which is formed here by the under surface of the sphenoid : in this way it becomes possible for nerves and vessels passing through the foramen to turn down on the outer wall or to cross below the roof and reach the septum, and thus to supply both sides of the nasal cavity.

The extreme inner part only of the back of the maxilla forms the front wall of the fossa ; the remainder makes the front wall of the zygomatic fossa, and the continuity of the two fossæ is very clear on this bone. The upper margin of the bone is the lower edge of the spheno-maxillary fissure, which opens from the orbit into the two fossæ : the extent to which the fissure opens into the spheno-maxillary fossa is really determined by the size of the spheno-maxillary area on the great wing, as can be seen on comparison of different skulls, so varies with the depth of the fossa. The fissure is directed downwards and outwards, and thus the maxillary nerve reaches its level by running outwards as well as forwards, without an upward direction : when the fossa is shallow the nerve may pass from the spheno-maxillary into the zygomatic fossa before reaching the level of the fissure.

The direction and construction of the posterior palatine canal can be understood by articulating the palate and maxilla, and the situation and relations of the tuberosity of the palate should be thoroughly investigated at the same time.

To sum up :—The fossa is connected *with the cranial cavity* by the foramen rotundum, with the *region of the naso-pharynx and Eustachian tube* (see p. 225) by the Vidian and pterygo-palatine canals, with the *nasal fossa* by the spheno-palatine foramen, with the *mouth cavity* by the posterior palatine canal (from which minute foramina also reach the outer wall of the nose), with *the orbit* by the spheno-maxillary fissure, and with the *zygomatic fossa* through the pterygo-maxillary fissure.

The structures passing through the spheno-maxillary fissure are :—
Infraorbital nerve and vessels ;
Temporo-malar nerve ;
Branches from Meckel's ganglion ascending to the orbit ;
The inferior ophthalmic vein.

Passing through the pterygo-maxillary fissure are :—
Internal maxillary vessels ;
Posterior dental nerves and vessels ;
Infraorbital nerve (sometimes).

The spheno-maxillary fissure is closed externally by the malar, connecting the maxilla and the sphenoid : rarely the maxilla may meet the sphenoid directly here, behind the malar, and this is the only place where these two bones may come into contact.

ETHMOID.

This bone lies immediately below the ethmoidal notch in the frontal bone, which is therefore closed by its upper surface, is applied to the front aspect of the sphenoidal body, and rests below on and between the two maxillæ. It thus forms the walls of the upper portion of the nasal fossæ.

It consists of :—(a) A median *perpendicular plate* which forms part of the bony septum of the nose, articulating with the other elements of the septum. (b) An upper horizontal *cribriform plate*, that is seen in the notch of the frontal, and is so named because it is pierced by numerous foramina for the filaments of the olfactory nerves : the inner edges of the orbital plates of the frontal rest on the horizontal top of the bone.

Anatomy of Skeleton

The upper part of the vertical plate appears above the cribriform plate in one region to make the strong process of the *crista galli*. (c) A *lateral mass* on each side suspended from the edges of the cribriform plate and consisting of an intricate series of thin-walled air cells attached to the inner aspect of a thin lateral plate termed the *os planum* or *lamina papyracea*. It is this lateral mass that rests on the maxilla. The mass of air cells constitute what is termed the *labyrinth*.

The superior surface (Fig. 188) shows :—(a) A lateral area on each side articulating

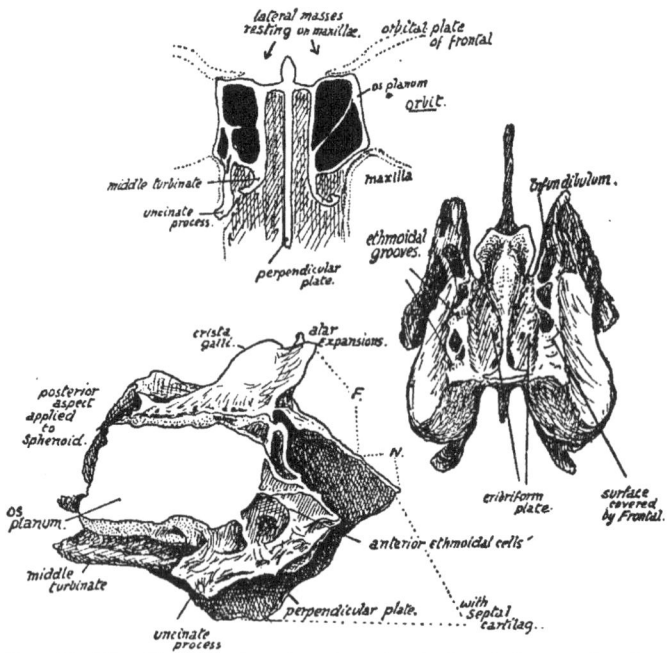

FIG. 188.—Ethmoid. Upper figure is a scheme showing how the bone consists of a central " perpendicular plate " connected above by a horizontal piece with the lateral masses which rest on the maxillæ, and how the outer walls of these lateral masses are formed by thin laminæ (os planum) which are in the inner walls of the orbits. The lateral masses are composed of air-cells, from the inner walls of which upper and middle turbinates are suspended and are thus connected with the roof. The orbital plates of the frontal rest on the horizontal piece of the ethmoid. On the right is the ethmoid seen from above, and on the left a view from the right side. *F.*, extent of articulation with frontal ; *N.*, with nasal bones.

with the orbital plate of the frontal. These surfaces usually exhibit broken air cells which are completed by the frontal. The anterior orbital or ethmoidal groove is seen running inwards and forwards across this surface, and the posterior groove lies a little behind it.

(b) The cribriform area supports the olfactory bulbs. It is covered by dura mater, under which the nasal nerves run forward from the anterior orbital grooves beside the crista galli to reach a notch on each side on the front border of the bone (sometimes completely closed in) in which they lie, between the ethmoid and the frontal : the anterior ethmoidal arteries run with them.

This surface is flush with the jugum sphenoidale behind, where it is notched to receive the " spine " of the sphenoid.

The *lower surface* of the plate forms part of the roof of the nasal fossa, and is therefore continuous in front with the small nasal areas on each side of the nasal spine of the frontal.

Vertical Plate.—The crista galli comes above the horizontal plate as a thick process to which the falx cerebri is attached, and from which strands of dura mater spread out over the neighbouring bones. In front of it the foramen cæcum of the frontal is placed, so that the crista forms its back wall and broadens out into two *alæ* to cover it.

The rest of the vertical plate is in the septum, and is drawn down below into an angle (Fig. 193), that fits in between the vomer, which articulates with its postero-inferior border, and the cartilage of the septum, that is fastened to its antero-inferior border. The upper and front part of the plate, just in front of the crista galli, comes in contact with the nasal spine of the frontal between the small nasal areas on that bone, and projects beyond this to support the two nasal bones. Behind, it comes up against the crest and top of the rostrum of the sphenoid. It is covered by mucous membrane and, deep to this, is grooved to carry olfactory fibres and their arachnoid coverings.

The *Lateral Mass* is applied to the front of the sphenoidal turbinal : it rests below on the inner and upper part of the maxilla, but its posterior part is separated from this bone by the vertical plate of the palate. The os planum forms the greater part of the inner wall of the orbital cavity : it does not, however, extend so far forward as the rest of the mass, and the deficiency is filled by the lachrymal bone, which thus articulates with the anterior margin of the os planum and partly covers in the exposed " labyrinth." The os planum is in a plane with the side wall of the sphenoidal body and turbinal, and the lachrymal connects it with the nasal process of the maxilla in front.

The *labyrinth* of the lateral mass consists of a number of air cells, with very thin walls, that are arranged in groups—*anterior, middle,* and *posterior*—which do not communicate directly. These cells are attached to the inner aspect of the os planum and project into the cavity of the nose. From their inner surfaces thin curved sheets of bone curve downwards and outwards and are known as the *upper* and *middle turbinate* bones : they are, of course, only portions of the ethmoid. A scheme of the arrangement is shown in Fig. 188. The middle turbinal comes down below the level of the lateral plate and can be recognised at once. The middle meatus of the nose lies below the middle turbinal, and the infundibulum runs up from the front part of this to the frontal sinus, and the position of the infundibulum can be recognised on the bone (Fig. 188) as a canal running upwards and forwards on the inner side of the mass that is exposed in front of the os planum, and continues below with the groove below the middle turbinal.

It is evident that the lachrymal must cover in the infundibulum and form its outer wall here if it is not separated from it by cells in this situation, which is the usual condition : many incomplete cells may be seen round it which will be completed by the lachrymal, frontal, and maxilla.

An *uncinate process* passes downwards and backwards from the front and lower part of the mass, outside the middle turbinal (Fig. 188), to articulate with the inferior turbinate bone and partly fill in the opening of the maxillary antrum, but this is not often seen complete on the disarticulated bone.

If the lateral mass is removed and looked at from the inner side the upper and lower turbinals are seen suspended from the inner wall of the ethmoidal cells : the superior meatus is between the two turbinals (see Fig. 194). Fig. 198 gives the appearance presented by the bones on section *in situ*. For further details of this region, see p. 244.

The lower walls of the middle ethmoidal cells form the part of the lateral mass that rests (Fig. 188) on the maxilla, and the middle turbinal is a direct continuation of these walls, so that when the walls become deficient behind the attached border of the middle turbinal obtains another articulation, with the *palate bone*, the vertical plate of which is here between it and the maxilla.

The ethmoid articulates with the *sphenoid* and its *turbinals* behind, the *palates* below, *maxillæ* below and also in front, *lachrymals* laterally in front, *frontal* above, and above and in front centrally, the two *nasals* and the *vomer* with its perpendicular plate, and the two *inferior turbinals* through its uncinate processes.

Development.

The nasal fossæ are at first enclosed, with the exception of their lower regions, by a cartilaginous capsule which is continuous over the roofs of the fossæ with the cartilaginous septum : cartilaginous supports grow from this into the turbinal masses and map out the air cells in part. Ossification starts on the surface of this capsule, just behind the lachrymal region, in the fifth month, and extends slowly into the cell walls and turbinates of the lateral mass, so that at birth the mass is practically ossified. The septum, however, is still cartilaginous, and the roof is fibrous, the cartilage having disappeared here. A centre appears in the membrane on each side of the upper part of the septal cartilage during the first year, and from this the bony plate is slowly formed during the next four or five years ; the crista galli is made by an early extension from these centres. The bony lateral mass in the meantime has extended its ossification into the roof and this joins with the completed septum.

LACHRYMAL.

Small shell-like bones placed near the front part of the inner orbital wall. Each bone articulates by its posterior margin with the os planum of the ethmoid, by its front border with the nasal process of the maxilla, rests below on that bone, and reaches the frontal above : its deep surface covers the antero-lateral aspect of the lateral mass of the ethmoid where the os planum is deficient, and forms outer walls here for the anterior ethmoidal cells and the infundibulum and a small part of the wall of the middle meatus, and a prolongation downwards of this surface, making the inner wall of the nasal duct, meets and articulates with an upper process of the inferior turbinate bone which forms the lower end of this inner bony wall. This downward process of the lachrymal also articulates by its posterior part with the uncinate process of the ethmoid, and to a very small extent helps to close in the opening of the antrum of Highmore.

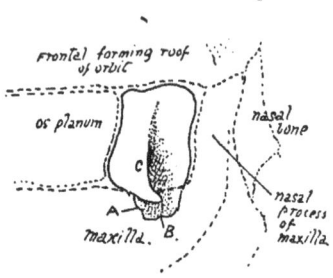

FIG. 189.—Orbital aspect of right lachrymal bone, with the adjoining bones indicated by interrupted lines. *A.*, process downwards forming inner wall of nasal duct and joining inferior turbinate ; *B.*, hamular process ; *C.*, lachrymal crest.

The deep surface shows impressions for the cells that are walled in by it.

The superficial, external, or *orbital surface*, on the whole slightly concave, has a posterior part level with the os planum, separated by a *lachrymal crest* from the deep fossa that holds the lachrymal sac and is continued down into the nasal duct. The crest ends below in the *hamular process*, turning forwards and outwards and resting on the maxilla, and making the outer margin of the opening. In rare instances the hamular process is found extending forward to the orbital margin; on the other hand, the process may be absent. The Tensor tarsi arises from the crest and passes forward to the outer side of the nasal sac, which it is supposed to compress on its contraction.

The bone is sometimes perforated, and is occasionally represented by several ossicles. Its complete absence has been recorded.

Ossification commences in membrane before the end of the third month, on the surface of the nasal capsule. It reaches the frontal within a few weeks, but does not extend back to articulate with the os planum until some months later, as the posterior or ethmoidal portion of the bone is the last to be formed.

VOMER.

A thin bony plate, composed of two fused laminæ, and situated in the lower part of the nasal septum. Its *upper end* is thickened to make two *alæ*, which are splayed

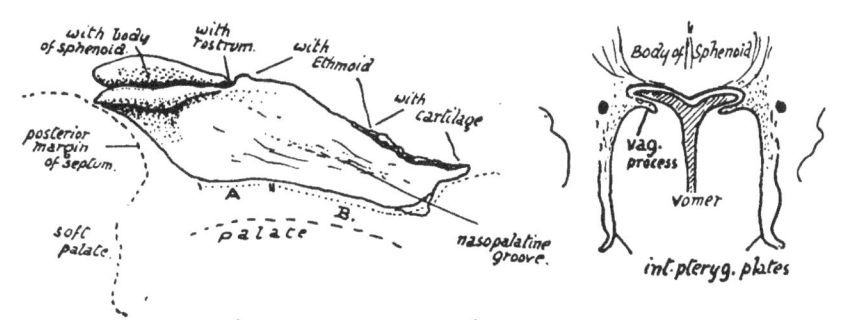

FIG. 190.—Vomer seen from the right, with outline of palate indicated by an interrupted line. *A.*, area of palate bones supporting Vomer; *B.*, region where it is carried on maxillæ. The lower front corner, shown as a separate piece, projects slightly into the anterior palatine foramen. The small figure shows schematically how the alæ articulate with the lower surface of the sphenoid and fit in between it and the vaginal processes of the internal pterygoid plates.

out under the body of the sphenoid and fit in between this and the vaginal plates of the internal pterygoid processes. The bone is covered by mucous membrane on both sides, and has, between it and the membrane, the naso-palatine nerve running forward and making the *naso-palatine groove* on it. Its *posterior margin*, thin but rounded, makes the free back edge of the bony septum, but this does not quite correspond with the edge of the complete septum, for the mucous membrane that forms the edge stands back from the bone (Fig. 190) for a little distance. The *lower border* rests on the septal crest made by the palate bones and maxillæ. The *antero-superior margin* supports the perpendicular plate of the ethmoid in its upper part, and is usually fused with it after middle life, but in its front part its two plates are distinct and enclose between them the margin of the septal cartilage (see Fig. 193).

242   Anatomy of Skeleton

Development.

A centre appears in membrane on each side of the septal part of the cartilaginous nasal capsule during the ninth week, and these join behind the cartilage to make the double bone in the course of a week or so. The bone thus formed is only a shallow trough in which the cartilaginous septum rests, and it separates this structure from the palate folds, and does not extend to the sphenoid or even into the " free edge " of the septum. It increases in size, but even at the beginning of the fourth month it is still in relation with the palate and has not effected any articulation with the sphenoid this occurs during this month. The two laminæ coalesce to a small extent in an upward and forward direction, the cartilage atrophying to a corresponding degree between them. The greater part of the depth of the bone, however, is the result of ossification in membrane extending downwards and backwards as the maxillary regions increase in height and the nasal fossæ enlarge, so that the free edge of the bone, which is at first practically horizontal, becomes gradually directed obliquely as the structure increases in depth.

It has been stated that the anterior end of the vomer is formed from an ossification which involves the vomerine cartilage in part; possibly this is associated with the small down-growth that projects from its front part into the upper opening of the anterior palatine foramen.

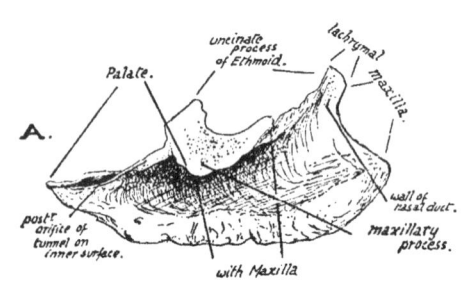

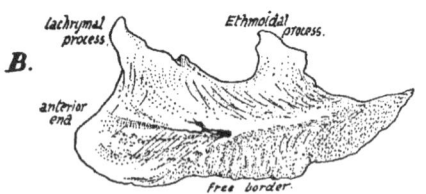

FIG. 191.—Right inferior turbinate bone, *A*, outer side, and *B*. inner or septal surface. To tell left from right, hold the bone with the pointed end behind, the thick roughened edge down, and the concavity externally.

INFERIOR TURBINATE BONE.

A bone projecting into the nasal cavity, attached to the maxilla and the other bones on the outer wall of the nose. It is covered by mucous membrane on both sides, and has its concave aspect looking toward the outer wall of the nose, and its convexity directed towards the septum.

It has :—(a) A *lower* or *free border*, thick and roughened by the vascular erectile tissue that covers it.

(b) A rounded *anterior end*, thick and covered by mucous membrane, which in the recent state is turned down off its front edge to form (Fig. 194) a fold passing towards the floor of the cavity.

(c) A pointed *posterior end*, lying against the vertical plate of the palate.

(d) An *upper edge*, by which it is attached. Connected with this border are three processes, *ethmoidal, lachrymal,* and *maxillary ;* the two first are directed upwards, but the maxillary process is turned down to fit into the lower part of the opening of the antrum. In front of the lachrymal process the bone articulates with the maxilla below the base of the nasal process ; the lachrymal process lies behind the nasal process and extends up to meet the lachrymal bone here, and, behind this, is in contact with

Skull and Hyoid 243

the uncinate process of the ethmoid. The ethmoidal and maxillary processes form a plate which fits into the lower part of the antral opening, so that the edge of the maxillary part articulates with the lower border of the foramen while the upper projection stands up in the opening to join the end of the uncinate process. The upper edge, behind this " plate," is attached across the vertical plate of the palate. These various articulations can be seen in Figs. 191 and 194, 3.

(e) A deeply-concave *outer surface*, widest in front, where it is also smoother; this part, just below the lachrymal process, forms the outer boundary of the lower portion and opening of the nasal duct.

(f) A convex and rough inner surface. A ridge in the posterior half, about midway up the bone, marks the position of a canal that carries a nerve and vessels from the posterior palatine set along the bone toward its front end.

Development.

In the early fœtus the lower edge of the nasal capsule, turned in, makes the projection of the lower turbinal: a centre appears on the surface of this cartilage in the fifth or sixth month or later and extends quickly. The bone frequently joins the ethmoid before middle life.

NASAL BONE.

A small bone, meeting its fellow in the middle line and forming the upper and front part of the roof of the nose, immediately below the frontal.

Each nasal bone has a thick rough *base* or upper end that articulates with the

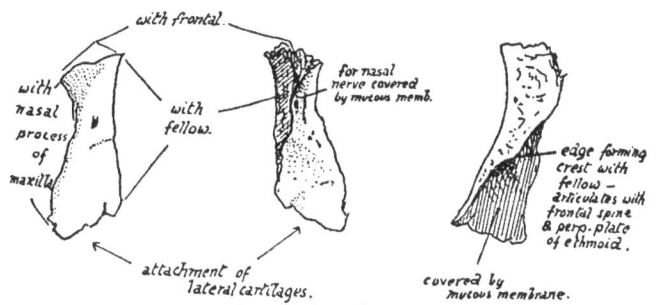

FIG. 192.—Right nasal bone, from the front, from behind, and from the inner side.

frontal, a thick rough *inner surface* that meets its fellow, a thinner *outer margin* that is supported by the nasal process of the maxilla, and a thin irregular *lower edge* that slopes upwards and inwards, and to which the cartilaginous structures of the side of the nose are fastened. The *front surface*, slightly convex from side to side, and concave in its upper part from above down, presents a vascular foramen and is covered by Pyramidalis nasi and skin: the *deep* or nasal surface, concave from side to side, is smooth and covered by mucous membrane, under which the nasal nerve runs down on the bone. The inner margin of this surface, where it meets its fellow, rests on the nasal spine of the frontal and on the perpendicular plate of the ethmoid in front of this (Fig. 193). If, as frequently occurs, this plate falls short of the lower margin of the bones, the septal cartilage articulates with the lower part of the median crest.

Anatomy of Skeleton

Development.

From a centre in membrane showing toward the end of the second month on the surface of the nasal capsule. The nasal nerve is separated from it by the cartilage, which does not completely disappear until after birth. A small second centre is said to form its inner part in cartilage, the notch in the lower edge marking the junction between the two portions.

NASAL FOSSÆ.

The structure of the lateral masses of the ethmoid can be understood best, perhaps, by a short study of the nasal fossa, which can be undertaken now with advantage.

The **Septum** (Fig. 193) is easy of comprehension : observe how the whole of its upper part is made by the perpendicular plate of the *ethmoid*, how this rests below and behind

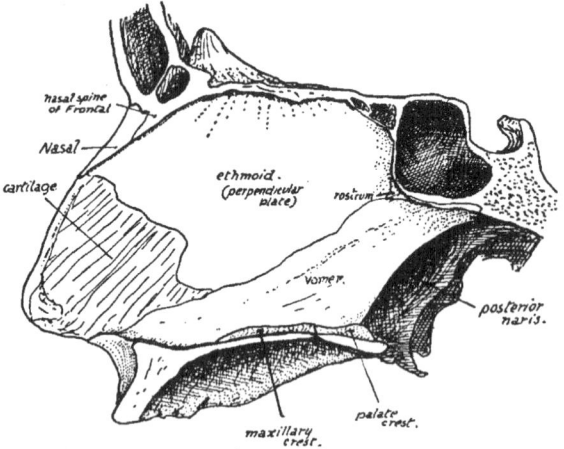

FIG. 193.—Nasal septum seen from the left side. The section is a little to the left of the middle line, so that the composition of the bony roof and floor is shown, and the anterior palatine canal is not opened up by it.

on the *vomer*, and how it projects forward far beyond the level of the rest of the bone to articulate with the nasal spine of the frontal and to carry the nasal bones in front of this. The other bones that enter into the formation of the septum are—the *frontal* spine, resting on the vertical plate and the crest (not always apparent) made in front of this by the meeting of the *nasal* bones ; the crest, made by the two *palate* bones and two *maxillæ* where they meet and support the vomer ; the rostrum of the *sphenoid*, fitting in between the ethmoid and vomer for a little distance. The remaining part of the septum is made by the *septal cartilage*, and on this lie the *vomerine* cartilage (*cartilage of Jacobson*) and the in-turned front part of the *cartilage of the aperture*.

The structures concerned in the building of the **outer wall**, however, are not so easily seen or understood. The student should endeavour first to obtain a clear idea of what constitutes the absolute outer wall of the fossa, irrespective of the projections and air cells that are attached to its nasal surface. Evidently the inner surface of the

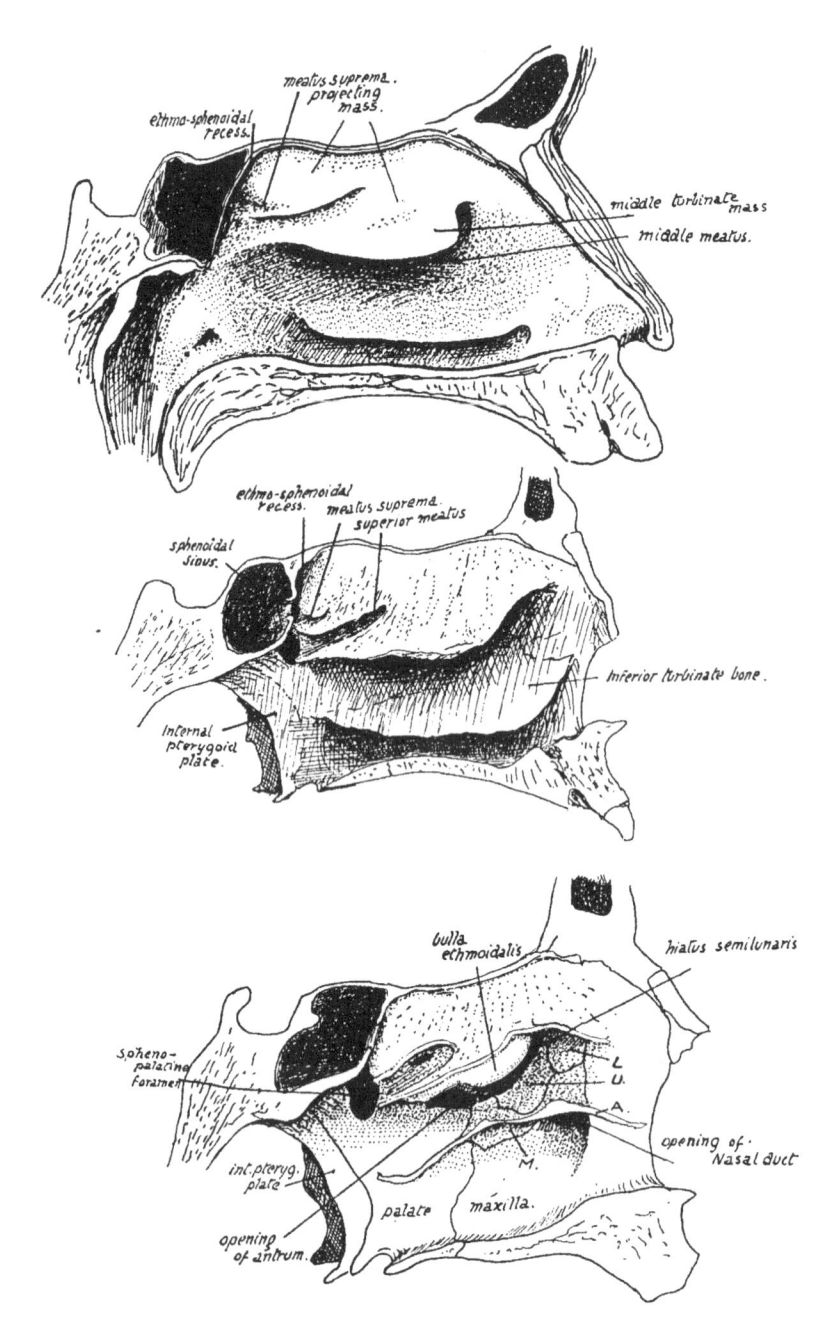

FIG. 194.

body of the maxilla and the vertical plate of the palate (applied to the maxilla) constitute the outer wall so far as they go, and separate the nasal fossa from the antrum of Highmore. But the nasal cavity extends higher than the jaw, and the outer wall above the level of the maxilla must separate the nasal and orbital cavities. Look, therefore, into the orbit and see how its inner wall is made (Fig. 195), and it can be seen that the thin *os planum* of the ethmoid forms the largest part of it, but is deficient in front, and here the lachrymal bone fills up the deficiency and lies between the os planum and the nasal process of the maxilla.

FIG. 194.—Outer wall of left nasal fossa. 1. With mucous membrane in position. 2. Mucous membrane removed. Notice the flat inner surface of the mass of the labyrinth, from which the upper and middle turbinals are suspended. 3. The free projections of the turbinals have been cut away, exposing the outer wall in the lower part of the fossa, but showing only the inner wall of the mass of the labyrinth in the upper portion of the fossa ; the lines of section of the turbinals is apparent. L. lachrymal ; U, unciform process of ethmoid ; A, cut edge of inferior turbinate ; M, its maxillary process turned down in the large antral opening, to articulate with its lower margin : the other processes are seen running up to articulate with U and L respectively.

We can say, then, that the upper part of the outer wall is made by the os planum, lachrymal, and nasal process of maxilla, while the lower and greater part is made by the body of the maxilla and vertical plate of palate. The lachrymal comes down for some distance internal to the maxilla to separate the nasal cavity from the nasal duct, and the uncinate process of the ethmoid also comes down, behind the lachrymal, on the inner side of the maxilla and closes in the front part of the antral opening. Another element, a separate *inferior turbinate bone*, is applied to the lower division of the wall, and a part of this turns down into the lower part of the antral opening and closes this : behind this the inferior turbinate is fixed to the vertical plate of the palate, and in front of the position of the antral opening the attached border of the inferior turbinate lies across the maxilla nearly as far forward as the anterior nasal margin. The line of the turbinate thus lies a little below the uncinate process and the lachrymal, and it gives short processes upwards to articulate with these (and still further close in the antral opening), so that these processes of the turbinate can be looked on as entering into the composition of the outer wall.

FIG. 195.—Scheme of the position of the lateral mass or labyrinth and the arrangement of turbinals. The inferior turbinate hangs from the outer wall below the ethmoid, but the others are parts of that bone ; one is shown hanging from the inner wall of the labyrinth. The whole labyrinth forms a mass in the upper part of the nose, so that the plane of attachment of the upper two turbinals is nearer the septum than that of the inferior turbinal, although the outer wall of the nasal cavity is further from the septum in the upper than in the lower part.

(The third diagram in Fig. 194 will give some idea of these arrangements on the outer wall : the line of the attached part of the inferior turbinate, which has been cut away, is shown at A, and its processes are seen joining the lachrymal (L.) and uncinate process (U.). The maxillary process, turned down, is seen at M.)

The anterior process of the inferior turbinal, by articulating with the lachrymal, completes the inner wall of the nasal duct and separates it from the nose, so that its lower opening must be below the level of the inferior turbinate.

The upper and middle turbinate bones are part of the ethmoid.

246   Anatomy of Skeleton

If the outer wall of the cavity with its soft parts (Fig. 194) is studied, it is seen that the inferior turbinal body is separated by the middle meatus from the middle turbinal, which is the lowest part of a mass that projects into the nasal fossa towards the septum evidently this mass lies between the outer wall (os planum) and the upper and middle

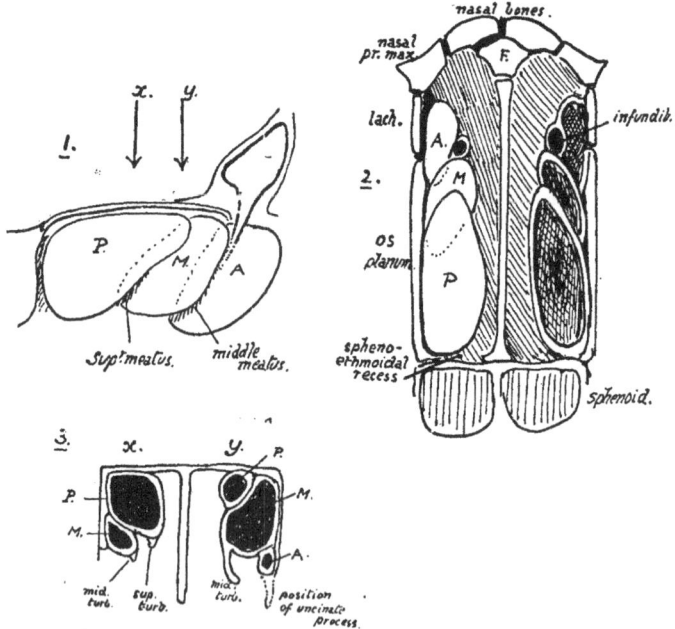

FIG. 196.—Schemes of the arrangements of the groups of cells which constitute the labyrinth. 1. Groups of left mass seen from the right (inner or septal side). *P.*, the posterior group, is on the whole above and internal to the middle group, *M.*, as well as behind it, and this group is again above and internal to the anterior group, *A.*; *i.e.*, each group overlaps the one in front of it internally and above and thus grooves may be considered to exist along the lower margins of *P.* and *M.*, directed forwards and upwards and due to the prominence of the more posterior group. Such grooves make the upper and middle meatuses, and the corresponding turbinals are suspended over them from the prominent overhanging margins. The frontal sinus and infundibulum may be considered an upgrowth from the groove between *M.* and *A.*, and thus it is separated from the outer wall by *A*. Sections through 1, about the levels of the arrows, would cut the groups somewhat as shown in 3, in which the position of the uncinate process of the ethmoid is also indicated, showing how the anterior group can extend into it downwards. 2 is the same scheme of arrangement seen from above, showing the relations of the groups to the outer wall (os planum and lachrymal). Compare this with the next figure and the one after that.

turbinates, which are simply dependent folds on its inner or septal aspect. On removing the mucous membrane (No. 2) the underlying bony structure presents the same appearance of turbinate bones as comparatively small folds or projections suspended from the inner or superficial aspect of a deeper mass; the general surface exposed is flat, marked by the mucous covering, and the middle turbinate bone hangs from its lower border and is thus held away from the outer wall of the nose by the depth of the intervening mass.

Skull and Hyoid

By cutting away the dependent turbinals the underlying mass is left by itself (No. 3), and the groove on its back part, which was partly covered by the upper turbinal, is seen to have an opening in it, while the removal of the middle turbinal has exposed the rounded lower and front part of the mass (*bulla ethmoidalis*) and, below this, the contracted opening of the antrum and a curved groove (*hiatus semilunaris*) leading up to the opening of the frontal sinus. The mass whose limits are thus roughly made out is the **labyrinth** of the lateral mass of the ethmoid, and consists of air cells with thin bony walls. A scheme of the relations between outer wall, labyrinth, and turbinal is shown in Fig. 195.

But the cavity represented in the labyrinth in this figure is purely schematic; there are really a number of cavities, separated by walls that are very variable in position individually, but they may be divided into groups. There are three such groups, *anterior*, *middle*, and *posterior*, and the air cells belonging to any one of these groups do not communicate with those of the other groups, and not always with each other: the cells of each group, however, open together—those of the posterior group in the upper meatus (Fig. 194), those of the middle and anterior groups in the middle meatus. The posterior ethmoidal cells are *behind*, *above*, and *internal* to the middle cells, and these are again *behind* and *internal* to the anterior group: these last are therefore the cells exposed on the outer aspect of the lateral mass where the os planum is deficient in front, and are covered in by the lachrymal.

The general relations of these groups to each other are shown schematically in Fig. 196: it must be again remembered that each group is not a single cavity, but is composed of a variable number of cells. The schematic position of the three groups is shown on section in the three planes, and it is clear that a transverse section through the front part of the mass will show anterior ethmoidal cells chiefly, with middle cells perhaps internal to them, and hiatus semilunaris situated between the two groups: a section across the middle of the mass may have only middle cells, with middle turbinal hanging from their inner and lower part; one further back goes through posterior cells, with the middle group perhaps represented externally and below, separated from the posterior group by the groove of the upper meatus, over which the upper turbinal hangs down from the inner wall of the posterior group. Two sections of the region are also shown (Fig. 197) to compare with the schemes and exhibit the inner subdivisions of these cell-groups.

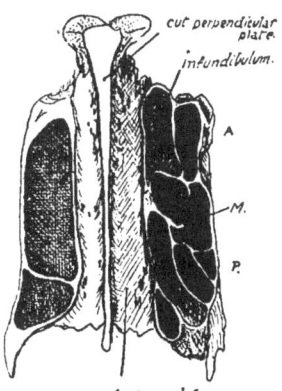

FIG. 197.—Sections at different levels through the lateral masses of an ethmoid, *seen from below*. On the right side of the figure the section runs at a lower level; A., anterior cells on outer side of infundibulum; M., four cells of middle group, the binder ones lying outside the posterior group which is represented by the inner cells at and behind the level of P. On the other side the section runs just below the roof and only opens cells of the posterior group, which are very large in this specimen. Observe that the back part of the posterior group may be partially separated, when the groove between it and the rest of the group becomes the meatus suprema and the fourth turbinal hangs from its projecting side. See Fig. 194.

The group of posterior cells is placed in front of the sphenoidal turbinate, and the recess (Fig. 194) between them is known as the *spheno-ethmoidal recess*: the opening of the sphenoidal sinus in the sphenoidal turbinate is thus in the recess. The posterior group may show a partial subdivision by a groove into which one or more of the cells

248   Anatomy of Skeleton

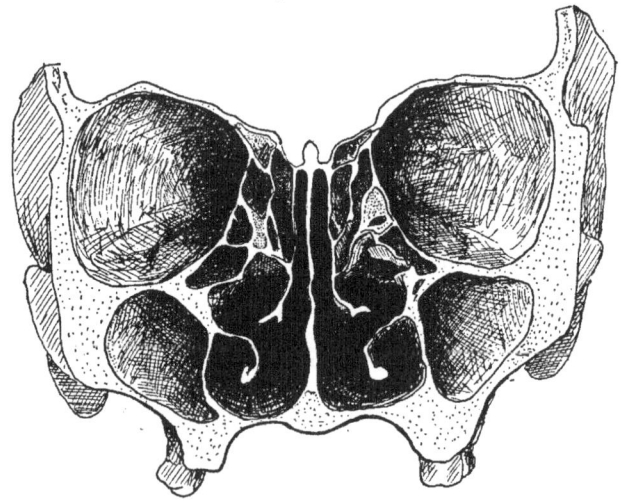

FIG. 198.—Section through an Adult Skull.

opens the groove is then termed the *meatus suprema*, and its upper boundary may even form a small " supreme " turbinal. The specimens figured in the first two drawings in Fig. 194 show such grooves.

FIG. 199.—Scheme of the roof of the front part of the nasal cavity. The inferior surface of the cribriform plate forms the roof as far forward as the front of the ethmoidal notch; in front of this is the roof area of the frontal: mucous membrane eXtends from this on to the posterior surface of the nasal bone (roof) and nasal process of maXilla (side wall). The nasal nerve passes between the ethmoid and frontal, to lie between the mucous membrane and the roof area of the frontal, and so on to the nasal bone.

The middle ethmoidal cells—as can be seen in the horizontal section in Fig. 197—are the smallest, and may only be represented by one cavity overlapped above and internally by the large posterior group. The anterior group seems to be the most variable in its development : the cells may, as in the specimen shown in Fig. 188, extend down into the uncinate process and be completed there by the maxilla and inferior turbinate, or they may tend to extend upwards between the infundibulum and the lachrymal and have their outer walls formed by the last-named bone. Further extensions of these cells are very common, involving the surrounding bones, so that cells of this group may have maxilla, lachrymal, inferior turbinate, or frontal forming part of their walls.

It is usual to see some incomplete cells on the upper aspect of the lateral mass, to be completed by the frontal, and such open recesses belong, as a rule, to the posterior group (Fig. 188), with the exception of the most anterior one, just behind the infundibulum, which is one of the middle series.

Open cells on the lower aspect of the lateral mass belong to the middle group and are closed by the maxilla and palate ; thus the ethmoidal surface of the orbital process of the palate shows cell walls, and the middle turbinate articulates with its lower and inner part.

The **roof** of the nasal fossa is highest in the ethmoidal region and slopes down in front of and behind this. It is made (Fig. 193) from behind forwards by the *sphenoid*, the *cribriform plate of the ethmoid, nasal area of frontal, nasal bone*, and finally by the junction between the cartilaginous alæ and septum. The student should follow the formation of the roof with care, especially in the region of the frontal bone : the nasal area of this bone is situated beside the nasal spine (see Fig. 199). The nasal nerve comes through the roof of the cavity between the frontal and ethmoidal parts of the roof, so that the nerve runs down below the frontal and nasal bones, to emerge once more between this last bone and the cartilaginous part of the roof. The olfactory fibres pierce the ethmoidal (cribriform) part of the roof.

The **floor,** made by the horizontal processes of palate and maxilla, is concave from side to side, and also in some degree from before backwards, owing to the higher level of the premaxillary region ; the curve of the floor is continuous with that of the inferior meatus at the side. The general level of the cavities of the nose and the maxillary sinus is about the same, but the front part of the nasal floor may be rather higher than the front part of the floor of the sinus.

The **posterior nares** are the hinder openings of the fossæ ; they are separated from each other by the posterior edge of the vomer. Each opening has the *vomer* as its inner boundary, the *internal pterygoid plate* for its outer wall, the *horizontal plate of the palate* below, and the *body of the sphenoid* above. But it must be noticed that the *ala of the vomer* and the *inner pterygoid plate* meet below the sphenoid and are therefore concerned in forming the *roof* of the opening.

The **anterior opening** is a common one for both fossæ unless the septal cartilage is in place. It is often termed the " pyriform aperture " owing to its shape, and is bounded laterally and below by the maxillæ and by the nasals above ; its lower edge is produced centrally into the (anterior) *nasal spine*.

A slight ridge may be visible just behind the margin at the infero-lateral angle, more apparent when the proper margin turns down, as it sometimes does, to be lost on the front surface of the premaxilla ; the ridge thus seen lies over the course of the anterior dental (incisive) nerve, and here minute branches can pierce the bone to supply the front end of the inferior meatus. The fossette (variable in size and definition) between the ridge and the true margin is seen well in lower races. The superolateral edges have the lateral cartilages attached to them by fibrous tissue. The lower edge is free and covered by the tissues of the lower margin of the narial aperture.

The width of the aperture varies considerably in different races ; compared with the height it is least in Europeans and greatest in certain Negro races. It is proportionately broader in the infant than in the adult, owing to the shallowness of the maxilla at birth.

Sphenoidal Sinus.—Variable in extent in different individuals and usually on different sides in any one individual. The variations, and the situations of the different incomplete bony septa found within the cavities, may be largely explained by reference to developmental points. There are paired centres for the presphenoid, situated in front of paired centres for the basisphenoid : the centres for the lingulæ lie outside these last and separate them from the inner parts of the great wings, which, however, come into relation, in front of the lingulæ, with the presphenoidal centres. Each sinus, growing back from the spheno-ethmoidal region, comes first into relation with the presphenoidal part and hollows this out, the condensed bone between the two ossific areas remaining as a median vertical septum. If now each cavity extends further back, equally on the two sides, deeper sinuses result : these involve the postsphenoid centres, and, being equal growths, still maintain a median antero-posterior septum in their deeper parts, formed by the condensation between the basal ossifications. This equal extension is uncommon, however, for usually the further extension is unequal : one sinus may, for example, involve the postsphenoid of its own side and then extend to the postsphenoid of the other side, so that the other sinus remains small, confined to the presphenoidal region, and the septum between them, median in front, swings to one side further back, where it corresponds with the condensation between the pre- and post-sphenoid. One or both of the " presphenoidal " sinuses may extend laterally (instead of, or in addition to, the backward growth) and so involve the neighbouring part of the great wing : in this case the backward and the lateral extensions are separated by a partial septum corresponding with the condensation of the lingula, and this septum, directed obliquely forwards and inwards, may be very prominent if the region of the lingula has not been invaded, or more or less destroyed if such an invasion has taken place. A septum of this sort lies under the carotid groove, as does also the outer part of a transversely directed septum associated with the condensation between pre- and post-sphenoids, and these might be called *carotid buttresses* within the cavity : they are usually better marked above than below and are situated of course in the deeper parts of the cavities at their sides. Various combinations of these extensions may be seen in different individuals, but in general, it may be recognised that there is a septum, median in front but as a rule directed to one or other side behind, with lateral and posterior loculi more or less separated by a carotid buttress : the deep and anterior parts of the cavities may show a partial separation by an incomplete transverse septum marking the plane of condensation between pre- and post-sphenoids, and this, like the deflected inter-sinus septum, may run into a carotid buttress. Exceptionally the sinus may extend (a) into the great wing, reaching as far, it may be, as the groove for the maxillary nerve, and even invading the base of the outer pterygoid plate, passing over the Vidian nerve, (b) into the roots of the small wing, the optic nerve and ophthalmic artery lying in a prominent bony canal projecting into the cavity, or (c) into the basi-occiput, from which it is nearly always separated by the condensation between this bone and the basisphenoid.

The sinus begins its growth about the third or fourth month of intra-uterine life, is at birth a definite separate small cavity, in relation with the front aspect of the sphenoidal ossification, and enclosed by its own turbinate, grows fairly rapidly in the next few years, so that a noticeable cavity, invading the presphenoid, is present at five years of age, and takes on a more rapid extension at or before the age of ten, and again at puberty.

The frontal sinus develops as an upgrowth from one of the groups of cells (anterior ethmoidal) which form under cover of the overhanging anterior and upper end of the middle turbinal. The enlarging cell extends slowly upwards, the extension beginning about the middle of intra-uterine life. The growing cavity does not, as a rule, reach the frontal bone before birth, but invades that bone within the first year, and grows steadily up to the ninth or tenth year : after this it seems to take on a more rapid growth. The extensions are usually unequal, so that the septum between the cavities is deviated in its upper part, though generally more or less median below. In metopic skulls (p. 200) a sinus never transgresses the middle line. In other cases one sinus may completely overshadow the other, so that it seems at first as if no septum were present. The opening of the sinus depends on the site of its origin : if it is an enlarged member of the lateral group of anterior cells, it opens into the top of the hiatus semilunaris, from which these cells take origin as outpouchings, but, if from more medially situated cells, as is perhaps more frequently the case, the sinus does not open directly into the upper end of the hiatus.

The adult sinus is extremely variable in form and extent. In a very general way it might be described as roughly pyramidal, the apex being directed upwards, but it is often rounded or ovoid, and may present a partial subdivision : this subdivision is in some cases really due to the simultaneous upgrowth of two cells which have partly coalesced. The front wall of the sinus is the thickest wall, and contains diploic tissue, the posterior wall being made of compact bone. The floor shows a general slope downwards and inwards towards the opening in its hinder and inner part, but there is often a depressed fossa in front of this, and the floor is generally uneven. It lies over the inner part of the orbital roof, not often going further than, or even so far as, the line of the supra-orbital nerve, and covers the anterior ethmoidal cells internally. The amount of its backward extension, greatest internally, is very variable, but it seldom reaches the depth of an inch : an average measurement in this direction would be in the neighbourhood of $\frac{3}{4}$ inch.

The *maxillary sinus* (*antrum of Highmore, maxillary antrum*) is a cavity situated in the maxilla, of an irregular pyramidal shape, with its base inwards. Its front wall is fairly thick, and is made by the facial surface of the maxilla, its floor is formed by the thick alveolar portion of the bone, and its roof and inner back wall are thin—as is its inner wall, which is made of the inner lamina of the maxilla, overlaid by the lower end of the lachrymal, the uncinate process of the ethmoid, the downturned maxillary process of the inferior turbinate, and the vertical plate of the palate bone (see Fig. 194). Its opening, narrowed by the approximation of these bones, is a small slit between the uncinate process, palate, and turbinate : occasionally there is a double opening, the accessory ostium being usually placed below and behind the ordinary opening. The sinus begins to grow as an out-pouching of the side wall of the nose during the third month, and is present at birth as a small cavity (circ. 5 × 5 × 12 mm.) in the inner part of the upper jaw. This extends out as far as the line of the infra-orbital nerve during the first year, so that the situation of this nerve is indicated by a bony ridge in the roof of the cavity during and after the second year. The sinus grows with the bone, fairly quickly, up to the 8th or 9th year, and after this rather more slowly, corresponding with the slower eruption of teeth now occurring : after the later years of puberty, its form is only altered by the addition of a postero-inferior angle associated with the last molar development. These three indefinite stages of its growth seem to be indicated in its floor, where there are two ridges, as a rule, visible : one of these is in the premolar region, and the other in the molar region. Other ridges seen in the cavity include the infra-orbital ridge and the crest running downwards and forwards from this, containing the anterior and middle dental nerves. At birth the sinus is, of course, some distance above the level of the nasal floor, but its floor reaches this level about the eighth year, and after this usually lies below the nasal level to a small extent. As the teeth fall out with age, the floor rises and comes a little above the level of the nasal floor. The floor is closely related to the roots of the molars, and does not often reach teeth in front of the second premolar : the canine may, however, be in the front wall of large sinuses, and abnormal cavities may occasionally extend in front of this. A large posterior ethmoidal cell may project into the back and upper part of the cavity, even giving in some cases the appearance of the existence of a double sinus.

MALAR.

An irregular bone which forms the outer wall of the orbit, separating it from the temporal fossa, and rests on the maxilla below, making the prominence of the cheek : it sends a process backwards, articulating with the zygoma of the temporal to complete the arch. It thus can be described as possessing *orbital, temporal* and *facial surfaces* in addition to the articular areas. It can be considered as formed by the junction of two plates at an angle of about 70 degrees : the inner plate is curved to make a concave orbital surface ; the outer plate forms the facial prominence, and is prolonged back to meet the zygoma. The angle between the plates forms part of the temporal fossa above, but below and in front it rests on the maxilla.

The *orbital surface* is concave and has an outer vertical and a lower horizontal portion : the outer part completes the outer wall of the orbit and articulates above with the frontal and postero-internally by a serrated edge with the sphenoid ; the projection of this posterior edge is sometimes described as the *orbital process*. A small tubercle or process on the upper part of the orbital surface marks the attachment of the fascial band that acts as a check to the over-action of the outer Rectus, which is in relation with the surface. The horizontal portion forms part of the floor of the orbit, resting on the maxilla, and extends inwards as the *infraorbital* or maxillary process : it is in contact with the Inferior Oblique muscle, and the opening of the malar canal, for the temporo-malar nerve, is between it and the outer part. The inner border of the horizontal portion articulates with the maxilla, and is continuous with the inner sphenoidal border of the vertical part, but between the two articular areas is a small non-articular portion of the margin (X in Fig. 200) which closes the outer end of the spheno-maxillary fissure.

The *facial surface* is slightly convex, producing a *malar prominence*, owing to the backward inclination of its lower and outer part ; along its front border it is covered by the lip of the maxillary surface that receives the bone. The remainder of the surface is covered by fibres of Orbicularis palpebrarum, and, deep to this, gives origin to the two Zygomatici on the prominence.

A foramen on this surface transmits the malar division of the temporo-malar nerve.

Anatomy of Skeleton

The upper border forms a thick *orbital margin* at its junction with the orbital surface, and this margin is prolonged inwards on to the infraorbital process, making more than half of the lower margin of the orbit ; it is prolonged up to meet the frontal bone at the *frontal process*. The front margin has already been noticed as articular : the *lower* or *masseteric border* is rounded and leads back to the zygomatic or *temporal process*, which joins the zygoma by a rough surface, bevelled internally and directed obliquely upwards and forwards ; above this the *posterior border* gives attachment to the temporal fascia, and a strong band of this fascia is fastened to a prominent *marginal* or *malar tubercle* near the top (*a* in Fig. 200) of this border below the frontal process.

The temporal surface is the upper and outer part of the angle included between the two constituent plates of the bone, the lower and inner part being articular for the maxilla : the temporal surface is therefore deeply concave, with a *front wall* made by the orbital plate and a *back* or *outer wall* formed by the upper part of the facial plate,

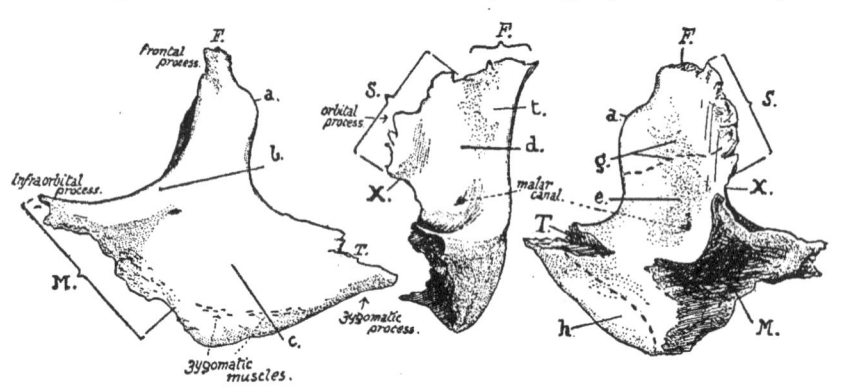

FIG. 200.—Left malar bone. Left figure, from the front: central figure, orbital aspect; right figure, posterior aspect. Articulating areas : *F*. with frontal ; *M*. with maXilla ; *S*. with sphenoid ; *T*. with temporal ; *a*. malar tubercle ; *b*. orbital margin ; *c*. facial surface covered by Orbicularis ; *d*. orbital surface ; *e*. temporal surface ; *g*. origin of Temporal muscle ; *h*. origin of Masseter ; *X*. non-articular border forming outer end of sphenomaXillary fissure ; *t*. tubercle for attachment of " check ligament " of Rectus eXternus.

and below and behind this by the inner surface of the zygomatic portion of this plate. The surface aids in the formation of the temporal fossa : the Temporal muscle arises from the upper part of its front wall ; below this it is in relation with the muscle and a movable pad of fat, and here the temporal division of the temporo-malar nerve has its exit and lies between the bone and the muscle : the lower and back part of this surface is in contact with Temporal muscle, and its lower portion gives origin to the front fibres of Masseter : this muscle arises only from the deep aspect of the zygomatic part and from the lower margin, not at all from the outer surface.

The malar canals or foramina vary in number ; there may be a common opening for the temporo-malar nerve with a double exit, or the nerve may enter the bone in two divisions, thus making four openings instead of three.

Development.

A bony centre appears in membrane outside and below the level of the eye at the end of the second month, and from this the ossification extends upwards and inwards, forwards and backwards. It is doubtful whether other centres are normally present, although they have certainly existed where the bone is found divided,* as is sometimes the case : possibly, however, a centre for the lower margin may be a normal occurrence.

The growing bone comes quickly into contact with the maxillary and temporal bones, but it does not reach the frontal until later. At birth the temporal surface has the appearance of a plate of bone applied to the remainder.

THE ORBIT.

Each orbit is a pyramidal-shaped space above the upper jaw. The upper surface of the *maxilla* makes the greater part of the *floor*, but the *malar* comes into the outer part, the *hamular process of the lachrymal* may be in the floor internally, and the *orbital process of the palate* is visible in the extreme back part of the floor at its junction with the inner wall (see Fig. 185).

The floor is separated from the *outer or posterior wall* by the spheno-maxillary fissure, except at its extreme outer end, where the malar is continued from the floor to the outer wall. The remaining and greater part of the outer wall is made by the *orbital surface of the great wing*. The outer wall is separated from the *roof* by the sphenoidal fissure in its back part, but outside this the great wing and the malar articulate with the frontal.

The *roof* is almost entirely formed by the *orbital plate of the frontal*, but the *small wing of the sphenoid* constitutes its most posterior portion. The *inner wall* is formed by the *body and turbinate of the sphenoid*, in front of this the *ethmoid*, and in front of this the *lachrymal* and the *nasal process of the maxilla* ; it has the anterior and posterior ethmoidal canals between it and the roof, above the os planum, and the nasal duct opens between it and the floor.

The *axis of the orbit* is directed backwards and inwards and slightly upwards, from the centre of the anterior opening to the optic foramen : the optic nerve lies practically in the line of the axis.

The *margins of the anterior opening* are made by several bones : above, the *frontal*, internally the *nasal process of the maxilla*, the *maxilla* and *malar* below, and the *malar* externally. Very rarely the hamular process of the lachrymal may be seen to reach the margin of the orbit. The margins are sharp except at the inner and upper angle, internal to the supraorbital foramen.

Owing to the obliquity of the axis the inner walls of the orbits are in nearly parallel planes, whereas the outer walls are in planes practically at right angles to each other. The depth of the fossa is about 2 inches.

The lachrymal sac lies in the lachrymal groove near the margin of the opening ; the Tensor tarsi arises from the crest behind the groove and passes forward to the outer side of the sac to join the Orbicularis palpebrarum, and the tendo oculi arises from the nasal process in front of the groove and passes out across the middle of the sac.

LOWER JAW OR MANDIBLE.

The only bone in the skull (with the exception of the tympanic ossicles) that is capable of separate movement : it carries the teeth opposing those of the upper jaw, gives insertion to the muscles of mastication, and origin to muscles of the tongue and floor of mouth and some muscles of expression.

It consists of two halves which are strongly joined in the middle line at the *sym-*

---

* This is rare, but may occur through a suture which divides the bone horizontally into an upper (larger) and lower (smaller) part.

physis. Each half has a *ramus* and a *body*, joined at an angle that varies with age and dentition. The term *angle of the jaw*, however, is applied to the prominence formed by the junction of the lower border of the body with the back margin of the ramus.

The *ramus* ends above in two processes—the *coronoid process* in front and the articular *head* behind, supported by the narrower *neck*. The two processes are separated by the *coronoid* or *sigmoid notch*. The *front* margin of the ramus is sharp, whereas the *posterior margin* is blunt and rounded. The *outer* surface is roughened by the insertion of the Masseter. On the *inner* side it presents (Fig. 202) the *inferior dental foramen*, whose inner lip is prolonged forward and upward into the *lingula :* leading downward and forward from the foramen toward the body is the *mylo-hyoid groove*. A rough area below and behind the foramen marks the insertion of the Internal Pterygoid.

The *body* has a thick rounded lower part or *splenium* supporting the *alveolar portion*, which contains the teeth. On the *inner* side is the continuation of the *mylo-hyoid groove*, and above this the *mylo-hyoid ridge* or *internal oblique line*, for attachment of Mylo-hyoid (Fig. 202).

Below the back part of these is the *submaxillary fossa*, and above their front part is the *sublingual fossa :* these are in relation with the corresponding glands. In front of the submaxillary fossa, and below the front part of the mylo-hyoid ridge, is a *digastric depression* or fossa for anterior belly of Digastric, which may, however, present itself in the form of a slightly elevated roughness. The mylo-hyoid line on each side nearly reaches the symphysis, and, above its level and on each side of the symphysis, are the *genial tubercles* for Genio-hyoids and Genio-glossi : these are described as forming two pairs, upper and lower, but usually it is difficult to distinguish them in this way, and there may be four tubercles, or a pair of them, or a vertical ridge representing them more or less fused.

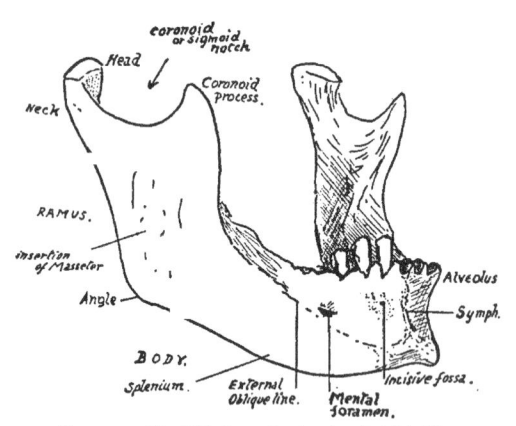

FIG. 01.—Mandible from the front and right side.

On its *outer* aspect the body presents, from before backward, the vertical ridge of the *symphysis*, ending below in the *mental protuberance*, the *incisive fossa*, the *mental foramen*, and the *external oblique line*, leading upwards and backwards to become continuous with the front edge of the coronoid process.

The body of the bone is mainly concerned with the support of the structures in the mouth.

The Mylo-hyoid forms the muscular floor of the mouth, and thus the mylo-hyoid ridge or internal oblique line can be taken to divide the inner surface of the body of the bone into an upper (or buccal) area and a lower (or cervical) portion. These two parts can only come into relation with structures that lie on the corresponding surfaces of the Mylo-hyoid : thus the continuation of the ridge forward would pass below the genial tubercles, because these are for " buccal " structures.

Owing to the obliquity of its line of origin the mylo-hyoid muscle, passing to the body of the hyoid, has its posterior fibres and " free edge " more vertical than horizontal, whereas its front fibres are horizontal. The Genio-hyoids lie on these front fibres, supporting the Genio-glossi, and the sublingual gland lies further out and comes into contact with the bone. Thus we have the sublingual fossa *outside* the genial tubercles and *above* the mylo hyoid ridge, and the mucous membrane, which covers the alveolar part of the bone, and turns in to form the floor of the mouth, is separated by the gland from the muscle here.

But the submaxillary gland comes into relation with the bone in a different way. Its " deep " part lies further back than the sublingual, up against the side of the tongue (Hyo-glossus), but here the longer and more vertical fibres of the mylo-hyoid are between it and the bone, so that the two are not in contact ; but this portion of the gland is continuous round the free edge of the muscle with the " superficial part," which is situated in the angle between the mylo-hyoid and the jaw (Fig. 203). Thus, although the deep part of the gland does not touch the " buccal " part of the bone, the superficial part comes into relation with the " cervical " portion, therefore *below* the Mylo-hyoid ridge. This piece of the gland lies in the digastric triangle, behind the

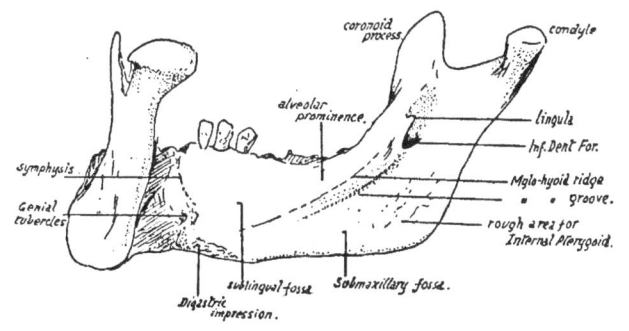

FIG. 202.—Inner aspect of right half of mandible.

anterior belly of the Digastric, so that the insertion of this muscle is further forward than the submaxillary fossa, but not on the symphysis.

Behind the position of the submaxillary fossa the gland extends back towards the angle of the jaw : it is still under cover of the bone, but is separated from it here by the lower part of the insertion of the Internal Pterygoid.

On the outer surface (Fig. 204) the *alveolus* is covered by mucous membrane, which is reflected from the jaw on to the cheeks and lip. Thus the membrane-covered surface extends to the origin of the Buccinator in the cheek region and is bounded in the region of the lip by the origin of muscles running up to the Orbicularis oris : these are the Depressor anguli oris below and behind the mental foramen, and the Depressor labii inferioris in front of this. Further forward, but not shown in the figure, the lower incisive slip of the Orbicularis arises from the incisive fossa. The muscle shown in the figure as arising under the mucous membrane in this fossa has its fibres directed down to be inserted into the skin of the chin ; therefore when it contracts it pulls up the skin and so lifts the lip, and is therefore termed Levator menti or Lev. lab. inferioris.

The remainder of the outer surface of the body is covered by the superficial tissues of the face. The deep fascia of the neck is attached to the lower border, and the Platysma, lying on this, gets some insertion into the bone from the chin outwards. In

256　　　　　　　　　Anatomy of Skeleton

the posterior part of the region, however, the muscle is carried on over the facial vessels, and usually blends here with the Risorius muscle.

The mucous membrane on the outside of the alveolus belongs to the alveolo-labial or extra-dental sulcus, and it can be seen (Fig. 204) on the bone that the sulcus becomes shallower as it is traced backwards: the Buccinator is attached opposite the molars, nearer the line of the teeth than the muscles further forward.

Behind the last molar the mucous membrane becomes continuous with that on the inner side of the jaw by a narrow strip crossing the extreme back part of the alveolus. This strip lies between the last tooth and the attachment to the bone of the pterygo-mandibular ligament, which passes downwards and outwards to this point from the bony region formed by the lower end of the internal pterygoid plate and the tuberosity of the palate.

The Buccinator arises from the front aspect of the ligament, and its line of origin extends on to the jaw (Fig. 204) forwards and outwards from the point of attachment: the upper Constrictor reaches the jaw along the back aspect of the ligament, obtains a few fibres of origin from the bone behind and internal to the attachment, and then passes along the mucous membrane to the tongue. The diagram in Fig. 206 shows the arrangement of the structures in this region, and the path of the lingual nerve is seen below, behind, and outside the fibres of the Constrictor, and then below the mucous membrane to reach the tongue.

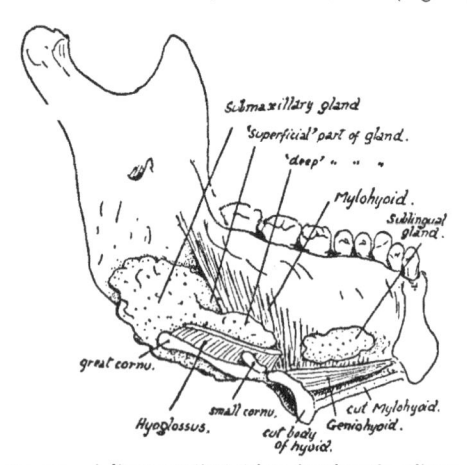

FIG. 203.—A diagrammatic sketch to show how the salivary glands of the floor of the mouth come into relation with the lower jaw. The mass of the tongue is supposed to have been removed, the hyoid divided through the body, and the left half of the mandible and Mylohyoid viewed from the inner side. The major portion of the submaxillary gland is seen to lie on the outer side of the Hyoglossus, which has been cut across near its origin, and the posterior part of the gland comes into relation with the Internal Pterygoid insertion; the front part of the gland reaches the edge of Mylo-hyoid and passes partly below and external to it and partly above and internal to it. It is the former of these two parts, the "superficial part," which alone lies in contact with the bone; the "deep part" lies in the sulcus between Hyoglossus and Mylo-hyoid. The sublingual gland, on the other hand, touches the bone further forward, above the origin of the muscle.

FIG. 204.—Attachments and relations of mandible. Upper figure shows the inner side of the right half of the bone. The parallel lines indicate the regions in relation with the inner and outer Pterygoid muscles; between the former and the bone several structures are seen (compare Fig. 207), but the external Pterygoid is in direct contact with the jaw and the Temporal insertion, the masseteric vessels alone intervening between it and the bone. The outer aspect (lower figure) shows the ramus covered by Masseter with the exception of a small area above and behind where the parotid lies on the bone. The body is divided by the Buccinator and oblique line into an upper and front part covered by mucous membrane, and a lower and back part covered by facial muscles, deep to which the facial vessels cross the bone and pass on to the Buccinator. The arrows mark the borders of the Temporal tendon; observe the extent of its insertion.

The same diagram also shows that the ramus of the jaw with its attached muscles is placed in contact with the side wall of the pharynx. The Internal Pterygoid is the deepest of these muscles, and is therefore the one in immediate contact with the

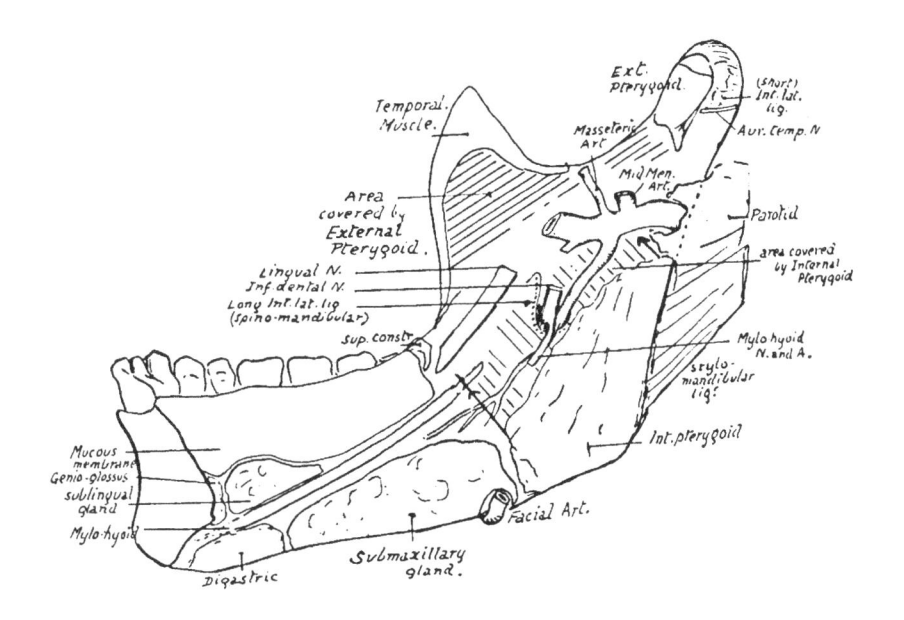

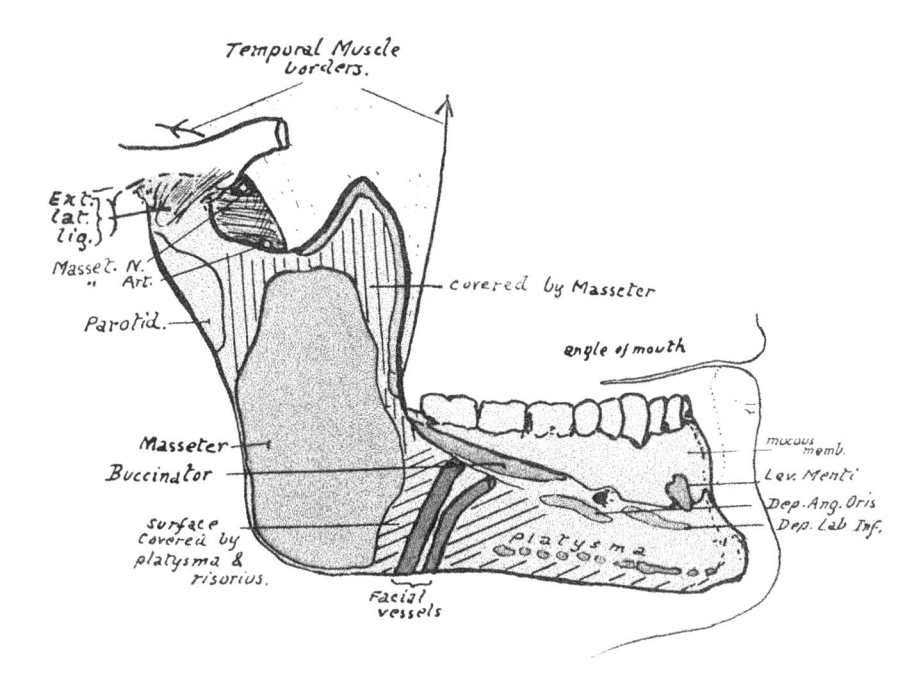

Fig. 204.

Skull and Hyoid

pharyngeal structures. Thus at its origin it is in contact with the internal pterygoid plate and the Tensor palati behind this (Fig. 205), which separates it from the sinus of Morgagni and the structures passing through this : below these it is in direct relation with the upper Constrictor and, a little lower, with the muscles running on the Constrictor to the pharyngeal wall, *i.e.*, Stylo-glossus and (deeper) Stylo-pharyngeus : near the angle it is in contact with the submaxillary gland, which—with the head in the natural position—separates it from Digastric and Stylo-hyoid, and from the external carotid artery where this turns upwards, outwards and backwards from the wall of the pharynx, along the Stylo-hyoid.

The insertion of the Internal Pterygoid (Fig. 204) is by mixed muscular and

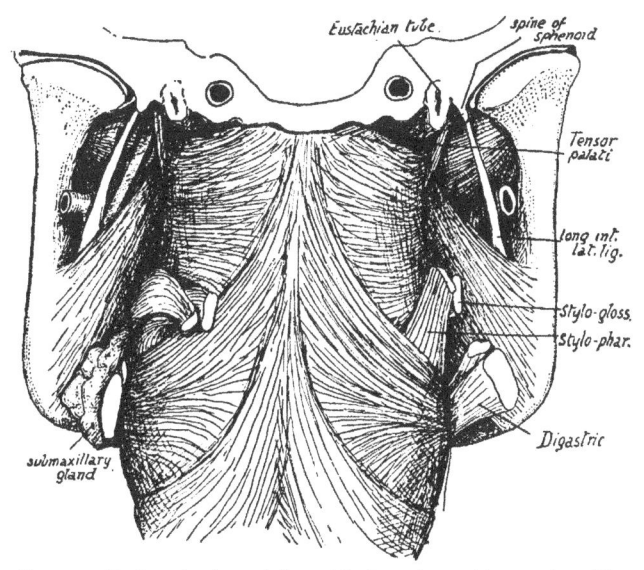

FIG. 205.—To show the deep relations of the lower jaw and its muscles. The pharynx, with the jaw in position, is seen from behind. On the right the Digastric, Stylo-hyoid, Stylo-pharyngeus, and Stylo-glossus are cut short on the left the last two muscles are turned back, and the submaxillary gland is shown between the internal pterygoid and Digastric and Stylo-hyoid, and above them. The parotid, with a quantity of loose tissue, occupies the interval above and behind the Pterygoideus internus.

tendinous fibres, as can be inferred from the roughnesses in its area : it extends high along the back part of the inner surface of the ramus, nearly reaching the neck, and here the parotid turns round the back of the bone and comes into relation with its inner aspect. The fibres of the muscle pass upwards, forwards and inwards to their origin, so that a rapidly-increasing interval occurs between them and the ramus. In the top of this interval the External Pterygoid separates the inner muscle from the plane of the jaw, and below this the bone has in relation with it, and between it and the Internal Pterygoid, the internal maxillary artery and some of its branches, the lingual and inferior dental nerves, the internal lateral ligament, the mylo-hyoid nerve and vessels, and frequently a small process of the parotid gland (Fig. 204).

The small masseteric artery runs to the coronoid notch between the bone and the

Anatomy of Skeleton

External Pterygoid, and passes immediately behind the Temporal tendon : its course *below* the outer Pterygoid is thus different from that of the masseteric nerve, which is *above* the muscle, between it and the base of the skull (see Fig. 182), and is therefore not a relation of the mandible.

The Temporal insertion is on the edge of the coronoid process, from the bottom of the notch almost to the alveolus ; it only extends slightly on to the surfaces of the process, usually more so on the inner than on the outer side.

The roughness of the masseteric area indicates that it also has tendinous fibres at its insertion. Observe that the area of insertion extends very high ; this is for the attachment of the more vertical posterior fibres, which have little play in the normal opening of the mouth and can therefore be fastened nearer their origin. The fibres that have most movement and more powerful action arise from the malar further forward and are directed downwards and backwards to the lower part of the ramus.

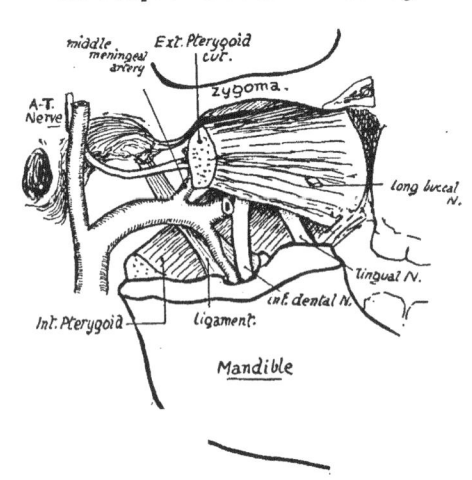

FIG. 207.—To show the structures lying in relation with the lower jaw above the level of the inferior dental foramen. For comparison with Figs. 204 and 206.

The upper part of the posterior margin of the ramus slopes back into the neck. Thus a triangular area of bone is exposed behind the level of the margin of the Masseter, and the parotid gland comes into relation with the bone here, being otherwise separated from its outer surface by the muscle. The gland also lies in contact with the border of the ramus, turning round it deeply to come into relation with the inner aspect and with the Internal Pterygoid, and the lower and deep part of its capsule is thickened to form the " stylo-mandibular ligament " which is attached to the lower part of the posterior margin.

The temporo-mandibular joint is a complex one, divided by an interarticular fibro-cartilage into two distinct cavities. The outer edge of the neck of the bone presents a ridge for the attachment of the external lateral ligament, the fibres of which are directed downwards and backwards from the eminentia articularis (Fig. 204). On the inner side another ridge marks the short internal ligament ; both these are thickened parts of the capsule. The capsule is very thin behind, but thicker in front where it receives some fibres of the External Pterygoid : it is attached all round the periphery of the fibro-cartilage, so that the menisco-mandibular and menisco-squamous cavities are quite distinct.

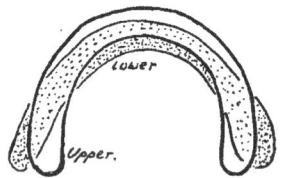

FIG. 208.—To illustrate the different curves of the dental arcades in upper and lower jaws ; lower arcade is shaded.

The fibro-cartilage fits on the articular eminence and in the articular cavity so that its upper surface is concavo-convex, while its lower surface is deeply concave to rest on the condyle of the mandible. The head of the mandible can plainly only

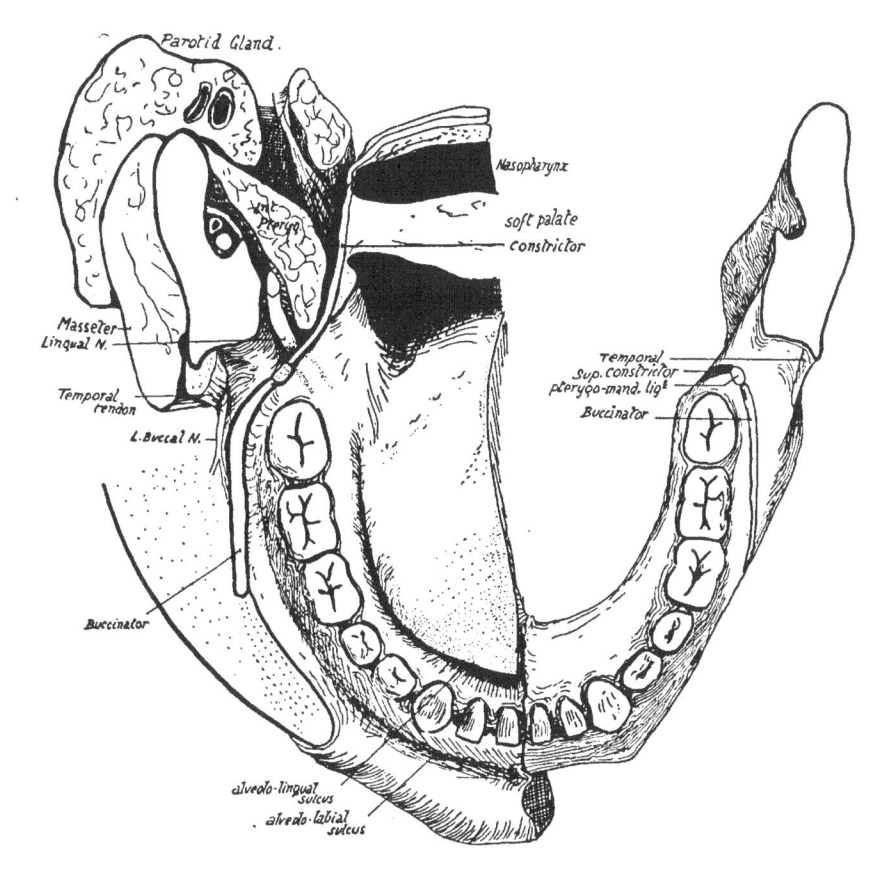

FIG. 206.—Diagram to show the relation of the mucous membrane and the wall of the pharynx to the mandible and its muscles.

Skull and Hyoid

rotate in the lower cavity, whereas the menisco-squamous joint is evidently only capable of a *sliding* movement.

A little consideration will show that it would be impossible to open the jaw wide if it rotated on an axis going through the temporal articulation, for the parotid and Sterno-mastoid would prevent any large movement of the angle backwards: this difficulty is overcome by putting the axis of rotation further down, when the jaw can be widely opened with half the amount of movement backwards of the angle but with a proportionate movement forward of the upper end. In such action the fixed point or axis of rotation is at the attachment of the long internal ligament: the foramen for the entering nerve and vessels is associated with this point, so that they enter the bone at its point of least movement. Some are inclined to place the position of the axis of movement lower still, at the attachment of the stylo-mandibular ligament. The External Pterygoid * starts the movement of opening by its contraction, thus drawing forward the fibro-cartilage as well as the head of the bone, so that the latter always has a hollow cavity in which it can turn: the forward movement of the bone with its depression (caused by its head being carried on to the eminentia articularis) tends to tighten the long ligament, and further action of the outer Pterygoid can only pull the upper end of the bone forward on the axis passing through the lingula. The long internal lateral ligament is attached to the lingula and the whole length of the adjoining margin of the foramen; it is fastened above to the spine of the sphenoid, which is on the inner border of the glenoid cavity. The short internal ligament is attached to the base of the sphenoidal spine, and the line of

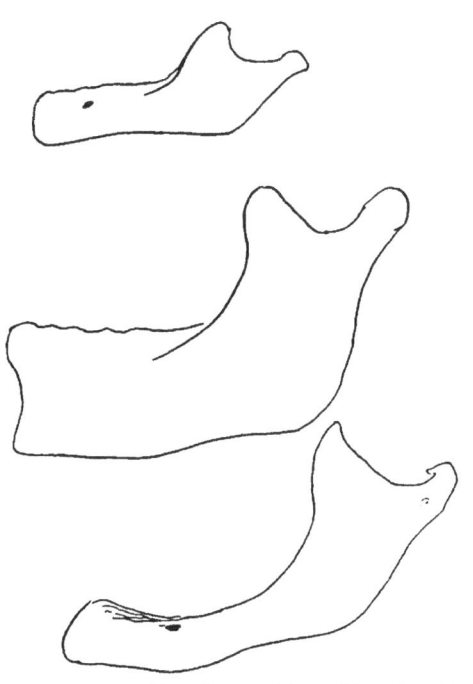

FIG. 209.—Types of jaws from a child, an adult, and an old person. The higher position of the mental foramen in the child is owing to the imperfect development of the alveolar part of the jaw, and in the old bone to absorption of the alveolus following loss of teeth. Notice also the difference between the angles; the angle becomes a necessity, as the mandible is depressed in consequence of the increasing height of the maxilla, for the purpose of keeping the margins of the two bones parallel, and hence the sharpest angle goes with the full dentition.

the rest of the capsule runs round the articular surface on the squama of the temporal. The auriculo-temporal nerve runs back between these two ligaments (Fig. 182) The condyle projects markedly inwards, and the neck exhibits a corresponding widening from side to side as it passes up to support the head: the wide part is concave in front, gives attachment to the front part of the capsule and, below this, to the

---

* The burden is thrown on this muscle only to show the movements of the joint; it must be understood that many other muscles are concerned.

260   Anatomy of Skeleton

External Pterygoid. The hollow below the neck *internally* is separated from the long ligament by an interval in which lie, between the bone and ligament, from above down (Fig. 207), the auriculo-temporal nerve, origin of middle meningeal, internal maxillary and sometimes a process of parotid, the inferior dental artery, the inferior dental nerve. The mylo-hyoid nerve and vessels must pierce the ligament to reach the groove, for the ligament is fastened to the whole length of the margin of the foramen.

Looking at the bone as a whole, it is noticeable that its alveolar part is more sharply curved than the rest of the body, with the result that its hinder portion forms a bracket-like projection on the buccal aspect of the bone. The dental arcade forms a parabolic curve, thus differing from the semi-ellipse of the upper arcade. The result is reflected in the disposition of the teeth : the parabolic curve being narrower in front than the ellipse, and broader further back, the teeth in the lower jaw will tend to be behind the others anteriorly and external to them behind. So we find the direction of the lower front teeth to be upward and forward while the back teeth are directed upwards and inwards to meet their opposers, and the same tendency can be observed in the direction of the sockets. The comparative shortness of the jaw also has effect in causing the direction of the front teeth.

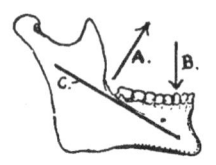

FIG. 210.—The muscles closing the jaw pull in the direction *A* and are resisted by the upper jaw acting in direction *B;* this would tend to open out the angle between ramus and body, an effect counteracted by the strong bar *C*.

The inferior dental canal runs, inside the bone, between the splenial and alveolar portions. Openings exist for vessels and nerves between the canal and the dental sockets, and the mental foramen carries the mental vessels and nerve to the surface. This foramen is situated as a rule below the jugum between the two bicuspids, about half-way between the upper and lower margins of the body ; but this level is only true in the adult tooth-bearing jaw, for, as shown in Fig. 209, it is different at other periods of life. In the same figure the differences between the angles formed by the axes of ramus and body are apparent, and the probable explanation of the occurrence of the angle is given.

The general build of the bone is in accord with the principle of resistance to the forces tending to destroy its form. Thus, with powerful muscles acting on its ramus and lifting the long arm of the body against the upper jaw, it is evident that the tendency of such action would be to open out the angle between body and ramus, and we find a thick and strong bar thrown downwards and forwards to subtend the angle and render it secure (Fig. 210). Secondarily there is the thickening of the splenium, which resists the downward push of the alveolar teeth.

The mandible is composed of a thick outer shell of dense bone with a small amount of cancellous tissue contained within it, so that its body is cut with difficulty by bone forceps : it resists decay longer than other parts of the skeleton, which probably accounts for it being often found as the sole representative of the bones of the face or skull in specimens from strata of past geological periods. Such ancient jaws show that the presence of a mental prominence has been a human characteristic for ages, but the earliest known jaws—of the early Pleistocene or late Pliocene periods—do not possess the prominence. It is accounted for by assuming that a recession of the alveolar part has taken place in man at the same time that the splenial portion was comparatively well preserved, perhaps in connection with specialisation of the tongue and floor of the mouth in association with the function of speech.

The direction of the condyles is peculiar. Their long axes are directed inwards and backwards, so differing from those of the glenoid cavities; they also differ in that they are not directed downwards as well as inwards. Such differences in the disposition of the bony surfaces can only be adjusted by the intervening meniscus. The direction of the axis of the condyle inwards and backwards seems to be associated with the side-to-side movement of the jaw; when the jaw is moved, for instance, to the right, the movement is accomplished by the left External Pterygoid and the right Digastric, and the bone apparently rotates round a centre passing vertically through the hyoid or behind this, the left condyle moving out and the right one in. From this point of view the condyles might be considered to lie in the line of the periphery of the circle in which the jaw moves: it is evident that they could not be directed inwards and forwards, unless the centre of rotation lay far behind the jaw, and the angle moved backwards.

A large part of the body of the bone is accessible to digital examination. Thus praetically the whole of the alveolus can be directly examined by the finger in the mouth: it may be pointed out here that the line of the closed mouth is above the lower teeth and its angle is about as far out as the first bicuspid. The outer aspect and lower border of the splenial portion are not thickly covered and can be palpated, and the facial artery can be compressed here against the bone. The inner aspect of this part is inaccessible.

The ramus, however, is thickly covered over nearly its whole extent, and can only be palpated from the outside through Masseter and parotid, although the angle is fairly clear. But the front border of the ramus can be investigated from the mouth: it can be understood from Fig. 206 that the finger, passed back in the alveolo-labial sulcus or along the cheek, can feel and examine the Temporal tendon and the ridges leading to the coronoid process, *through the plane of the Buccinator*, and it is possible also to feel the pterygo-mandibular ligament in this plane if the structure is put on the stretch by opening the mouth. The student can familiarise himself with these observations by examining his own mouth, and he will also be able to find his lingual nerve where it lies on the bone under the mucous membrane behind and below the last molar.

Development.

The bone develops in the first or mandibular visceral arch from an ossification commencing in membrane to the outer side of Meckel's cartilage during the sixth week; thus there is only one centre for each half of the bone. The centre first shows in the lateral part of the bone, and, extending forward, grows round Meckel's cartilage and encloses it in a groove which closes in later: it also spreads slowly upwards and backwards to make the ramus and its processes. The condyle and coronoid are indicated at the beginning of the third month and are practically formed by the middle of this month: a cartilaginous change is visible in the mesenchyme of the condylar, coronoid, and angular regions before the spreading ossification reaches them.

The bone is in two halves *at birth*, and does not fuse until a year or more has elapsed.

Meckel's cartilage atrophies in its greater part, but its front portion, taken into the bone, is ossified, and probably therefore represented in the bone between the mental foramen and the symphysis, and is possibly responsible for the prominence of the chin.

THE DEVELOPMENT OF THE SKULL AS A WHOLE.

The base of the skull is preformed in cartilage which begins to make its appearance in the latter half of the second month: the vault is developed in membrane, as are the orbital plates of the frontal bone.

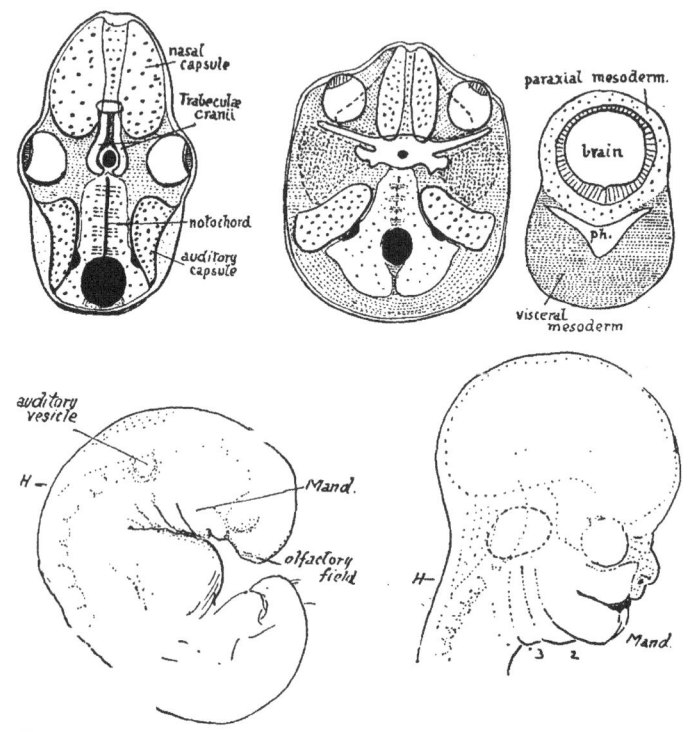

FIG. 211.—The first figure represents what might be considered a typical (lower) vertebrate skull-base. The hinder portion is made by bones related with somites beside the notochord, hence is "parachordal," and eXtends forward to the pituitary foramen. In front of this are the paired "trabeculæ cranii," enclosing the foramen behind, and connected by an intertrabecula in front. Special sense-capsules make up the rest of the cartilaginous base: of these the nasal capsules eXtend forward below and in front of the trabeculæ, only their upper and posterior parts really coming into the floor of the proper cranial cavity : the eye probably had a cartilaginous capsule originally, but this is not found now, only the fibrous sclerotic possibly representing it. The auditory capsules, containing the inner ear, are partly basal, beside the parachordal region, but are mainly in the lateral walls. All these structures are formed in " paraXial " mesoderm, which immediately surrounds the neural tube. In the human skull (second figure) the area of the base is much increased, especially in a lateral direction, as the result of brain growth. The auditory capsules in consequence become definitely basal, placed on each side of the parachordal region (which corresponds with the same region in the lower type, although it is developed dorsal to the notochord) while the parachordal cartilage shows backward eXtensions meeting behind the foramen magnum. The cartilage surrounding the pituitary foramen may be trabecular, but there is no definite indication of this. In front of this the nasal capsules come into the base, but, as eXplained in the teXt, they only get into this position secondarily. The solid cartilaginous septum between them is continuous in formation with the pre-pituitary base, and may possibly be trabecular or intertrabecular. The eyes have moved forward, probably on account of the development of the temporo-sphenoidal region of the brain, and a process of cartilage (spheno-ethmoidal plate) is thrown over each from the nasal capsule, as indicated by the interrupted line. A wide area of paraXial mesoderm is thus left between the eye and the auditory capsule : in this a cartilaginous orbitosphenoid eXtends out behind the eye, and another shorter process behind this, also continuous with the central basal chondrification, marks the base of the future alisphenoid. The remaining paraXial tissue in this area does not ossify or chondrify, but probably forms the dura mater between the orbitosphenoid and the tentorium. The greater part of the alisphenoid is formed in membrane from (visceral) mesoderm deep to the paraXial plane, and comes up into place secondarily from below ; it is indicated in the figure by darker stippling. The third figure shows the general relations of the visceral mesoderm, in the floor of the pharynX, to the paraXial mesoderm round the neural tube. Observe that the visceral tissue comes round the pharynX, to form a support for the paraXial. With the widening of the brain a corresponding widening and thickening of the visceral mass takes place in support. The relation between the two layers is only present at first behind the pituitary level, but, later, eXtensions (maXillary processes) from the first visceral arch grow forward to support the eyes and overhanging brain, etc. The lower figures show the formation of the face from the embryonic condition ; *H* is the hind-brain. Notice how the maXillary process from the mandibular arch (*mand.*) has grown forward below the eye and is applied to the side of the nasal capsule, which has been formed round the olfactory field.

Skull and Hyoid

In lower vertebrates the basal part of the brain case is divisible into a posterior portion, the *parachordal bars*, and an anterior part composed of the *trabeculæ cranii* (Fig. 211). In the human skull there are no signs of such cartilaginous bars, although there is some indefinite suggestion of the conditions in the precartilaginous state, but the portion of the base between the foramen magnum and the pituitary fossa can nevertheless be termed the *parachordal* part, because the notochord runs in relation with it as far as the dorsum sellæ. In front of this, however, there are no indications of trabecular structure whatever, and this front part of the base of the skull, moreover, is possibly of a quite different morphological value, so that it should not be called the "trabecular part," but might be termed with more propriety the *prechordal* part.

The notochord does not lie altogether in the parachordal part of the base (see Fig. 216), but below it, in relation with the roof of the pharynx, for the greater portion of its course here; but, in spite of this, the basiocciput and basisphenoid may be looked on as developing in connection with the mesodermal somites that lie on each side of the notochord. Four pairs of somites are taken up in this way into the skull base: if there are any others cephalad to these they are lost in the condensation round the end of the notochord, and give no separate indication of their existence as distinct structures at any time in development, so far as has been observed at present.

Before chondrification begins the mesenchyme of the base forms a continuous condensation in the future basioccipital and sphenoidal regions, and from the latter extends out somewhat at the sides, while from the former situation the condensation shows a tendency to extend back round the neural tube towards the region of the future post-occipital.

Cartilaginous change first occurs in the basiocciput and basisphenoid, spreading later from this into the mesenchyme of the root of the alisphenoid, orbito-sphenoid, and post-occipital. At the same time another element becomes apparent in the base: this is the *periotic capsule*, developed as a chondrification round the otocyst, which has come into position beside the parachordal part of the base, behind the alisphenoid. Thus, when the process has gone on for a little time the base would present the various distinct parts shown in the scheme in Fig. 211, where the periotic capsules are seen as separate elements, the small and great wings as small projections from the sphenoidal central mass, and the post-occipital as two cartilaginous sheets spreading round the foramen magnum to meet behind it. A little later these different parts unite and form a mass of cartilage continuous at the sides and in front and behind with the thinner *membranous* covering of the sides of the brain. The cartilaginous portion or *chondrocranium* made in this way includes the basi-, ex-, and post- occipital, the basisphenoid and basal parts of the great wings, the orbitosphenoids, and the petrous temporal. The bones that correspond with these regions therefore *ossify in cartilage*: the *membrane bones* in continuity with them include the orbital plates of the frontal, the greater part of the alisphenoids, the squamous temporal, and the occipital bone above the superior curved line. The nasal capsule (cartilaginous) is also present, in front of the sphenoid, but it is possibly not in the same morphological plane as the rest of the base and will be considered separately later.

While these modifications are proceeding in the immediate coverings of the brain, additional *visceral* elements are being added on the under aspect of the base. These are derived from the visceral arches that constitute the floor of the primitive pharynx: it is evident (Fig. 211) that the outer and upper parts of the mesoderm of these arches must come into relation with the ventro-lateral part of the covering of the brain, which lies in close association with the roof of the pharynx. In this way the upper ends of certain of the *cartilaginous* bars formed in these arches are taken up into the skull, while other bones are developed *in membrane* in the arches and form the facial portion of the skeleton ; these two sets of structures constitute together the *visceral part* of the skull. It is only in the first three arches that a dorsal moiety of each bar is formed, and in the case of the third bar this quickly disappears, so that only *the upper ends of the first two bars are finally represented in the skull ;* these form (see Fig. 172) the ossicles of the ear and the styloid process.

The facial skeleton is developed secondarily from the mesenchyme of the mandibular (first) arch. This arch gives a *maxillary process* forward on each side of the mouth cavity, below the anterior part of the cranial capsule and the eye, and on the sides of the nasal region : the arrangement is shown in Fig. 211, which also indicates the general formation of the face by the arch and its maxillary process. The lower jaw ossifies in the mesenchyme of the *first arch itself*, and in its *maxillary process* are formed the maxilla, palate, internal pterygoid plate, and malar.

The **nasal capsule** is the cartilaginous wall of the nasal fossæ, surrounding them above and at the sides and giving a cartilaginous partition between them. Its anterior part is undoubtedly developed round the early olfactory pit, but there is some reason to suppose that its hinder part is formed from the maxillary mesenchyme, a view which, *inter alia*, renders intelligible the formation of the palate bone on the inner side of the capsule while the maxilla is developed external to it.

The early nasal pits are placed just above the fronto-nasal process, and these small cavities gradually extend upwards and backwards, their cartilaginous capsules forming round them as they extend, so that ultimately they come into their final positions, and the capsules are flush with the presphenoid above. Then the spheno-ethmoidal plates grow out along the orbitosphenoidal processes. The back of the cartilaginous *septum*, however, is associated as an early condensation with the central part of the basal cartilage, and the value of this portion may differ from that of the rest. When formed, the capsule surrounds the cavities, and has projections and spaces within it which map out the ethmoid, while the lower edges of its side walls turn in and fo m the basis of the inferior turbinals.

It would be better, perhaps, for us at present to look on the nasal capsule simply as a *sense-capsule* developed round an organ of special sense, without regard to the possible values of its different parts, and to speak of the ethmoid, lachrymal, nasal, and vomer as formed in connection with the capsule, the first preformed *in cartilage* and the others ossified in *membrane* on the surface of the cartilage. Perhaps the premaxilla—if such a bone has a separate existence (see p. 235) in the human skull—should be placed in the same class, as it is developed in the *fronto-nasal process* made by the fusion of the inner walls of the two olfactory pits.*

By looking on the nasal capsule as a special sense-organ we put it for the moment on the same level as the periotic capsule and can thus bring its upper surface, the part that shows between the orbital plates of the frontal, into the conception of the cartilaginous base, making it continuous with the prechordal portion of this base. The capsule of the eye does not become cartilaginous in the human skull, and is probably represented by the sclerotic.

* It has been suggested that the maxillary mesenchyme invades the frontal-nasal process from the sides and provides the tissues of the region, with the exception of the portion of the process exposed behind as the " incisive papilla." Although this view explains the nervous supply, etc., of the parts, its embryological support is so doubtful that it would be wiser to look on the " premaxilla " as being possibly in line with some other bones round the nasal capsule, without regard to its more detailed morphological value as indicated in development.

Skull and Hyoid·

We find, then, that the skull can be said to be made up of bones which may be divided into classes according to their source of origin. Thus we have :—

(*a*) **Bones of the Cranial Capsule.**

(*b*) **Special Sense Capsules.**—The ethmoid and petrous in this group can perhaps be added as a subdivision to (*a*), but the vomer, nasal, and lachrymal may be placed in this class as formed in membrane on the capsule, and the premaxilla may perhaps be put provisionally with them.

(*c*) **Visceral Skeleton.**—This includes (1) membrane bones formed in the arches, mandible, maxilla, malar, palate, internal pterygoid plate ; (2) bones formed in cartilaginous bars of the arches—styloid process, ossicles of ear.

(*d*) **The Tympanic Plate.**—The value of this is uncertain, and, although it is often referred to as homologous with certain ossifications found in lower types of jaw, it would be safer from our present point of view to place it simply in a class by itself without further remark on its morphology.

The bones of the *cranial capsule* can be further subdivided into cartilage and membrane bones : the *membrane bones* are the frontal, parietal, squamous temporal, and interparietal part of occipital, with a large part of the great wing of sphenoid. The chondro-cranium underlies that part of the brain that expands least, and the membrane bones cover the portion that grows most rapidly.

The cerebral nerves pass out of the embryonic skull between the various parts of the chondro-cranium, and the foramina of exit in the adult skull may be divided into groups on this basis. Thus we get foramina (*a*) between various distinct parts of the skull, as the jugular between petrous and occipital ; (*b*) between different parts of what is one bone (in the adult), as the anterior condyloid foramen between basi- and ex- occipital. A class occurs as a subdivision of (*a*), the foramen being formed by ossification extending from one of the bordering elements round the issuing structure : thus we may get the foramen ovale, for instance, as a secondary enclosure of a nerve originally passing between alisphenoid and periotic capsule, or in some cases, as in the canalis innominatus of the sphenoid, the enclosure may only take place occasionally.

We have seen that the skull can be broadly divided into a dorsal and lateral group of bones developed in relation with the capsule of the brain, and a ventral skeleton of a visceral origin.

So far the division, given above, is founded simply on the embryonic development and is therefore unassailable. But when we come to inquire into the remoter values of the parts concerned, and to compare the human skull with those of lower types, the situation becomes more complicated and bristles with difficulties.

So far as the mammals are concerned, the comparison is easy and straightforward, but below this class the difficulties begin. In all classes of vertebrate skulls the uncertainties are not so much with regard to the bones of the brain capsule, nor with reference to the greater part of the visceral skeleton, but are mainly associated with the comparative values of the prechordal and nasal capsular regions—in fact, they may in general be said to centre round the nose and roof of the mouth.

It would be out of place in the present work to enter on a consideration of the questions involved, and it will be sufficient to mention that the membrane bones are generally thought to represent dermal structures which, originally placed superficial to the coverings of the brain, as this expanded in evolution came to occupy a deeper situation to fill the call for increased covering : thus in the Elasmobranch fishes the brain case is a continuous cartilaginous one, but as we proceed further up the vertebrate scale we find the cartilage becoming confined to a relatively lessening basal area, while its place on the vertex is taken by membrane bones presumably derived from more superficial tissues. A similar view is taken of the formation of the jaws, but here the question is complicated by the failure of the human bones to show centres of ossification corresponding in number with the presumed dermal bones concerned : thus each half of the lower jaw, for example, has only one centre instead of a number in agreement with the number of elements (angular, splenial, etc.) that are supposed to have sunk into the deeper tissues of the mandibular arch to make the bone. It seems doubtful, moreover, whether certain cartilage bones have not been replaced by membrane bones which from their positions could not have come from the surface.

The chondro-cranium is replaced by the *osseous cranium*. The times of appearance of the ossific centres vary in different bones ; but two main periods of activity in this direction are to be noticed, the first about the end of the second month, or beginning of the third month, and the second about the fifth month. All the bones formed in the cartilaginous base, excepting the special sense capsules, begin to ossify in the first period, as do also all the membrane bones of the vault ; the lingula begins a little later, in the fourth month. The bones formed in the special sense capsules, on the other hand, belong to the second period, including the ethmoid, inferior turbinal, sphenoidal turbinal, and petrous centres : the centre for the vertical plate of the ethmoid is even later, occurring after birth. The lachrymal, nasal and vomer, membrane bones lying against the nasal capsule, belong to the first period, as also does the tympanic bone, a membrane bone applied to the periotic capsule in part. In the visceral skeleton there is greater variation, the mandible showing centres in the fifth week, before the centres of the maxillary process, the derivatives of which belong to the first period ; this includes the maxilla, palate, internal pterygoid plate, and malar.[*] But the visceral bones further back are much later than the second period, the tympano-hyal not showing a centre until a month or so before birth and the stylo-hyal not until some months after that event : the upper ends of these bars, however, come into the second period by commencing to ossify as the small bones of the middle ear in the fifth month.

Summing up this account of the ossification, it may be said that all the membrane bones begin to ossify in the first period, the cartilaginous cranial bones partly in the first and partly in the second (special sense capsules), and the cartilaginous visceral bones considerably later.

So it comes about that, save in the perpendicular plate of ethmoid and the styloid process, all the bones of the skull are in process of formation for some time before birth, although the various elements concerned have not joined with each other in all cases to make the bones as we see them in the older skull. Thus the bony post-, ex-, and basi- occiput are distinct ; the basiocciput is separated by cartilage from the basi-sphenoid, this partially from the presphenoid, the squamous from the petrous temporal, and the two halves of the lower jaw are not joined in the middle line. In addition the membrane bones of the cranial vault, though they grow rapidly, do not meet each other over the quickly-growing brain, and at birth are separated by lines and areas of fibrous tissue where the periosteal covering of the bones becomes directly continuous with the dura mater lying immediately underneath.[†] The largest of these inter-osseous fibrous areas occur at the angles of the parietal bone : the areas are termed *fontanelles*, and these larger ones are the *anterior* and *posterior median* at the bregma

---

[*] There is some reason to suppose that the lachrymal and vomer may be formed in tissues belonging to the maxillary process where this is applied to the capsule, when their ossification periods would come into line with the other " maxillary " bones. The outer pterygoid plate is ossified by extension from the sphenoid; it is looked on as a muscular process, and thus as not having a morphological value equal to or as great as that of the inner plate.

[†] Before the sutures close the bones of the vault are thickest at their centres and thin away into the fibrous tissue at their edges. After the edges come into contact they begin to increase in thickness and ultimately become the thickest parts. At the same time the central parts of the bones become less prominent.

and *lambda* respectively, and the *anterior* and *posterior lateral* at the *pterion* and *asterion* (Fig. 212). The bregmatic fontanelle is four-sided, between the two parietals and two halves of the frontal ; the lambdoid is triangular, between occipital and parietals ; and the two lateral intervals are irregular in shape, that at the asterion sometimes extending back for a little distance between the post- and supra- occipital.

The lateral fontanelles are as a rule practically, although not actually, closed at birth, and the posterior median interval closes shortly after : the bregmatic fontanelle, however, remains open and palpable from the surface till about the end of the second year, and the observation of its condition is of some clinical value. The various bones of the vault come into complete contact by the age of four or five ; in the case of the frontal the two halves are partly joined at that age : occasionally a bone fails to meet its neighbour and the interval is filled by an *accessory* centre, constituting a *Wormian bone.* Such *true* Wormian bones are to be distinguished from *false* Wormian bones, which are normal centres that have abnormally failed to join the bone with which they should be fused.

Fontanelles are not confined to the regions mentioned above, for membranous intervals sometimes occur elsewhere wherever parts of the bone do not come together adequately at their usual time : thus a *metopic fontanelle* may be seen at the root of the nose at the lower end of the metopic suture, an *occipital* one may occur at the hinder margin of the foramen magnum between the halves of the post-occipital, or a *parietal* (*sagittal*) fontanelle may mark the site of the parietal foramen.

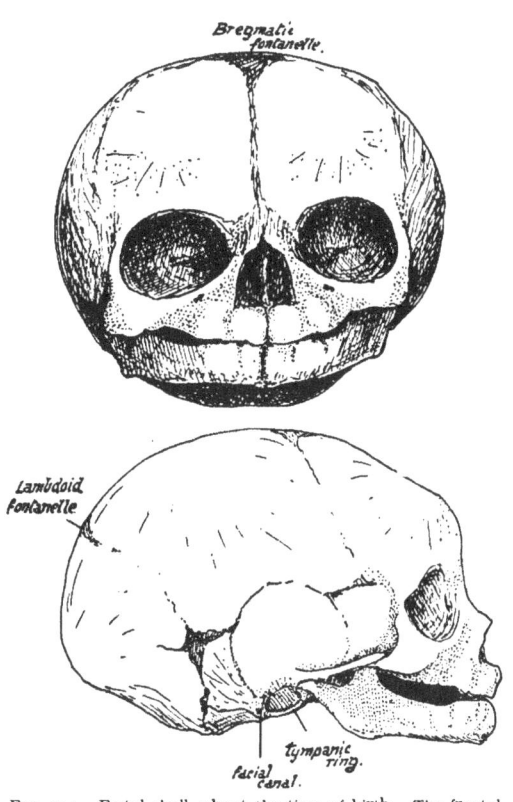

FIG. 212.—Fœtal skull, about the time of birth. The frontal region is artificially depressed to some extent.

There are many things to be noticed about the skull at birth in addition to the fontanelles. In Fig. 212 the skull at birth is shown, and it plainly differs from the adult structure. The most striking thing about the fœtal skull is the comparative smallness of the face and largeness of the cranium, but another point of distinction can be found by measurement : a line drawn through the condyles practically divides the fœtal base into two equal halves, whereas in the adult the part that lies in front of the line is proportionately much increased and gives a ratio of 1·7 : 1 instead of 1 : 1, as in the new-born skull. We have therefore to deal with an excessive growth *in length in the front part of the skull* during the change to the adult type, as well as with a *vertical increase*, both affecting the anterior portion of the skull ; the former is mainly due to cranial growth in front and the latter mainly to facial increase, but the face increases in depth also with the added cranial length. When the facial growth occurs without the proportionate cranial growth the adult skull presents the appearance of *prognathism* that characterises the heads of the lower races.

The characters of the face in the new-born skull depend almost entirely on the immature condition of the maxillæ and mandible. Each maxilla (Fig. 212) is flattened vertically, owing on the one hand to the almost complete absence of the alveolar process and on the other to the small size of the antrum, which only occupies the inner part of the bone to a small extent, leaving the remainder more or less flat. Whatever prominence there may be in the maxilla is due to the tooth germs embedded in the

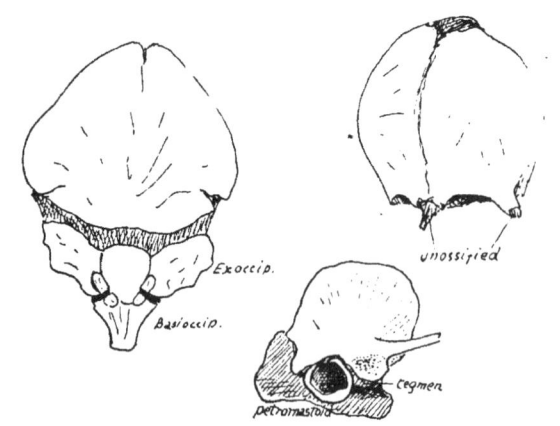

Fig. 213.—Occipital, frontal, and temporal at birth.

bone, but these are properly supra-alveolar, as can be seen by comparing the skull with an adult one.

The results of this shallowness of the maxillæ can be well seen in the nasal fossæ. These are much broader compared with their height than in the adult, and the various bones that make their walls are correspondingly modified : thus the palate has a short but broad vertical plate, the internal pterygoid plate is similarly modified, and the vomer shows a decreased vertical height while its breadth is proportionately greater, and thus it forms a gutter that holds the septal cartilage in it. On the other hand, the ethmoid and lachrymals lie above the level of the maxillary body and are not affected by its subsequent growth, so they present at birth no essential differences, save in size, from the adult conditions. The malar, owing to the want of development of the alveolar region, overhangs the opening of the mouth on each side and may ·be felt easily through the mucous membrane. Another result of the alveolar deficiency is seen in the hard palate : it is much less curved than in the adult—in fact it is nearly flat, for in the full-grown bone the concavity is largely due to the presence of the alveolar eminences.* The posterior

* In old toothless jaws with loss of alveolar processes there is a return to the flattened palate.

Skull and Hyoid

choanæ, like the anterior openings, are proportionately less high than in the adult, and their plane is more nearly horizontal.

The mandible is in two halves joined by fibrous tissue. Each half shows a short ramus making a very large angle (175 degrees) with the body. The body is poorly developed both in its alveolus and its splenial portion, and presents prominences that mark the position of the dental sacs in the bone. The coronoid process is higher with reference to the condyle than in the adult.

The growth of the alveolar regions in both jaws depends on dentition, and thus its full development in each is not reached before the twentieth year or thereabouts, although the temporary dentition brings about a rapid growth in the first five or six years of life. The space required by the teeth and the increasing height of the maxillæ push the mandible down, and in this way lead (Fig. 209) to the lessening of the angle between ramus and body in the lower jaw, in order to keep the teeth in apposition; the angle at the first dentition is about 140 degrees and in adult jaws about 120 degrees. Thus the appearance of an angle in the mandible may to some extent be also looked on as a result of the later growth of the upper jaws. The body of the mandible increases in length proportionately in keeping with the antero-posterior growth of the face, and in this way provides room for the three permanent molars.

The smallness of the maxilla is reflected in the orbit, where the position of the two main fissures is decidedly lower, and the spheno-maxillary fissure is also much wider, than in the adult: as the maxilla grows the fissure is narrowed, and at the same time the malar region is pushed and rotated into a more external position, thus raising the levels of the fissures.

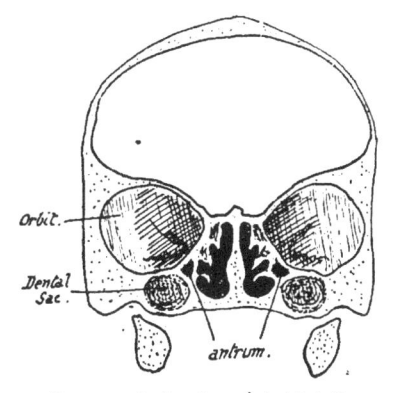

FIG. 214.—Section through fœtal skull.

The frontal bone, in two halves, is seen above the face: each half shows a marked frontal eminence from which ridges radiate on the surface of the bone.

Among numerous other minor details of the front aspect it may be pointed out that the lower edges of the nasal bones are straight and not notched or serrated as in the adult, and that the lachrymal may be somewhat rotated so that its surface looks forwards as well as outwards.

The *lateral view* of the new-born skull shows well-marked parietal eminences, where the skull is broadest. At a lower level the junction of bones at the asterion and pterion may be more or less incomplete (see previous reference to fontanelles). The temporal squama is not fused with the underlying petrous, nor is there any appearance of a mastoid process: the tympanic plate is represented by the tympanic ring, so that the membrane is practically on the surface of the skull.* The ring is fused with the squama in part, and its circumference is not much less than that of the adult membrane, so that this membrane is proportionately larger in the newly-born. The want of development of the mastoid and tympanic plate leaves the stylo-mastoid foramen

---

* This does not mean that the membrane is visible in the living infant from the surface, for the outer fibrous and cartilaginous meatus is of proper proportionate length and thus conceals the membrane.

more on the outer than the basal aspect : when these structures grow it is overlapped by them externally and becomes a basal opening. The Digastric arises from the outer surface just behind the foramen. The zygoma is not joined with the malar and does not possess complete roots. The styloid process is cartilaginous.

A *posterior view* shows the prominent occipital protuberance, from which fine ridges radiate out into the upper part of the bone ; a partial separation of this upper part from the lower may be seen at the sides, or a sagittal slit between the two halves of the supra-occipital above. These slits are not always present, but when they are they run into the postero-lateral and lambdoid fontanelles respectively. The occipital bone at birth is in four pieces joined by cartilage.

The junction between the bones in the new-born skull may be summed up as follows :—

(a) Fusion of half bones across mid-line : all cartilage bones and no membrane bones.

(b) Fusion of elements making single bones : has occurred in membrane bones, but not at all or only to some extent in cartilage bones.

The skull grows rapidly during the first six or seven years of life : the growth is largely in the facial region, the volume of which compared with that of the cranium has been estimated to be 1 : 8 at birth and to have risen to 1 : 4 at five years of age. The second dentition is accompanied by another period of increased development, so that by the twenty-first year the ratio between face and cranium has risen to 1 : 2. This second growth period brings the air sinuses particularly into evidence, so that the expression may be considerably modified by it.

The adult skull has a cranial capacity varying between 1,000 and 1,800 c.c., with an average of about 1,400 c.c. In Europeans the skull is usually over 1,450 c.c. in capacity, and such skulls are termed *megacephalic*. *Mesocephalic* skulls, such as the Chinese, have a capacity between 1,350 and 1,450 c.c., while *microcephalic* skulls, such as are found in lower races, have a capacity below 1,350 c.c.*

The adult human skull is characterised by its relatively enormous capacity with a jaw arch comparatively small. The large brain development that brings about these human characters also leads to an increased width of the middle region of the skull, with shifting forward of the lateral eyes and corresponding narrowing of the nose : this last condition, however, is no doubt associated with the general retrogression of the facial region. The character of the human foramen magnum shows complete adaptation to the erect position, the plane of the foramen being practically horizontal, whereas in anthropoid apes it is rather oblique and in quadrupeds it looks directly backwards. The human skull is almost perfectly balanced on the vertebral column, and its muscular attachments for support are proportionately weak. Other evidence of increased brain development can be seen in the delayed fusion of the various elements that cover in the cranial cavity † : in this connection it may be pointed out that the cranial elements have increased in relative size and development, while some of the facial bones, such as the lachrymal and palate, show signs of retrogression when compared with those seen in skulls of lower animals.

Many of the facts mentioned above can be used as data when endeavouring to determine approximately the age of a skull at the time of death. Thus the age of young skulls can be determined with fair accuracy from the number of erupted teeth : the presence of well-formed air sinuses puts the age after puberty ; absence of fusion between basisphenoid and basioccipital places it

---

* For further skull measurements the student must consult works on anthropology.

† The separation of the bones over the vault no doubt enables the increase in size of the head to take place more rapidly, but probably it is not an essential factor in the increase. The greatest proportionate increase in size takes place of course during fœtal life, but a more gradual growth goes on, with a marked increase in weight, from the first few years of life until adult age : as old age approaches the skull decreases somewhat in size and loses about two-fifths of its weight. The increase in size of the cranial bones, especially after they have met at the sutures, is brought about as in the other bones by addition to the surface with corresponding removal of bony tissue from their deep aspect : as growth proceeds in this way the parietal and frontal eminences become less marked.

before twenty-five ; the styloid process is not joined to the temporal until middle life ; the vomer and mesethmoid are synostosed after forty-five ; and the skull in old age is edentulous, with wasted alveolar processes, and is lighter and has thinner bones with large air sinuses.

It must be remembered that female skulls are thinner and lighter than male : they are shorter in proportion to their breadth : they are also smaller, with a capacity about one-tenth less, but this is in proportion to the smaller bulk of the female body.* The male skull is rougher and more ridged, with more prominent mastoid processes, zygomatic arches, and occipital protuberance : the superciliary ridges are more rounded and prominent, but the parietal eminences are less marked than in the female skull. In other words, the female skull preserves somewhat an appearance of immaturity. The tympanic plate is said to be more developed in the male and shows a sharp border, while in the other sex the edge is more rounded. The female palate is narrower than the male. There is little sexual difference before puberty.

In comparison of skulls of different races it is always advisable, when possible, to compare the measurements of male skulls. For such measurements and indices the student must consult works dealing with anthropometry, but a few may be given here as they are in common use :—

(a) Maximum length—male, about 8 inches or less ; female, about $\frac{1}{2}$ inch less.

(b) Maximum circumference (varies between $17\frac{1}{4}$ inches and $21\frac{1}{4}$ inches). In Europeans —male, about $20\frac{1}{2}$ inches ; female, about $19\frac{1}{2}$ inches.

(c) Cephalic index : $\frac{\text{max. breadth}}{\text{max. length}} \times 100 = $ ceph. ind.

Ceph. ind. of Europeans is 75 to 80 (mesaticephalic).
,, ,, some races is over 80 (brachycephalic).
,, ,, others is under 75 (dolichocephalic).

Usually rather larger in women than in men, owing to relatively greater breadth.
(All these measurements must of course be made between the same points if they are to have any value in comparing different skulls, and these points are not always similar in the measurements given by individual observers.)

(d) The Facial Angle.—This estimates the amount of prognathism or projection of the face bones. It is obtained in various ways, depending in general on a basal skull line contrasted with a line drawn from the nasion to a projecting point on the upper jaw. The contained angle is the one required. In older measurements the opposite angle was taken, so that the angle increased as prognathism decreased.

It may not be out of place to point out here that the anthropoid skull is most like the human type in its young state, diverging from it as it develops. The skull of a chimpanzee is distinguished from that of man by the sloping plane of its foramen magnum already mentioned, the smallness of its condyles, the diastema in the canine region of its jaws, its large intermaxillary bone, jaws, teeth, and petrous bone, the want of a crista galli and flatness of its (usually fused) nasals, and the presence of a well-marked post-glenoid process. In this last character the negro skull resembles it to some extent.

HYOID.

A U-shaped bone situated in the ventral floor of the pharynx below the base of the tongue and in front of the epiglottis. It presents a median unpaired *body*, a long *large cornu* on each side, and a small nodule, not always bony, the *small cornu*, situated above the junction of body and great cornu.

The body is concave on its *deep or posterior surface*, lodging here the infra-hyoid bursa : the bursa lies in front of the thyro-hyoid membrane, which is therefore attached to the upper margin of the posterior surface. The *anterior surface* is convex and divided by a *transverse horizontal ridge* into an upper and lower portion : the surface gives attachment to muscles of the tongue and floor of the mouth and along its lower edge to infra-hyoid muscles (Omo-hyoid, Thyro-hyoid and Sterno-hyoid).

The *greater cornu*, flattened, has usually an *upper surface* that gives origin to Hyo-glossus and Middle Constrictor, a *lower surface* that has Thyro-hyoid and thyro-

* The female skull is *relatively* larger : compared in weight with the rest of the skeleton it gives a ratio 1 : 6. Male ratio 1 : 8.

272   Anatomy of Skeleton

hyoid membrane attached to it, a prominent *outer margin* which affords insertion to Stylo-hyoid and Digastric, a smoother *inner edge* that is in relation with the mucous membrane of the pharynx, and an enlarged extremity which is fastened to the thyroid cartilage by the thyro-hyoid ligament.

The *lesser cornu* is fastened by fibrous tissue to the body and great cornu, and by the stylo-hyoid ligament to the styloid process. It affords origin to some fibres of the Middle Constrictor and the Chondro-glossus.

The hyoid bone is an ossification of the ventral portions of the cartilaginous bars of the second and third pharyngeal arches: the dorsal end of the second bar is fused with the petro-mastoid and forms the styloid process, while the intermediate part is the stylo-hyoid ligament; the upper end of the third bar is only indicated for a short period as a condensation of cells that ultimately disappears entirely. The lesser cornua (*cerato-hyals*) are parts of the second bars, while each greater cornu (*thyro-hyal*) represents the third bar of its side. The body is sometimes termed the *basihyal*, but this name should more properly be applied to the transverse ridge on the body, which probably alone represents the ventral junction of the two bars of the second arches (Fig. 217). The more complete development of the second bar causes the small cornua to appear longer at one stage of their formation than the great cornua (Fig. 218), but it must of course be remembered that the contrast is mainly in appearance only, because there is no bone in the young specimen, and in the adult the large second bar is as fully represented, although by more sharply differentiated bony and ligamentous tissues.

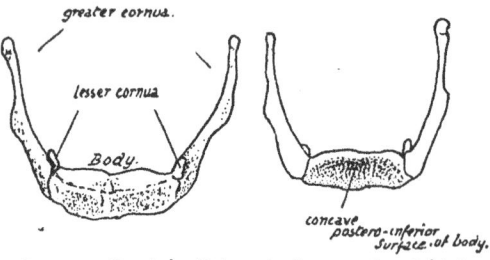

FIG. 215.—Female hyoid from the front and from behind.

The relation of the thyroid (median) down-growth to the hyoid is of some interest and importance; it is frequently said to lie in or even behind the bone, but this is probably not a true statement of the case. In the early period, when the anlage of the hyoid is first recognisable, the thyro-glossal duct can be found passing altogether ventral to the rudiment of the bone. This is shown in Fig. 216. At a later stage the epithelial rod that runs down to the thyroid gland is still ventral to the hyoid but is in close contact with it, and is in fact folded somewhat round its lower edge (Fig. 216), so that a potential thyroid tube might remain behind the plane of the body although it has not got there by growing down behind the hyoid. This is the condition found in all normal embryos and young fœtuses, and the occurrence of thyroid cysts * or fistulæ are persistences of this earlier state.

The bursa that lies behind the *body* of the bone is median in position, and for this reason is probably not derived from the thyro-glossal duct: on each side of it the hollow in the bone is filled by fibro-fatty tissue. The line of attachment of the thyro-hyoid membrane can usually be made out above the hollow, and in the central part some fibrous tissue connects the bone with the epiglottis. The front surface is practically altogether muscular, and the transverse ridge is plainly not a secondary marking on

---

* Such cysts are usually unilateral and—so far at least as I have had opportunity of examining them—even when they seem to extend up behind the bone the proper relation may be made out with a little trouble and the continuity of the bone established behind them.

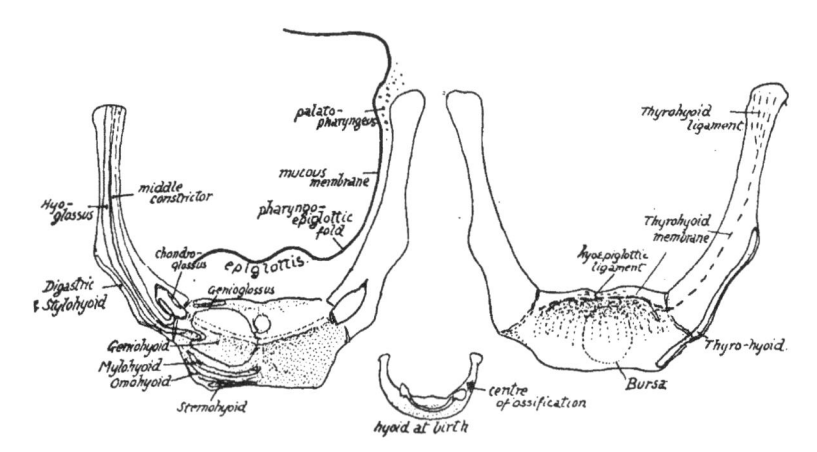

FIG. 217.—Attachments to hyoid. Observe the area occupied by insertion of Geniohyoid, and how the Hyoglossus encroaches on this area on the body, being also attached to the whole length of the great cornu outside the middle Constrictor. Genioglossus has only a very small area behind Geniohyoid, and Mylohyoid has a linear insertion just above the infra-hyoid muscles. Digastric and Stylohyoid are on the lower edge of the great cornu and do not reach the body, but Thyrohyoid, on the deep side of the edge, reaches also the edge of the body. Ligamentous hyo-epiglottic fibres are attached above the thyrohyoid membrane. Chondroglossus, middle Constrictor, and stylohyoid lig. are attached to the small cornu. The central figure shows the hyoid at birth, cartilaginous but for a centre in one great cornu: observe the cartilaginous bar joining the lesser cornua. In the left-hand figure the relation of mucous membrane and Palato-pharyngeus to the bone is indicated.

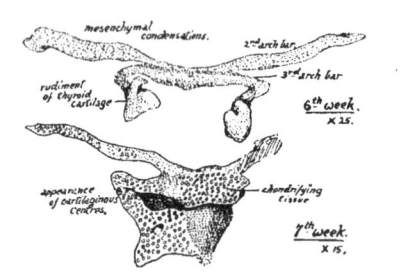

FIG. 218.—Early conditions of the hyoid and thyroid cartilages: from models.

Skull and Hyoid

the bone, in that it does not define any muscular area, but (Fig. 217) is covered by one of these on each side : the ridge connects the regions of the lesser cornua and probably marks the basal part of the second bar. It is sometimes partly separated from the rest of the body. The median vertical ridge, however, may be looked on as a secondary line where it lies below the horizontal ridge ; but the part above this ridge, though possibly in part secondary, may have some atavistic value, for it sometimes shows a short process that has been homologised with the " entoglossal process " found in some lower animals.

The *greater cornu* presents a prominent *angle*, particularly in male bones, at its outer and front part : this is made by the attachment of the expansions of Digastric and Stylo-hyoid.

The lower and back margin has the thyro-hyoid membrane attached to its whole length. . The upper surface has the Middle Constrictor arising from its length, and

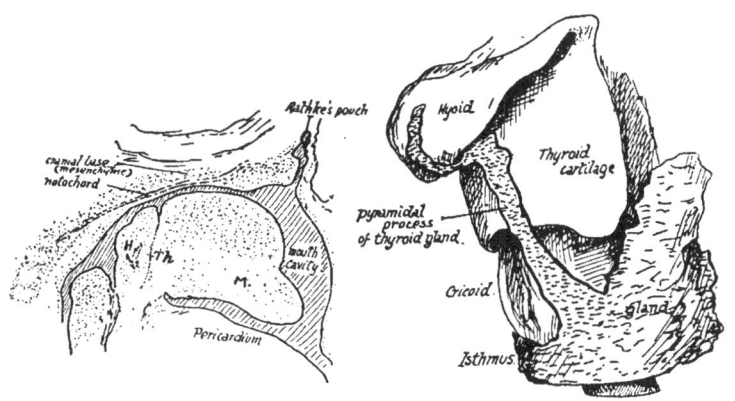

FIG. 216.—1, sagittal section through the region of the mouth in an embryo of the fifth week, showing the track of the median thyroid downgrowth ventral to the hyoid rudiment. *M.* mandibular arch. 2, from a model exhibiting the thyroid stem in position as normally seen, partly wrapped round the hyoid, but properly on the ventral aspect. Early in third month.

the fibres of the muscle cover the extremity of the cornu as they turn down into the wall of the pharynx : all the origin of the Constrictor is hidden by the Hyo-glossus, which also arises from the length of the cornu and extends on to the body for a little distance, lying outside the lesser cornu. The *lesser cornu* and the stylo-hyoid ligament are in the plane of the Middle Constrictor, which arises from them : the Digastric and Stylo-hyoid are superficial to the Hyo-glossus and therefore cannot be attached to the small cornu. Any fibres arising from the small cornu must be deep to Hyo-glossus, and thus, beside the Constrictor, we find the Chondro-glossus arising from the inner aspect and passing up to the deep aspect of the Hyo-glossus.

The muscular areas on the bone are shown in Fig. 217 and a description given of the limits of the areas: observe especially that Hyo-glossus reaches the body, Thyro-hyoid does the same, Genio-glossus has a very small insertion, and Genio-hyoid has the largest insertion on the body.

The bone lies in the level of the second to third cervical vertebræ, and about on a level with the lower margin of the lower jaw when the head is held in the natural position. The body can be felt with difficulty from the surface owing to the covering of muscle fibres, and the same is true of the greater cornu; the small cornu is not palpable. The cornua can be felt from within the pharynx (Fig. 219), but the body is rather hidden from this aspect by the epiglottis: the great cornu lies under the pharyngo-epiglottic fold and the mucous membrane in the upper part of the pyriform fossa, the floor of which rests (below the hyoid) on the thyro-hyoid membrane and upper part of the thyroid ala.

Development.

The mesenchymal structure shown in Fig. 218 is chondrified from centres that appear in it in the fifth or sixth week. A single median cartilaginous centre is found

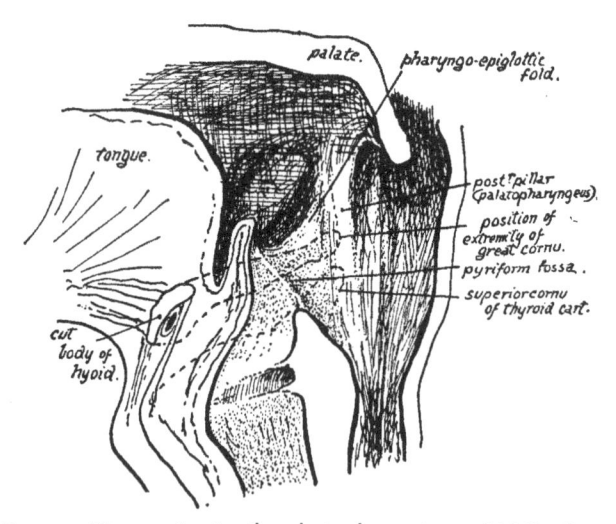

FIG. 219.—Diagram of section through the pharynx in a sagittal direction to show the position of the great cornu and upper cornu of thyroid (interrupted lines) to the folds and fossæ in that region. The cavity is shown opened out. Fibres of Palatopharyngeus run down on the inner side of the extremity of the greater cornu, under the mucous membrane.

early in the back of the rudiment of the body. Other chondral centres appear (as seen in Fig. 218) in the bars of the second and third arches. The whole structure is chondrified from these centres during the third and fourth months, and at birth (Fig. 217) bony centres have just appeared near the ventral ends of the great cornua. The body is ossified from a single (? paired) centre showing shortly afterwards, but the small cornua apparently do not commence to ossify until just before puberty. The extremity of each great cornu remains cartilaginous until nearly middle life: epiphyses have been described for these ends.

The body and great cornua may fuse after middle life; the small cornu may join the greater, or in rare cases may fuse with the body.

INDEX

The names of parts of bones are not given separately in the index, except in special cases, being included in the names of the bones of which they are parts.

The names of separate vessels, muscles, nerves and ligamentous bands are given grouped under the titles arteries, veins, muscles, nerves, *and* ligaments.

ABDOMINAL wall, attached to ilium, 134
Acetabulum, 118, 125, 127
Acromial angle, 69, 74
Adductor region, 129
   tubercle, 138, 150
Age of skulls, 270
Air-cells
 frontal, 200
 mastoid, 217
 maxillary, 229
 nasal, 247
 sphenoid, 219
Alæ of sacrum, 39
Alcock's canal, 133
Alveoli, formation of dental, 232
Alveolo-labial sulcus, 256
Anal fascia, 132
Angle of jaw, 254, 259
  Louis, 57, 60
 formed opposite third rib, 59
Angle or angles:
 ankle movement, 176
 astragalus, neck, 177
 facial, 270
 feet, 187
 femur, neck, 138, 142
 humero-trochlear, 89
 jaw, 260, 268
 lumbo-sacral, 40
 pelvic, inlet and outlet, 122
 pubic arch, 122
 sternum with vertebral column, 59
Annulus fibrosus, 11
Ano-coccygeal ligament, 45
Anterior angles of ribs, 48, 50
  common ligament, 14
Anthropoid foot, 187
  foramen magnum, 270
  sacrum, 42
  skull, 271
Antrum of Highmore, 231, 250
   development, 233
  mastoid, 217
Aortic impression, 28
Arches of foot, 173
Arnold's nerve, 211
Artery or arteries:
 "acromial rete," 74
 acromio-thoracic axis, 76
 articular of knee, 148, 160, 162
 axillary, 65, 68
 carotid, external, 257
  internal, 211, 225
 circumflex, anterior, 85
  external, 127, 146
  internal, 130, 146
  posterior, 83

Artery or arteries—*continued.*
 deep palmar arch, 108, 113
 dorsalis pedis, 171
   communicating, 184
   scapulæ, 73
 external circumflex, 127, 146
 facial, 256
 ilio-lumbar, 41
 inferior profunda, 88
 intercostal, 28, 51
 internal circumflex, 130, 146
  mammary, 56, 57, 59
  maxillary, 236
  pudic, 129, 133
 interosseous, anterior, 92, 99
   posterior, 92, 98
 masseteric, 258
 middle meningeal, 202, 216
 nutrient of femur, 146
 obliterated hypogastric, 131
 obturator, 130
 occipital, 207, 212
 ophthalmic, 196
 perforating, 146
 peroneal, 167
  anterior, 162, 167
 popliteal, 146
 posterior interosseous, 92, 99
  palatine, 231, 235
 radial, 96
 sacral, lateral, 42
  middle, 41
 superior epigastric, 56
  profunda, 86
 supraorbital, 200
 suprascapular, 68, 72, 76
 tympanic, 216
 ulnar, 95, 97
 vertebral, 25, 26
Articular masses, cervical, morphology, 21
 processes, cervical, 25
  affecting transverse process, 25
  inhibit rotation, 34
Articulations, neural, 25
  neuro-central, 25
  varieties, 1
Asterion, 198
Astragalus, 175
Atlas, 16, 24
 anterior tubercle, 24
 at birth, 37
 oblique ligament of, 26
 venous notch, 26
Atmospheric pressure in hip, 140
Auricular point, 198
  surface, 120
Axilla, 77, 85

Axis, at birth, 24, 37
  description, 17, 22
Axis of orbit, 253
  pelvic cavity, 123

Back, 28
Ball of toe, 184
Base of skull, 191
Basi-hyal, 215, 271
Bertin, bones of, 227
Bicipital tubercle, 70, 74
Bigelow's ligament, 126, 140
Bone, composition, 2
  varieties, 2
Bones, growth, 4
  moulding, 5
  ossification, 3
  parts of, 2
  structure, 2
  varieties, 2
  vascular supply, 4
Brachycephaly, 270
Brain, relation to skull, 204, 207
Bregma, 198
Brim of pelvis, 121
Bulla ethmoidalis, 247
Bursæ (named or in relation with):
  acromion, 67, 74
  adductor magnus, 144
  anterior annular ligament of ankle, 178
  axis, 17
  biceps humeri, 84, 97
  brachialis anticus, 95
  calcaneum, 179
  carpus, 104
  clavicle, 68
  coracoid, 74, 78
  crucial ligaments, 149
  crureus, 145
  deltoid, 84
  digital, 115
  extensors at wrist, 99
    of carpus, 113
  femur, 143, 144, 145, 149
  flexors, at wrist, 104
    in fingers, 110, 115
    radial, 109
    ulnar, 97
  gluteal, 143, 144
  hyoid, 272
  infrapatellar, 153, 160
  intertubercular, 84
  ischium, 128
  metacarpus, 113
  oblique ligament, 97
  obturator internus, 128, 130
  odontoid, 21
  olecranon, 95
  peronei, 169, 179, 180
  phalanges of hand, 115
  popliteal, 157
  psoas, 126
  radius, 97, 99
  sartorius, 163
  semimembranosus, 162
  subacromial, 66, 74, 83
  subcoracoid, 72, 75
  subcrureal, 144, 145
  subdeltoid, 83
  subpsoas, 126

Bursæ (named or in relation with)—continued.
  subscapular, 72, 82
  supinator brevis, 97
  suprapinatus, 66, 73
  tendo Achillis, 179
  tensor palati, 225
  tibialis anticus, 182
    posticus, 156
  trapezium, 109
  trapezius, 73
  triceps, 95
  trochanters, 143, 144
  ulna, 95, 97
  unciform, 175

Calcanean process of cuboid, 181
Calcar femorale, 144
Canalis innominatus, 221, 265
Carotid sheath, 25
  tubercle, 21
Carpal concavity, 104
Carpus, 101, 103, 105
  added elements, 116
  ossification, 116
Cartilages, 1
Cartilaginous skull, 262
Cavernous sinus, 225
Centres of ossification, 3
Centrum, 11, 28
Cerato-hyals, 215, 271
Cerebellum, 196, 207
Cerebral cavity, 194
Cerebrum and frontal, 201
  occipital, 207
  parietal, 204
  temporal, 214, 216
Cervical nerves, 21
  rib, 16
  vertebræ, 15, 20
Chassaignac's tubercle, 21
Choanæ, 268
Chondro-cranium, 262, 265
Clavicle, 65
  comparative, 69
  morphology, 68
  on sternum, 58
  ossification, 69
Cleft palate, 235
Coccyx, 44
  ossification, 45
Colles' fascia, 132
Conjoint tendon, 134
Conjugate of pelvis, 123
Cornua, coccyx, 44
  sacrum, 38
Coronal suture, 189, 203
Coronoid process, attachments, 95
Costal arch, 47, 55
  cartilages, 47, 55, 58
    first, 56
  element in vertebræ, 13
    cervical, 16, 21
    lumbar, 34
    sacrum, 40, 43
  facets, 26, 29, 30
Costo-coracoid membrane, 55, 65, 68, 76
  transverse bar, 13, 25
Cotyloid fossa, 126
  notch, 118
Cranial capacity, 270

Index

Cranial sinuses, 196
  skeleton, 189
Cranio-pharyngeal canals, 227
Cranium, 189
Crista colli (rib), 45, 50
Crus penis, 132
Cuboid, 180
Cuneiform, foot, 181
  hand, 102, 108
Cupola of nasal capsule, 228
Curved lines on ilium, 127
Curves in ribs, 46
  spine, 13

DEEP transverse fascia (leg), 164, 167
Dens, 17, 23
Dermal bones in skull, 265
Development of acetabulum, 126
  astragalus, 177, 186
  carpus, 116
  cervical articulations, 24
  clavicle, 69
  femur, 150
  fibula, 169
  humerus, 89
  ligamentum teres, 126
  limbs, 63
  metacarpus, 116
  metatarsus, 186
  os innominatum, 137
  palate, 232
  patella, 154
  phalanges, foot, 187
  hand, 11
  radius, 100
  ribs, 55
  sacrum, 43
  scapula, 79
  skull, 262
  sternum, 59
  tarsus, 186
  tibia, 169
  transverse tubercles, 35
  ulna, 100
  vertebræ, 36
Diameters of pelvis, 123
Diaphragma sellæ, 219
Diaphysis, 4
Digital fossa, femur, 138, 144
  fibula, 157
Diploe, 196
Dissections studied with bones, 7
Dolichocephaly, 270
Dolichohierism, 42
Dorsal vertebræ, 26, 28
Dorsum sellæ, 194, 219
Ductus cochleæ, 213
  endolymphaticus, 212
Dura mater, with sphenoid, 221
  temporal, 214

EAR region, 209
Early skull, 262
Ectochondral growth, 3
Emarginated patella, 152
Endochondral growth, 3
Endognathion, 234
Endothoracic fascia, 51
Entoglossal process, 273

Epiglottis, 274
Epihyal, 215
Epiotic, 218
Epiphyseal lines, 3
  plate, 11
Epiphyses, 3
Epipubis, 137
Episternal bar, 59
Ethmoid, 237
  ossification, 240
Ethmoidal cells, 247
  notch, 200
  process, 242
Eustachian spine, 225
  tube, 208
  development, 217, 222, 226
Examination of carpus, 110
  femur, 150
  fibula, 168
  humerus, 88
  os innominatum, 137
  radius, 100
  ribs, 49, 50
Exognathion, 234
Extradental sulcus, 256

FACE, 228
Facets on ribs, 52
Facial skeleton, 188, 196, 264
Factors in moulding bones, 5
Falciform edge on ischium, 123, 132
  process of sciatic ligament, 132
False pelvis, 121
Falx cerebri, 195, 201, 207
Femoral sheath, 131
  tubercle, 138, 143
Femur, 138
  articulations, 140, 148, 158
  ossification, 151
Fenestra ovalis, 217
  rotunda, 217
Fibrous attachments, effects on bones, 6
Fibula, 154, 164
  articulations, 157, 167
  ossification, 169
Filum terminale, 42, 44
Fingers, 115
  in examining bones, 6
Floor of pelvis, 120, 131
Fœtal skull, 267
Fontanelle, metopic, 202
Fontanelles, 266
Foot, 169
  ligamentous outer part, 174
  position of, 187
Foramen lacerum, 193, 195, 225
  magnum, 191, 202, 270
Foramina in skull, classification, 265
Forearm, 92
Forehead, 199
Fossæ of skull, 194
Fovea femoris, 141
  humeri, 82
Frontal, 198
  fœtal, 269
  ossification, 202
  sinus, 200, 250
  development, 202
Fronto-nasal process, 232
  parietal suture, 189

GASSERIAN ganglion, 221
Genial tubercles, 254
Gimbernat's ligament, 120, 130
Glabella, 198
Glaserian fissure, formation, 209
    structures in, 216
Gluteal aponeurosis, 127
    planes, 127
    region, 127
Gorilla, sacral index, 42
Great sciatic ligament, 122, 127, 141
    morphology, 135
    notch, 136
Growing ends of bones, 4

HAMULAR process, lachrymal, 242, 253
    pterygoid, 224
Hand, 101
Hard palate, 235, 268
Haversian gland, 126
Head, 188
Hiatus sacralis, 38, 42
    semilunaris, 247
Hip and shoulder contrasted, 141
His, canal of, 272
Homologies of pelvic and pectoral regions, 136
Homology between human and animal pelves, 136
Huguier, canal of, 216
Human characters in pelvis, 136
Humerus, 80
    ossification, 89
Hüschke, foramen of, 218
Hyoid, 271
    embryonic, 272
    ossification, 274
Hypochordal bar, 37
Hypoischium, 134, 137
Hypothenar eminence, 104, 108
Hypotrochanteric fossa, 145

ILIAC crest, 133
    fascia, 130
    fossa and region, 130
Ilio-pectineal eminence, 118
    line, 118, 125
    trochanteric band, 127, 143
Illustrations, use of, 8
Incisive papilla, 235
    suture, 234, 235
Incisura ethmoidalis, 200
Infantile foot, 187
Inferior petrosal sinus, 208, 213, 225
    turbinate, 242
    ossification, 243
Infraorbital suture, 233
Infrapatellar pad, 153, 160
Infratemporal fossa, 229
Inion, 198
Injuries, effect of blood supply in, 4
    of acromion, 73
    clavicle, 67, 76
    growing ends, 4
    vault of skull, 4
Innominate bone, 118, 137
    ossification, 137
Intercondylic region, 149
Intercostal membranes, 52, 56
Intermuscular septa, femur, 146

Intermuscular septa, humerus, 81, 87
Internal brachial ligament, 87
Interosseous membrane, forearm, 92
    leg, 156, 165
Interparietal, 208
    suture, 189
Intersegmental position of vertebræ, 37
Intertrochanteric lines, 144
Intervertebral discs, 9
    development, 37
    in sacrum, 42
    foramen, 10
    notches, 10
Intrajugular process, 206
Ischio-capsular band, 127
    pubic region, 129
    rectal fossa, 132
Ischium, 118
Iter chordæ (anterius et posterius), 217

JACOBSON'S cartilage, 244
    nerve, 211
Joints:
    ankle, 167
    carpo-metacarpal, 102, 111
    costo-vertebral, 51, 52, 53
    elbow, 94
    hip, 126, 140
    interphalangeal, 115
    jaw, 215, 257
    knee, 149, 153, 158
    metacarpo-phalangeal, 114
    metatarso-phalangeal, 185
    midcarpal, 103
    midtarsal, 170, 179
    occipito-atlo-axoid, 18
    sacro-iliac, 40, 125
    shoulder, 81
    sterno-clavicular, 66
    vertebral, 14
Jugular foramen, 193, 206, 211
    notch, 58
    tubercle, 206
    vein and atlas, 25
Jugum sphenoidale, 195, 219

KERCKRING, ossicle of, 208
Knee-joint, 149, 153, 158

LABYRINTH of ethmoid, 238, 239, 246
Lachrymal bone, 238, 239, 269
    ossification, 240
    crest, 239
    gland, 200
    process, 241, 245
    sac, 240, 251
Lambda, 198
Lambdoid suture, 189, 203
Lamina papyracea, 237
Laminæ in cervical region, 22
    markings on, 22
Lateral curve in spinal column, 14
    mass, atlas, 238
    ethmoid, 245
    sinus, 195, 196, 207, 208
Levatores longiores, 51
Ligaments:
    acromio-clavicular, 67, 74

Ligaments—*continued.*
  alaria, 149, 154
  of ankle, 164, 185
  annular, foot, 164—178
    hand, 100, 105, 106, 107
  anterior common, 20, 33, 41, 44
  astragalo-scaphoid, 176
  atlo-axoid, 17, 20
  axial, 17
  Bigelow's, 126
  calcaneo-navicular, 177, 179, 181
  carpal, 103, 105
  chondro-sternal, 56, 58
  coraco-acromial, 74, 76
    clavicular, 67, 76, 78
    humeral, 76, 80, 82
  costo-lumbar, 33
    transverse, 51, 52, 53
  cotyloid, 126
  crucial, 148, 158
  cruciate, 88
  cruciform, 17
  cubo-navicular, 174, 181
  of elbow, 86, 88, 93, 94
    foot, 174
  Gimbernat's, 119, 130
  gleno-humeral, 72, 82
  glenoid, 75, 82
  great sciatic, 40, 44, 122, 127
  of hip, 140
  hyo-epiglottic, 273
  ilio-femoral, 126, 140, 141
    lumbar, 33, 41, 134
  interclavicular, 57
  interclinoid, 217, 220, 225
  interlaminar, 14
  intermetatarsal, 184
  internal brachial, 87
  interosseous, arm, 98
    leg, 157, 165
  interspinous, 15
  intertransverse, 15
  of jaw, 257
    knee, 149, 158
  lateral of ankle, 164, 167, 176
    elbow, 87, 88, 92
    jaw, 216, 258
    knee, 149, 160, 162, 163, 164
    wrist, 103
  mandibular, 258
  Meckelian, 216
  metacarpo-phalangeal, 114
  mucosum, 149, 154
  nuchæ, 22, 207
  oblique, 92, 98
  occipito-odontoid, 17, 18, 19
  orbicular, 92, 96, 97
  palpebral, 200
  patellar, 151, 159
  piso-uncinate, 105
  plantar, 174, 180
  Poupart's, 119, 129
  pterygo-spinous, 224
  pubofemoral, 128, 140
  radio-ulnar, 92
  rhomboid, 56, 66, 68
  sacro-iliac, 40, 125
  sciatic, 41, 44, 122, 126, 130
  of shoulder, 82
    sole, 174
  spinal, 14, 17, 22

Ligaments—*continued.*
  spino-glenoid, 73
  stellate, 28, 51
  sternoclavicular, 65
  stylo-hyoid, 271
  stylo-mandibular, 258
  subflava, 14, 22, 31, 34
  subpubic, 122
  supraorbital, 200
  suprascapular, 75
  temporo-mandibular, 215
  teres, 126, 141
  tibio-fibular, 151, 160
  transverse, of ankle, 164
    axial, 17
    hip, 126
    humeral, 82, 84
    metatarsal, 185
  triangular, 132
  vertebral, 14, 17, 22
  Winslow's, 149, 158
Limb-girdles, 62
Limbs, formation of, 62, 63, 64
  homology in, 62, 78
  morphology of, 62
  parallelism between, 62
  position of, 64
  rotation of, 64
Linea alba, 57
  aspera, 151
  suprema, 207
Loins, 33
Longitudinal sinus, 201, 203, 207
Lower extremity, 118
  jaw, 254
Lucas, groove of, 223
Lumbar vertebræ, 32
  ossification, 35
  vessels, 33
Lumbo-sacral angle, 13, 40

MALAR, 251
  ossification, 253
Mammillary processes or tubercles, 32 35
  attachments, 34
  on sacrum, 39
Mandible, 253
  ancient, 260
  differences with age, 260
  digital examination of, 261
  ossification, 261, 265
Marrow, red and yellow, 2
  in thoracic bones, 57
Maxilla, 218, 236, 267
  ossification, 232
Maxillary process, 264
  of inferior turbinate, 232, 242, 245
  sinus, 250
Measurements and proportions. *See also under* ANGLES.
  acetabular parts, 125
  acetabulo-pubic, 135
  astragalus, articular, 176
  bases of skulls, fœtal and adult, 267
  cephalic index, 270
  cervical bodies, 15
  cranial capacity, 270
    measurements, 270
  dorsal bodies, 28

Measurements and proportions—*continued*.
  face, during skull-growth, 269
  femur, length, 139
         neck, 142
  mandible, 268
  orbit, 252
  ribs, proportion of parts, 52
  sacral index, 42
  skull, growth and weight, 270
         ratio in sexes, 270
  thoracic index, 47
         measurements, 47
  upper limb, 64
Meatus, external, 216
         in infant, 269
         internal, 215, 217
Meatuses of nose, 248
Meckel's cartilage, 208, 216, 261
Median eye, 203
Megacephaly, 270
Membrana sacciformis, 92
           tectoria, 19
Membrane bones, 3
           in skull, 272, 275, 277
Meniscus, jaw, 215
          knee, 158
          sterno-clavicular, 66
          wrist, 95
Mental prominence, 254, 259, 261
Mesaticephaly, 270
Mesocephaly, 270
Mesognathion, 234
Mesoscapula, 79
Metacarpals, 110
  distinction from metatarsals, 105
  ossification, 116
Metacarpus, 105, 109
Metacromion, 74
Metatarsus, 170, 175, 183
Metopic suture, 190, 200, 202
Microcephaly, 270
Mirror relation between limbs, 65
Morphological neck, 83
Mouth, movement, 258
       position of, 271
       region, 254, 268
Muscle-markings on bones, 6
Muscle or muscles:
  abductor metatarsi quinti, 185
  adductors, thigh, 119, 128, 146
           thumb, 113
  anconeus, 88, 95
  azygos uvulæ, 234
  biceps, arm, 75, 80, 82, 84, 97
         thigh, 134, 146, 160, 162
  brachialis anticus, 86, 95
  buccinator, 229, 256
  chondroglossus, 273
  coccygeus, 41, 132
  complexus, 207
  compressor urethræ, 132
  constrictors, 224, 256, 271
  coraco-brachialis, 85
  corrugator supercilii, 200
  deltoid, 66, 68, 71, 73, 78, 85
  diaphragm, 33, 56, 58
  digastric, 208, 254, 257, 271
  erector spinæ, 40, 48, 52, 55, 119, 133, 134
  extensor brevis digitorum, 177
           pollicis, 99

Muscle or muscles—*continued*.
  extensor carpi radialis brevior, 99, 111
                         longior, 86
                 ulnaris, 97, 115
           of foot, 162
           indicis, 99
           longus digitorum, 162, 165, 167
                 pollicis, 99
           ossis metacarpi pollicis, 99, 109
  external oblique, 49, 54, 56
  facial, 218, 251, 255
  flexor accessorius, 179
         carpi radialis, 106, 109
               ulnaris, 95, 97, 108
         longus digitorum, 163
               hallucis, 165, 176, 179
               pollicis, 91
         profundus digitorum, 74
         sublimis digitorum, 87, 98, 115
  gastrocnemius, 148, 159, 179
  genial, 254, 273
  gluteal, 127, 140, 143, 145
  gluteus maximus, 41, 44, 127
  gracilis, 129, 163
  hyo-glossus, 255, 271
  hypothenar, 104, 107, 110
  iliacus, 129, 144
  ilio-costalis, 51, 54
  inferior oblique, 23, 25
  infrahyoid, 271
  infraspinatus, 71, 82, 83
  intercostals, 51, 52, 54
  internal oblique, 55
  interosseous, 110, 112
  latissimus dorsi, 49, 55, 76, 83
  levator anguli scapulæ, 23, 25, 77
          ani, 120, 133
          costæ, 52, 53, 55
          palati, 213, 225
  longissimus dorsi, 34, 41, 52
  masseter, 215, 252, 254
  multifidus spinæ, 35, 41
  mylo-hyoid, 255
  obliqui abdominis (pelv.), 130
  obturator externus, 128, 142, 144
            internus, 120, 131, 132, 143
  occipito-frontalis, 198, 202, 216
  omo-hyoid, 77, 271
  orbicularis oris, 228, 256
              palpebrarum, 198, 251
  orbital, 200, 223, 229, 251
  palatopharyngeus, 274
  pectineus, 130, 140, 146
  pectoralis major, 49, 53, 55, 57, 66, 68, 84
              minor, 53, 76
  perineal, 132
  peroneus brevis, 169, 180, 185
           longus, 162, 164, 172, 179, 182, 183
           tertius, 167, 185
  plantaris, 158, 179
  platysma, 68, 256
  popliteus, 140, 160, 162
  post-vertebral, 22, 30
  prevertebral, 20
  pronator quadratus, 96, 98
           teres, 87, 98
  psoas magnus, 33, 34, 41, 129, 140, 144
        parvus, 131
  pterygoids, 221, 223, 233, 254, 257
  pyriformis, 41, 128, 143

Index

Muscle or muscles—*continued.*
  quadratus lumborum, 34, 55, 133
  quadriceps, 145, 151
  recti antici, 20, 206
    postici, 207
  rectus abdominis, 49, 56, 58, 134
    femoris, 126, 139
  rhomboids, 50, 70, 77
  sartorius, 134, 158
  scalenus anticus, 20, 49
    medius, 23, 51
    pleuralis, 53
    posticus, 49, 55
  semimembranosus, 134, 159
  semitendinosus, 136, 159
  serrati postici, 48, 54, 55
  serratus magnus, 48, 49, 55, 70, 78
  of sole, 179
  soleus, 163, 164, 166, 179
  sphincter ani, 45
  splenius, 23, 25, 207, 216
  sterno-hyoid, 59, 64, 67
    mastoid, 58, 67, 207, 216, 258
  sterno-thyroid, 56, 58
  stylo-glossus, 215, 257
    hyoid, 215, 257, 271
    pharyngeus, 215, 256
  subclavius, 56, 66, 67
  subscapularis, 49, 71, 76, 82, 83, 89
  superior oblique, 207
  supinator brevis, 95, 98
    longus, 86, 99
  supraspinatus, 67, 72, 73, 81, 82, 83
  temporal, 200, 202, 216, 223, 251, 258
  tensor fasciæ, 127, 134
    palati, 217, 221, 225, 237, 257
    tarsi, 241, 253
  teres muscles, 73, 81, 83, 84
  thenar, 104, 106, 108
  tibialis anticus, 162, 178, 182, 183
    posticus, 165, 168, 173, 177
  transversalis, 34, 54, 55, 134
  trapezius, 66, 68, 71, 73, 207
  triangularis sterni, 56, 59
  triceps, 69, 72, 83, 86
  vasti, 145, 152, 160, 163
Muscles, covering thorax, 48
  effect on bones, 5
Myrtiform fossa, 229

Nares, 250, 268
Nasal aperture, 250, 268
  bone, 243
    ossification, 243
    fœtal, 269
  capsule, 240, 243, 262, 264
    bones in relation with, 227, 235, 240
  fossa, 244
    floor, 248
    fœtal, 277
    relation to antrum, 250
    roof, 249
Nasion, 198
Neck, 20, 207
Nerve or nerves
  anterior tibial, 167
  Arnold's, 211
  auriculo-temporal, 222
  carotid plexus, 211, 224
  cerebral, 275

Nerve or nerves—*continued.*
  cervical, 21, 23
  chorda tympani, 216, 222
  circumflex, 83, 86
  coccygeal, 41, 44
  dental, 230, 250, 258
  descending cervical, 68
  digital, 114
  external popliteal, 167
  facial, 212
  glosso-pharyngeal, 211
  gluteal, 129
  great occipital, 23
    superficial petrosal, 214, 224
  ilio-hypogastric, 33
    inguinal, 33
  intercostal, 51, 56
  lingual, 256, 257
  mandibular, 221
  masseteric, 215, 222, 258
  maxillary, 221, 225, 237
  Meckel's ganglion, 237
  musculo-cutaneous, arm, 87
    leg, 167
    spiral, 86
  mylo-hyoid, 148, 258
  nasal, 238
  naso-palatine, 231, 241
  obturator, 131
  to obturator internus, 128
  olfactory, 238
  ophthalmic, 221, 225
  optic, 195, 219
  perineal, 133
  posterior interosseous, 99
    palatine, 234
  to quadratus femoris, 128
  sacral, 41
  sciatic, 127, 135
  sixth, 214, 225
  spheno-palatine, 236
  spinal accessory, 25, 211
  suboccipital, 25
  supraorbital, 200
  suprascapular, 76
  temporal, 222
  temporo-malar, 237, 250, 253
  ulnar, 86, 88, 96, 97
  vagus, 211
Nerves in psoas, 33
Neural canal, 19
  groove, 21
Neuro-central lip, 21
  suture, 11
  synovial cavity, 21
Norma basalis, 191
  frontalis, 189
  lateralis, 189, 190
  occipitalis, 189
  verticalis, 189
Notochord, 37
Nucleus pulposus, 11, 37
Nutrient foramina, 4

Oblique ligament (Cooper), 94
  line, analysis, 163
Occipital, 204
  at birth, 269
  ossification, 208
Odontoid process, 17, 23

Index

Olecranon, attachments, 95
Olfactory pit, 264
Omo-hyoid fascia, 59, 68, 77
Opisthotic, 218
Orbit, 253
    fœtal, 268
Orbital process (malar), 251
Os acetabuli, 125
    calcis, 177
    epiptericum, 227
    innominatum, 118, 124
    interparietale, 227
    magnum, 102
    planum, 229, 246
    trigonum, 176
Osseous cranium, 265
Ossicle of Kerckring, 208
Ossicles of ear, development, 209, 210
Ossification, centres of, 3
    earlier in women, 100, 116, 117, 187
    of atlas and axis, 37
        cervical vertebræ, 36
        clavicle, 68
        coccyx, 44
        femur, 150
        fibula, 169
        humerus, 89
        lumbar vertebræ, 36
        metatarsus, 186
        os innominatum, 137
        patella, 154
        phalanges, foot, 187
        radius, 100
        rib, 55
        sacrum, 43
        scapula, 78
        sternum, 60
        tarsus, 186
        tibia, 169
        ulna, 100
    process of, 3
Osteoblasts, 3
Osteoclasts, 3
Otocyst, 208, 212

PACCHIONIAN bodies, 201, 203
Palate bone, 233, 235, 239
    ossification, 234
    fœtal, 268
    folds, 232
    hard, 235
Palm, structure of, 104
Parachordals, 262
Parietal, 202
    eminence, fœtal, 269
    ossification, 202
    pelvic fascia, 41, 127
Parieto-occipital suture, 189, 203
Paroccipital process, 206
Parotid, 258
Patella, 152
    ossification, 154
Pectoral girdle, 62, 64, 78
Pelvic axis, 123
    cavity, 121
    fascia, 41, 131
    floor, 41, 44
    girdle, 62, 78
    inlet and outlet, 121

Pelvis, 118, 121
    of reptiles, 136
    sexual differences, 122, 136
Penis, relation to pubis, 135
Perineal region, 132
Periods in skull ossification, 265
Periosteum, 2
Periotic capsule, 262
Peritoneum, with innominate, 131
    sacrum, 41
Peroneal septa, 162, 167
    tubercle, 179
Phalanges, foot, 170, 185
    hand, 105, 115
    ossification, 117
Pharyngeal derivatives in skull, 263
    tonsil, 206
    tubercle, 206
Pharyngo-epiglottic fold, 274
Pharynx and hyoid, 271, 273
    lateral recess, 211
Pisiform, 102, 108
Pituitary fossa, 194, 219
    gland, 194, 227
Plantar ligaments, 174
Platyhierism, 42
Pleistocene or pliocene jaws, 260
Pleura on ribs, 47, 50, 52, 54, 56
    sternum, 50
Popliteal notch, 155
Postaxial and preaxial borders, 64
Posterior common ligament, 14
    intercostal membrane, 50
    nasal spine, 235
Post-meatal spine, 217
Postscapula, 79
Poupart's ligament, 119, 130
Preauricular groove, 136
Prechordal part of skull, 262
Preinterparietals, 208
Premaxilla, 229, 232, 235, 264
Premaxillary suture, 232
Prepubis, 137
Prescapula, 79
Pretracheal fascia, 59
Prevertebral muscles, 20
Primary ridges on bones, 6
Prognathism, 267
Prominence of fourth lumbar in women, 13
Promontory of sacrum, 40
Pronator ridge, 96
Pro-otic, 218
Proportions. *See under* MEASUREMENTS.
Pterion, 198
Pterotic, 218
Pterygo-maxillary fissure, 236
    palatine canal, 225
Pterygoid fossa, 224
    region, 221, 257
Pubic part of fascia lata, 130
Pubis, 118
Pulmonary groove, 49
Pyriform aperture, 250
    fossa, 274

QUADRATE of birds, 264
    tubercle, 138, 144

RACIAL differences in nares, 251
    skull, 270

Index

Radius, 89, 97
    axis of movement, 94
    ossification, 100
Rathke's pouch, 227
Region of foot, 171
    hip, 140
    knee, 149
    leg, 160
    spine, 11
    thigh, 145
Reid's base line, 198
Retinacula of femur, 142
" Retinacula patellæ," 153
Ribs, 46, 51
    ossification, 57
Riolan, bones of, 217
Rolandic fissure and parietal, 204
Rosenmüller, fossa of, 211
Rostrum, 220, 244
Rotation of fœtal head, 126
    radius, 95
    scapula, 78
    vertebræ inhibited, 34
Round ligament of uterus, 131

SACRAL canal, 38, 41, 42
    groove, 38
    index, 42
    notch, 38
Sacrum, 37, 42, 43
Sagittal suture, 189, 202
Scalene tubercle, 49
Scaphoid, foot, 181
    hand, 102, 105
Scapula, 69
    in relation with thorax, 50, 77
    ossification, 79
Scapular index, 78
    rotators, 77
Sciatic foramen, 122
Sclerotome, 37
Semicircular canals, 217
Semilunar bone, 102, 106
    cartilages, 149, 158
Sense capsules, skull, 274
Septum nasi, 243
    edge, 241, 243
Sesamoids in foot, 184
    great toe, 172, 183
    hand, 114
    peroneus longus, 180
Sexual differences, coccyx, 44
    femur, 141
    foot, 178
    sacrum, 40, 42
    skull, 270
    sternum, 59
Shape of bones, factors in, 5
Shoulder, 73, 80, 82
Sibson's fascia, 55
Sinus tarsi, 171, 175, 177
    tympani, 217
Skeleton, functions, 1
    study of, 5
Skull, 188
    bones, 198
    development, 261
    growth of, 266
    movements on column, 18
Specimens, value of for comparison, 7

Spheno-ethmoidal recess, 247
    maxillary fissure, 223, 236, 251
        fœtal, 268
    fossa, 236
    palatine foramen, 234, 236
Sphenoid, 227, 219
    ossification, 226
Sphenoidal turbinal, 219, 226, 253
    sinus, 249
Spinal canal, 9, 15, 22
    column, 9, 11, 12, 13, 14
    cord, 15
    foramen, 9, 28, 32
Spinous processes, 14, 32
Spiral line, femur, 144
Sternal plates, 60
Sternum, 57
    ossification, 61
Straight sinus, 207
Structure of bones, 2
    femur, 145
    fibula, 169
    foot, 172
    hard palate, 232, 235
    labyrinth of ethmoid, 238, 246
    mandible, 260
    nasal fossæ, 244
    orbit, 253
    os innominatum, 118, 125, 137
    sacrum, 46
    scapula, 77
    skull, 188, 196, 262
    spheno-maxillary fossa, 236
    sternum, 58
    tarsal bones, 187
    temporal, 208
    vertebræ, 10, 36
Study of skeleton, 5
Stylo-hyal, 215
Subacromial bursa, 66
Subcostal angle, 47, 55
Sublingual gland, 255
Submaxillary gland, 255, 256
Subpsoas bursa, 126
Subpubic arch, 122
Subsagittal suture of Pozzi, 202
Subscapular fossa, formation of, 72
Supracondylar process, humerus, 87
Supracostal groove, 53
Surface texture of bones, 6
Sustentaculum tali, 178
Symphysis pubis, 134

TABLES of skull, 196
Tarso-metatarsal line, 171, 184
Tarsus, 169, 171, 175
Teeth, effect on antrum, 251
    shape of jaw, 260
    relations between jaws, 231, 259
    to lower jaw, 259
    upper jaw, 231
Temporal, 208
    fossa, 191, 200, 202, 223, 252
    ossification, 218
Tendo-oculi, 231, 254
Tentorium cerebelli, 195, 207, 214
Theca of fingers, 115
Thenar eminence, 104, 106
Thoracic index, 48
Thorax, 46, 48

Thumb, movements in, 102, 114
Thyreo-glossal duct, 272
   hyal, 271
   hyoid membrane, 272
Thyroid foramen, 119, 129, 136
   gland, cysts, with hyoid, 272
Tibia, 154, 160
   ossification, 169
Torcular Herophili, 207
Torus occipitalis transversus, 207
   palatinus, 235
Trabeculæ cranii, 262
Transverse processes, on last two dorsal, 39
   lumbar, 33
   tubercles on cervical, 24, 25
   width, 14
Trapezium, 102, 108
Trapezoid, 102, 109
Triangular fibro-cartilage, wrist, 92, 100
Trochlear fossette, 200
   surface, femur, 139, 149
True pelvis, 121
Tubercle and thoracic index, 48
Tubercles on transverse processes, 35
   cervical, 16, 20
   relations of muscles and nerves, 16, 20
Tuberosity of ischium, 133
   maxilla, 230
   palate, 233
Tubo-tympanic recess, 208
Turbinate, ethmoidal, 239, 245
   inferior, 242, 245
   maxillo-, 242, 245
   sphenoidal, 253
Tympanic membrane, 269
   plate, 264, 269
   ring, 209
Tympano-hyal, 215, 217
Tympanum, 217
   development, 209

ULNA, 89, 94
   ossification, 100
   sexual differences, 97
Unciform, 102, 110
Uncinate process, ethmoid, 235, 239
   on ribs, 54

Upper arm, 84
   limb, 65

VAGINAL plates, pterygoid, 225, 241
Vas deferens, 131
Vascular spot, temporal, 216
Veddah foot, 187
Vein connecting cephalic and jugular, 68
Veins:
   diploic, 196, 207
   external iliac, 131
   jugular, 196, 211
   ophthalmic, 236
   spinal, 22, 28
   subclavian, 56
   suprascapular, 76
Venæ basis vertebræ, 22
Ventral bar of scapula, 71
Vertebræ, false, 11
   postsacral, 12
   presacral, 11
   true, 11
   typical, 9
Vertebral aponeurosis, 40, 51
   column, 9
   groove, 14
Vertebrarterial foramen, 16, 22
Vertebrate skull, 265
Vessels. *See under* ARTERIES.
Vidian canal, 225, 236
Visceral skeleton in hyoid, 271
   skull, 263, 264, 266
Vomer, 241
   ossification, 241
Vomerine cartilage, 244
   crest, 232

WEIGHT transference in pelvis, 124
Wormian bones, 189, 202, 276

XIPHOID cartilage, 58
   development, 69

ZYGOMATIC arch, 190, 216
   fossa, 229

University of Toronto Library

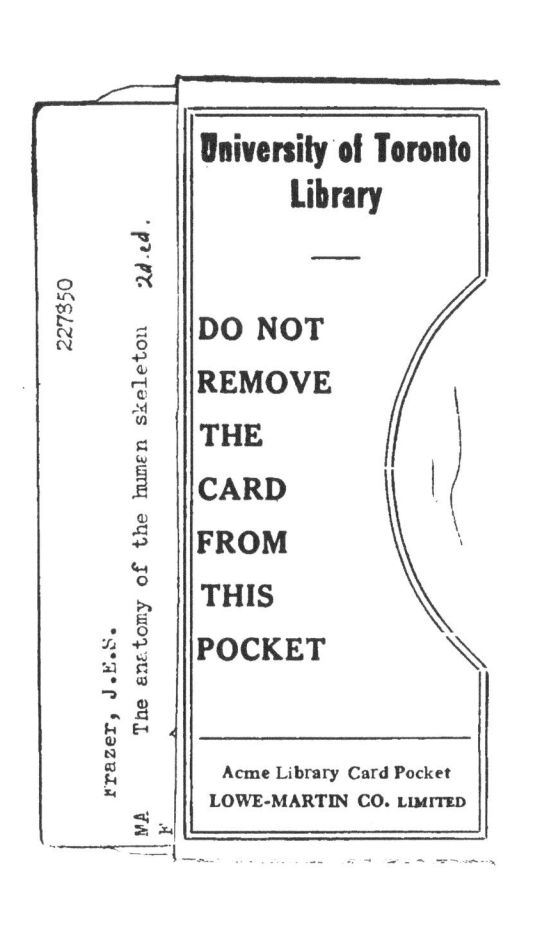

DO NOT
REMOVE
THE
CARD
FROM
THIS
POCKET

Acme Library Card Pocket
LOWE-MARTIN CO. LIMITED

227350

Frazer, J.E.S.
The anatomy of the human skeleton 2d.ed.

Lightning Source UK Ltd.
Milton Keynes UK
UKOW06f2148170416

272436UK00005B/95/P